Springer Tracts in Mechanical Engineering

Springer Tracts in Mechanical Engineering (STME) publishes the latest developments in Mechanical Engineering - quickly, informally and with high quality. The intent is to cover all the main branches of mechanical engineering, both theoretical and applied, including:

- Engineering Design
- Machinery and Machine Elements
- Mechanical Structures and Stress Analysis
- Automotive Engineering
- Engine Technology
- Aerospace Technology and Astronautics
- Nanotechnology and Microengineering
- Control, Robotics, Mechatronics
- MEMS
- Theoretical and Applied Mechanics
- Dynamical Systems, Control
- Fluids Mechanics
- Engineering Thermodynamics, Heat and Mass Transfer
- Manufacturing
- Precision Engineering, Instrumentation, Measurement
- Materials Engineering
- Tribology and Surface Technology

Within the scope of the series are monographs, professional books or graduate textbooks, edited volumes as well as outstanding PhD theses and books purposely devoted to support education in mechanical engineering at graduate and post-graduate levels.

Indexed by SCOPUS, zbMATH, SCImago.

Please check our Lecture Notes in Mechanical Engineering at http://www.springer.com/series/11236 if you are interested in conference proceedings.

To submit a proposal or for further inquiries, please contact the Springer Editor **in your region**:

Ms. Ella Zhang (China)
Email: ella.zhang@springernature.com
Priya Vyas (India)
Email: priya.vyas@springer.com
Dr. Leontina Di Cecco (All other countries)
Email: leontina.dicecco@springer.com

All books published in the series are submitted for consideration in Web of Science.

More information about this series at http://www.springer.com/series/11693

Xueyan Xing · Jinkun Liu

Vibration Control Methods of Mechanical Distributed Parameter Systems

清华大学出版社
TSINGHUA UNIVERSITY PRESS

Springer

Xueyan Xing
School of Automation Science
and Electrical Engineering
Beihang University
Beijing, China

Jinkun Liu
School of Automation Science
and Electrical Engineering
Beihang University
Beijing, China

ISSN 2195-9862 ISSN 2195-9870 (electronic)
Springer Tracts in Mechanical Engineering
ISBN 978-981-16-1531-3 ISBN 978-981-16-1532-0 (eBook)
https://doi.org/10.1007/978-981-16-1532-0

Jointly published with Tsinghua University Press
The print edition is not for sale in China. Customers from China please order the print book from: Tsinghua University Press.

This Springer imprint is published by the registered company Springer Nature Singapore Pte Ltd.
The registered company address is: 152 Beach Road, #21-01/04 Gateway East, Singapore 189721, Singapore

Preface

In modern industry, distributed parameter systems widely exist in various fields of machinery, medical treatment, transportation, etc., such that the research on modeling and control of distributed parameter systems has attracted extensive attention of researchers. Different from traditional lumped parameter systems, state variables of distributed parameter systems are not only related to time, but also closely related to the spatial position, which brings great difficulties to the modeling and control design of distributed parameter systems. In actual engineering, there are many systems that can be regarded as distributed parameter systems, such as flexible manipulators, overhead crane bridge beams, marine risers, aerial refueling hoses, elevator cables. However, these distributed parameter systems inevitably suffer from vibration under external disturbances, and the vibration is coupled with the operation movement, which greatly degrades the control performance of the system. How to build mathematical models of distributed parameter systems and design control laws to achieve vibration suppression of systems is a major difficulty in the current control field. In view of the above considerations, mechanical distributed parameter systems in industry, including flexible strings, flexible beams, mobile vehicle-mounted flexible manipulators, and overhead crane bridge beams, have been regarded as the main research objects of this book. Common control problems in actual engineering, such as actuator failures and output constraints, have been considered in this book and linear matrix inequality (LMI) technology is applied to the controller design. Control laws have been proposed based on partial differential equation (PDE) models to achieve system stabilization.

The PDE model avoids the control and observer spillover problems caused by the traditional truncated ordinary differential equation (ODE) model, which improves the actual control effect and increases the difficulty of control design. Since typical structures of distributed parameter systems are considered, the context in this book is representative, which lays foundation for control theory research of mechanical distributed parameter systems. This book provides a thorough ground in PDE modeling and controller design for flexible mechanical systems. The proposed boundary and distributed controllers are verified using MATLAB. In this book, simulation examples are used to illustrate the successful application of the theory. The main context of this book is presented as following:

A general introduction of modeling and control methods for distributed parameters systems is given in Chap. 1.

In Chap. 2, preliminaries including Hamilton principle, finite difference method, and several useful definitions and lemmas are provided for modeling and control design of distributed parameter control systems.

In Chap. 3, the LMI-based PDE boundary control strategy is developed for Euler-Bernoulli beam systems using radial basis function (RBF) neural networks. Since the control design conditions are given in the form of a set of LMIs, the feasibility of the designed controller is easy to be demonstrated. Feasible regions of control gains can be systematically and efficiently achieved by involving LMIs. The system stability can be guaranteed taking into account both parametric and boundary modeling uncertainties, which further increases the availability and engineering application range of the proposed robust adaptive neural network control.

In Chap. 4, the LMI-based boundary and distributed control schemes are developed and the effective control gains can be efficiently solved with a powerful MATLAB toolbox. The exponential stability of the closed-loop system is guaranteed in the presence of both parametric uncertainties and unknown distributed disturbances by the proposed control technique and disturbance observer.

In Chap. 5, considering the external disturbances in practical engineering, boundary and distributed observers are introduced to compensate the boundary and distributed disturbances on the string for ensuring the control effect. Based on the proposed boundary and distributed disturbance observers and LMI techniques which can be directly resolved with a MATLAB toolbox, boundary and distributed controllers are proposed through converting the stability condition to the equivalent LMI problem to find the feasible solutions for a set of LMIs.

In Chap. 6, a vehicle-mounted flexible manipulator system which is composed of a moving rigid vehicle, a flexible arm, and a payload is described by a coupled PDE model by Hamilton principle. Based on the original PDE model, the basic control law is presented to regulate the positions of vehicle and joint. Beyond that, the elastic vibration of the flexible manipulator has also been eliminated under the proposed control law. By extending LaSalle's invariance principle to the infinite dimensional PDE system, the asymptotic stability of the closed-loop system is proven.

In Chap. 7, based on the PDE model, an adaptive fault-tolerant control scheme is developed to restrain the elastic vibration of the flexible manipulator with actuation redundancy and boundary uncertainty using RBF neural networks. Positions of the mobile platform and joint can also be regulated under the proposed control laws.

In Chap. 8, a switching fault-tolerant control is proposed for a moving vehicle-mounted flexible manipulator with actuator failure and a barrier Lyapunov function (BLF) is introduced to prevent constraint transgression. Without actuator failure, both position regulation and vibration suppression can be guaranteed by the designed fault-free control laws. Moreover, on the basis of the BLF and proposed monitoring functions, state constraints are consistently met and control inputs continue functioning even when a portion of the actuators fail. Under the supervision of the proposed monitoring functions, fault detection and actuator switching can be performed simultaneously without requiring a priori knowledge of failure types.

In Chap. 9, the state-feedback control laws are proposed for the vehicle-mounted flexible manipulator system, under which both position regulation and vibration suppression can be achieved with prescribed performance. Tracking errors and their convergence rates can be regulated within certain pre-specified values.

In Chap. 10, the PDE model is established for the overhead crane bridge system considering the effect of the rotational inertia and shear deformation of the bridge beam. In the established PDE model, the main beam is regarded as an infinite-dimensional Timoshenko beam and the middle trolley is considered as a rigid body. Then the vibration control is proposed to stabilize the system.

In Chap. 11, a PDE model-based robust control is developed for the overhead crane bridge system. By applying proposed control laws, vibration and rotation of the system can be suppressed. Besides, output constraints of the middle trolley are effectively ensured with actuator faults and environmental disturbances.

In Chap. 12, the unknown control direction problem is resolved for the overhead crane bridge system with uncertain parameters and disturbances. The stability of the closed-loop system is demonstrated using Lyapunov's direct method.

Beijing, China
December 2020

Xueyan Xing
Jinkun Liu

Acknowledgements

We heartily appreciate our co-workers who have contributed to the collaborative studies of this book. We would also like to express our sincere appreciation to our colleagues who have contributed to the collaborative research.

Last but not least, we are deeply gratitude to our families for their invaluable loves, supports, and sacrifices over the years.

This work is supported by "Research for the PDE modeling and control theory of a two-link rigid-flexible manipulator in three-dimensional task space," funded by the National Natural Science Foundation of China (NSFC) under the Grant 61873296, China.

Beijing, China
December 2020

Xueyan Xing
Jinkun Liu

Contents

About the Authors

Xueyan Xing received the B.S. degree in automation from Northeastern University, Shenyang, China, in 2015, the M.S. degree in control engineering from the Harbin Institute of Technology, Harbin, China, in 2017, and the Ph.D. degree in control theory and control engineering from Beihang University in 2020. She is currently a research fellow in the department of engineering and design, University of Sussex in UK. Her current research interests include distributed parameter systems, vibration control, human-robot interaction and robot control.

Jinkun Liu was born on October 14, 1965, and he received BS, MS, and Ph.D. degrees from Northeastern University, Shenyang, China, in 1989, 1994, and 1997, respectively. He was a Postdoctoral Fellow in Zhejiang University from 1997 to 1999. He is currently a full professor in Beihang University, Beijing, P. R. China. His main research interest is adaptive boundary control for flexible manipulator. He has published more than 100 research papers and 10 books.

Chapter 1
Introduction

1.1 Background and Motivation

With the rapid development of today's economy and technology, mechanical systems are widely applied in the industry. As we know, the state of many industrial systems in actual engineering is generally not only related to time, but also related to space. In such case, the traditional lumped parameter model cannot fully express the dynamic characteristics of the system, and it is necessary to treat these systems as distributed parameter systems for modeling and controller design. Due to the wide application prospects of the distributed parameter system, it prevails in the aerospace industry, robotics industry, machinery manufacturing industry and many other fields [1–10]. In recent years, the modeling and control of distributed parameter systems have attracted widespread attention in the control field. The analysis theory and control method of the distributed parameter system can also be applied to systems with similar structures in other related cross fields to increase industrial productivity and improve social and economic benefits.

There are many common systems in actual engineering that can be regarded as distributed parameter systems, such as flexible manipulators, crane cables, satellite solar panels, overhead crane bridges, flexible ocean risers, and flexible flapping-wing aircraft wings. The following examples illustrate the application of several distributed parameter systems in high-tech engineering fields.

(1) Flexible maniulpators

With the increasing complexity and accuracy of industrial tasks, flexible robotic arm systems have been desperately required by different kinds of industrial fields. The flexible manipulator can be divided into two types: the immovable flexible manipulator fixed at the initial end and the mobile flexible manipulator with a moving platform. Existing related researches mostly assume that the flexible manipulator is fixed at the initial end without considering the moving carrier, which greatly weakens

X. Xing and J. Liu, *Vibration Control Methods of Mechanical Distributed Parameter Systems*, Springer Tracts in Mechanical Engineering,
https://doi.org/10.1007/978-981-16-1532-0_1

the nonlinearity and coupling of the system, and thus the established model is greatly simplified. However, since the traditional arm is fixed, the scope of its operation is severely restricted. Due to the movement of the carrier of the robot arm, the mobile flexible manipulator has a wider range of motion with larger operating space, which can greatly make up for the shortcomings of the fixed manipulator. Therefore, the research on dynamic modeling and control of mobile flexible manipulator with movable platform has important theoretical significance and practical application value.

(2) Flexible string

Flexible strings are commonly used in practical engineering, such as cables of overhead cranes, conveyer belts, and cables of elevators. In the process of picking and conveying heavy objects, the flexible string inevitably vibrates, which will seriously affect the positioning and control accuracy of the transportation of the load and even cause harmful effects to the system. Therefore, how to suppress the vibration of the string during operation is one of the most important problems to be solved urgently. In control design, the flexible string is always described by wave equations, which is one of typical structures of distributed parameter systems.

(3) Overhead crane bridge

In today's society, the overhead crane is widely applied in industry. The overhead crane is mainly composed of the main bridge beam, middle trolley, and cable. The payload is lifted or lowered by the cable after positioning the trolley. The payload can be transported to the ideal location by the moving trolley. In the process of transporting the payload in the actual project, the bridge beam inevitably vibrates due to the influence of the external disturbance and the inertia of the payload, which seriously affects the positioning and control accuracy of the payload. How to achieve the vibration suppression of the bridge beam is a vital research issue.

In actual engineering, systems with flexible structures and spatial vibrations are often regarded as distributed parameter systems. For such kind of systems, designing control laws based on lumped parameter models easily leads to low-accuracy control due to the neglect of spatial effects. As a result, distributed parameter models are used instead of lumped parameter models to improve the control performance. However, since distributed parameter systems usually vibrate during the operation, the residual vibration interacts with the system movement, and seriously affects the working state and operating life of the system. The vibration even makes the control system completely out of control and causes a safety accident.

Considering the adverse effects of the vibration of the system, developing control methods of distributed parameter systems has significant value. How to establish mathematical models for distributed parameter systems and eliminate system vibration has become a hot research direction and one of the key research issues.

1.2 Modeling Methods of Mechanical Distributed Parameter Systems

Since the distributed parameter system usually works in the infinite dimensional task space, infinite dimensional modes of the system need to be considered in the modeling process, which brings difficulties to the modeling of the system. At present, the mathematical models of distributed parameter systems are mainly divided into two types: finite-dimensional truncated models described by ordinary differential equations (ODEs) and the infinite-dimensional distributed parameter model described by partial differential equations (PDEs).

1.2.1 Modeling Method of Truncated ODE Model

Designing control laws based on truncated models needs to reduce the order of the system such that inessential system modes are ignored and only significant modes are remained in the truncated ODE model. The commonly used discretization methods of models are introduced below.

1. Assumed mode method

The assumed mode method is often used to discretize the continuous distributed parameter system [11–13]. It uses different boundary conditions of the system to obtain system mode functions. Then the solution of the distributed parameter system is approximated by using a linear combination of finite mode functions. The n-dimensional discrete model of the double-link flexible arm is established in [14] using the assumed mode method. A numerical analysis method is proposed for the vibration analysis of rectangular plates and stiffened panels in [15] using the assumed mode method. Dynamic analysis of stepped thickness rectangular plate structures is given in [16] based on the assumed mode method. The accuracy of the solution of the assumed mode method is usually related to the number of finite mode functions. Increasing the number of finite mode functions can improve the accuracy of the assumed mode method, but the amount of calculation will accordingly increase. It is necessary to properly select the number of mode functions according to the actual requirement.

2. Finite element method

The finite element method [17–19] discretizes the continuous distributed parameter system into finite elements, and takes the variable of the node of element and its first derivative as generalized coordinates. Then the shape function is constructed using the node of element and the dynamic equations of the continuous system can be obtained by integrating element matrices. The finite element method is used in [20] to study the dynamics of soft pneumatic actuators. The characteristic of the human middle ear, such as ligaments and tendons, is investigated in [21] based on the finite element method and a three-dimensional finite element model is established. The finite element method and the Lagrange method are combined in [22] to derive the model of a single-link flexible manipulator and a nonlinear feedback controller is proposed. A high-order trace finite element method is used in [23] to discretize continuous PDEs and obtain higher discretization precision. However, the structure of the dynamic model obtained by the finite element method and the solving procedure are usually complicated. In addition, the finite element method cannot be used to perform parametric analysis of the system, which limits its application in practical engineering.

3. Lumped parameter method

The finite segment method [24] and the lumped mass method [25] are the two most common methods in the lumped parameter method. The finite segment method discretizes the continuous distributed parameter system into a finite number of rigid segments and connects them through flexible connectors. The lumped mass method discretizes the distributed mass of the continuous distributed parameter system to several nodes and the model of distributed parameter system is then obtained by approximating the mass of the discrete system.

In [26], differential equations are established for cable-stayed bridges with shear lag analysis based on the principle of minimum potential energy and the finite segment method. The finite segment method is used in modeling of railroad track structures in [27]. It can be seen from [26] and [27] that the finite segment method introduces nonlinear coupling effects by replacing flexible components with rigid ones. Transient characteristics of belt conveyor is studied in [28] by lumped mass method and the dynamic characteristics of the midwater trawl system are analyzed by a two-dimensional lumped mass method in [29]. Compared with the finite segment method, the lumped mass method is simpler and easier to understand, but the accuracy of the obtained model is relatively low. As a result, the lumped mass method is usually used in engineering to obtain complex models of distributed parameter systems.

4. Galerkin method

The Galerkin method combines polynomial functions to make the weighted integral of the accumulation satisfy the given original equation, and then the approximate dynamic characteristics of the continuous distributed parameter system can be obtained by solving a set of linear equations [30–32]. In [33], the discontinuous Galerkin method is used for the viscoelastic wave modeling, which shortens the time and storage required by the traditional Galerkin method. Dynamics of three-dimensional fracture mechanics are studied in [34] combining both finite element and Galerkin method. However, what is obtained by Galerkin method is the approximate solution of the system, which cannot guarantee that all points of the system satisfy the original system equations. As a result, the Galerkin method still lacks modeling accuracy.

Assumed mode method, finite element method, lumped parameter method and Galerkin method establish approximate models of distributed parameter systems through order reduction and approximation, which makes the subsequent controller design become easier. However, these discretization methods ignore the influence of high-order modes of the continuous distributed parameter system, which may cause the control and observation spillover problems, and reduce the control accuracy. In order to enhance the control performance, it is necessary to consider higher-order modes in these methods and increase the order of the designed controller. However, this will increase the difficulty of control implementation and increase the amount of calculation.

1.2.2 Modeling Method of PDE Model

In order to avoid the problems of low control accuracy, spillover effects, and system instability caused by the above-mentioned truncated model, the PDE distributed parameter model has received extensive attention in the control field. The distributed parameter model is generally obtained by Lagrange method [35, 36], Hamilton principle method [37–40], Kane method [41], Newton–Euler method [42], and expressed by a series of coupled ODEs and PDEs.

Compared with the truncated discrete model, the continuous distributed parameter model does not need to be discretized for the controller design, which improves the control accuracy and control effect. In [43], the boundary control is designed for the PDE system based on the backstepping method. A linear boundary feedback control method is addressed in [44] to suppress the vibration of the axially moving flexible string system with geometric nonlinearity. In [45], a boundary feedback controller is proposed to exponentially stabilize the axially moving flexible string system. External disturbances and spatial changes of tension of the flexible string system are considered in PDE modeling and control design in [46].

Since the PDE model does not need to reduce the order of the designed controller, it largely avoids control spillover problem. However, the infinite-dimensional PDE distributed parameter model takes into account the higher-order modes of the system, which increases the difficulty of control design and stability analysis. At present, the types of PDE distributed parameter models are very limited. The research results of control methods designed based on the PDE distributed parameter model are still not sufficient compared with the ones based on the finite-dimensional ODE model.

1.3 Control Strategy of Mechanical Distributed Parameter Systems

According to the position of the controller in the distributed parameter system, the control strategy is usually divided into boundary control and distributed control. Boundary control means that the control input signal of the actuator acts on the end boundary of the distributed parameter system, and the system can be tracked or suppressed by controlling the boundary points of the system. Distributed control means that the control force acts at every point of the distributed parameter system to achieve the control goal.

1. Boundary control
   Boundary control is an effective control method and easy to implement for distributed parameter systems [47–49]. From the perspective of mathematical model expression, boundary control usually appears in boundary conditions in distributed parameter models. Since only a small number of boundary actuators and sensors are needed in the implementation of boundary control, the impact of actuators and sensors on the system is usually negligible. A boundary controller is designed for an Euler-Bernoulli beam with rigid base and free end in [50]. Lagrange method and assumed mode method are used to establish a dynamic model of a flexible manipulator rotating on a horizontal plane in [51], and a boundary controller is proposed at the end of the system to achieve stabilization of the closed-loop system. The elastic vibration of the moving beam system is suppressed by designing a boundary controller in [52]. Boundary control is an economical and highly feasible control method. In the research on distributed parameter systems in recent years, the boundary control has received much more attention and applications.
2. Distributed control
   Distributed control requires actuators and sensors distributed at every point of the continuous distributed parameter system [6, 53–55]. From the perspective of the mathematical model expression, distributed control is generally located in the infinite-dimensional governing equations in the PDE model of the distributed parameter system, rather than in boundary conditions as the boundary control. Theoretically, the implementation of the distributed control is on the basis of an infinite number of sensors and actuators. Consequently, the distributed control is a function that is related to both time and space. Distributed control usually uses smart materials, such as piezoelectric ceramics, as actuators. With the rapid development of technology and material science, the realization of distributed control has become possible. A fuzzy distributed control problem is investigated for a class of nonlinear distributed parameter systems in [56]. PDEs are used to model the dynamics of multi-agent systems and the distributed control is proposed to achieve tracking control in [57]. The saturated distributed control method is designed in [58] to stabilize a class of distributed parameter systems. A synthetic distributed output feedback controller is developed for controlling a first-order hyperbolic PDE system with two-time scales in [59],

where the control input, control output, and measurement output are all spatially distributed.

Compared with the boundary control, distributed control has a larger control range and can obtain better control effect. Although distributed control has a wider range of applications, but it requires distributed actuators and sensors which makes it more difficult to implement. In practical applications, considering the control effect and economic cost, an appropriate control strategy should be selected to control the distributed parameter system.

1.4 Stability Analysis Method of Mechanical Distributed Parameter Systems

1.4.1 Lyapunov Theory

In the field of automatic control, Lyapunov stability analysis theory has been widely used in various linear and nonlinear systems to analyze and prove the stability of the system [60–68].

In order to analyze the stability of the system, Lyapunov stability theory can be divided into two main methods, that is, the first Lyapunov method and Lyapunov's direct method. The first Lyapunov method has complicated steps. It needs to obtain the explicit solution of differential equations and use the characteristics of the solution to indirectly analyze the stability performance of the controlled system. The Lyapunov's direct method needs to construct an appropriate Lyapunov function and analyze the Lyapunov function to directly obtain the stability of the system, which avoids solving the differential equations. An adaptive boundary controller is designed for the PDE system with unknown parameters and the system stability is analyzed by the Lyapunov's direct method in [69]. A Lyapunov adaptive boundary control is proposed for parabolic PDEs with spatially varying coefficients in [70]. In [71] and [72], the research results of vibration control of beams with varying moving speed are given and the Lyapunov's direct method is used for the stability analysis. The Lyapunov's direct method is used for the control design and stability analysis of the axially moving string system in [73].

Compared with the first Lyapunov method, Lyapunov's direct method requires little background knowledge and simple mathematical tools, which thus has wider application. It is one of the most commonly used methods for stability analysis of distributed parameter systems in the control field.

1.4.2 Operator Semigroup Theory

As a powerful mathematical tool, operator semigroup theory is an important part of functional analysis [74–79] and widely used in engineering and mathematics fields, such as differential equations, quantum theory, approximation theory, and Markov processes. The operator semigroup method has also been extended to the control field and applied to the stability analysis of the PDE system. It is also used for proofing the existence and uniqueness of the solution of the closed-loop PDE system. In [80], the boundary control is proposed for two connected strings with joint anti-damping and semigroup theory is used to prove the exponential stability of the closed-loop system. The operator semigroup theory is used in [9] to prove the asymptotic stability of the axially moving string under spatiotemporally varying tension and proposed controller. In [81] and [82], exponentially stable observers are proposed for the distributed parameter systems, and the operator semigroup theory is employed to prove the existence and uniqueness of the solution of the systems.

Although the operator semigroup theory has high flexibility and a wide application range compared with the Lyapunov direct method, it is more logically rigorous and difficult to understand. The use of the operator semigroup theory requires strong mathematical foundation and knowledge of functional theory, which makes it difficult to master.

1.5 Typical Engineering Problems of Distributed Parameter System

1.5.1 Common Model Structure

In the existing research, distributed parameter systems can be described by two categories of models according to the order, namely the beam structure model and the string structure model. These two types of structures are introduced below.

1. Beam structure
   The beam structure model has a fourth-order partial derivative to the space, which belongs to the fourth-order PDE model. The two most widely used beam structures in actual engineering are Euler-Bernoulli beam structure [83–85] and Timoshenko beam structure [86, 87]. Euler-Bernoulli beam is an approximate model, in which the influence of moment of inertia and shear deformation is not considered due to its slender structure. Therefore, Euler-Bernoulli beam needs to satisfy the Kirchhoff assumption, that is, the beam section, which is perpendicular to the neutral axis when there is no deformation, is always perpendicular to the neutral axis. Compared with the Euler-Bernoulli beam model which is more suitable for beams with slender structures, the Timoshenko beam model considers the coupling of the moment of inertia and the shear deformation.

As a result, the dynamic characteristics of the Timoshenko beam model are more accurate than the Euler-Bernoulli beam model. In actual engineering, the distributed parameter system with beam structure has a wide range of applications, such as the flexible manipulators, main beams of the crane bridges, marine risers, etc.

2. String structure
   The string structure model [88–90] usually considers the influence of the system tension. Since a second-order partial derivative to the space exists in the string model, it belongs to the second-order PDE model. The distributed parameter system with string structure is also widely used in practical engineering, such as crane cables, conveyor belts, elevator cables, etc.

1.5.2 Common Engineering Control Problems

1. Output Constraints

In engineering, once the output signal of the control system exceeds the given range, the equipment may be seriously damaged and safety accidents may occur. For safety reasons, the output signal of the control system needs to be less than a certain value such that the safety accidents can be prevented and the control performance degradation caused by the violation of output constraints can also be avoided. Therefore, the problem of output constraints [91–95] is one of common control problems, which needs to be addressed in the control field.

2. Actuator Failure

With the rapid development of engineering technology, how to improve the safety and reliability of control systems has become one of the significant control problems. If the actuator of the control system fails, the designed control laws will not continue to guarantee the control effect, which will degrade the control performance and even cause a series of safety issues. In addition, due to changes of the external environment during operation, actuator failures are inevitable, and the failure modes and failure time are hard to predict. In such case, the fault-tolerant control is urgently needed to enable the system to maintain the ideal system performance even after the actuator fails. The fault-tolerant control is proposed for the flexible spacecraft in [96], which can realize attitude tracking in the case of actuator failure. The adaptive sliding mode fault-tolerant attitude tracking control is addressed in [97] for the flexible spacecraft subject to actuator saturation. The adaptive fault-tolerant control strategy is proposed in [98] to eliminate the vibration of the rigid-flexible coupling manipulator.

1.6 Outline of the Book

The general objectives of the book are to develop constructive and systematic PDE models and design control methods for mechanical distributed parameter systems. By investigating the characteristics of several different PDE models, control strategies are proposed to achieve the ideal performance for the concerned systems. The book starts with a brief introduction of modeling and control techniques for mechanical distributed parameter systems in this book.

In Chap. 2, necessary lemmas, definitions, and properties for the subsequent modeling, controller design, and stability analysis of distributed parameter systems are presented.

In Chap. 3, the boundary robust control is designed for the Euler-Bernoulli beam on the basis of radial basis function (RBF) neural networks based on the linear matrix inequality (LMI) technology. By solving the given LMIs, the feasible control gains can be achieved. Under the proposed control, the vibration of the Euler-Bernoulli beam can be restricted in the presence of parametric and system boundary uncertainties.

In Chap. 4, boundary and distributed control laws are designed for the Euler-Bernoulli beam. The control gains are expressed in the form of LMIs, which can be directly obtained using MATLAB. To deal with external unknown distributed disturbances, an observer is designed. Under the designed control and observer, the exponential stability of the disturbed Euler-Bernoulli beam with uncertain parameters is guaranteed.

In Chap. 5, the vibration control problem of the flexible string is investigated. Boundary and distributed observers are designed to observe environmental disturbances acting on the flexible string. Then, based on the proposed observers, LMI techniques are used to design boundary and distributed controllers for the vibration suppression of the system.

In Chap. 6, the PDE model is established for the vehicle-mounted flexible manipulator system with consideration of the vehicle, joint, flexible manipulator, and payload using Hamilton principle. Based on the PDE model, the control scheme is addressed to realize the position and angle regulation of the vehicle and joint, respectively. Besides, the vibration suppression of the payload and the soft link can also be realized.

In Chap. 7, based on the PDE model, an adaptive fault-tolerant control scheme is proposed to restrict the elastic vibration of the flexible manipulator with actuation redundancy. RBF neural networks are used to compensate the effect of system uncertainty. With the proposed control, the closed-loop system is asymptotically stable and its asymptotic stability is proven by extending LaSalle's invariance principle.

In Chap. 8, barrier Lyapunov function (BLF) is used to design the switching fault-tolerant control for the vehicle-mounted flexible manipulator, which realizes the vibration suppression of the flexible manipulator and the payload when actuators are fault-free. Besides, the position regulation of the vehicle and the angle adjustment of the joint can also be achieved with healthy actuators. When the actuator fails, the designed control law can make the output signals of the system still satisfy the output constraint.

In Chap. 9, the actual engineering requirement of control performance is considered and the control scheme is designed for the vehicle-mounted flexible manipulator such that the prescribed performance of the system can be realized under control while vibration suppression and position regulation are achieved. Under the proposed control, the convergence speed and output errors of the system are constrained within the given performance index function.

In Chap. 10, the PDE modeling of the overhead bridge crane system is investigated taking into account the dynamic coupling characteristics of the main bridge beam and middle trolley. The PDE model established in this chapter lays the foundation for the subsequent control design. Then the vibration control is proposed to achieve exponential stability of the closed-loop system.

In Chap. 11, the output constraints of the disturbed overhead crane bridge can be ensured under the designed control laws such that the vibration and rotation of the trolley are within the given bounds. With control, the vibration and rotation of the bridge beam can be effectively suppressed. The stability of the closed-loop system is proved using the Lyapunov's direct method.

In Chap. 12, the control laws are designed for the overhead crane bridge for stabilization subject to the unknown control direction, environmental disturbances, and uncertain system parameters. With the proposed control, the vibration and rotation of the system are restricted and finally converge to adjustable ranges.

References

1. S.S. Ge, W. He, B.V.E. How, Y.S. Choo, Boundary control of a coupled nonlinear flexible marine riser. IEEE Trans. Control Syst. Technol. **18**(5), 1080–1091 (2010)
2. Q.H. Ngo, K.S. Hong, Skew control of a quay container crane. J. Mech. Sci. Technol. **23**(12), 3332–3339 (2009)
3. A. Paranjape, S.J. Chung, M. Krstic, PDE boundary control for flexible articulated aircraft wings, in *Aiaa Guidance, Navigation, & Control Conference* (2013), pp. 6486–6500
4. A.A. Paranjape, S.J. Chung, J. Kim, Novel dihedral-based control of flapping-wing aircraft with application to perching. IEEE Trans. Rob. **29**(5), 1071–1084 (2013)
5. K.D. Do, J. Pan, Boundary control of transverse motion of marine risers with actuator dynamics. J. Sound Vib. **318**(4–5), 768–791 (2008)
6. D. Halim, S.O.R. Moheimani, Spatial resonant control of flexible structures-application to a piezoelectric laminate beam. IEEE Trans. Control Syst. Technol. **9**(1), 37–53 (2001)
7. K.S. Hong, J. Bentsman, Application of averaging method for integro-differential equations to model reference adaptive control of parabolic systems. Automatica **30**(9), 1415–1419 (1994)
8. L. Zhang, J. Liu, Observer-based partial differential equation boundary control for a flexible two-link manipulator in task space. IET Control Theory Appl. **6**(13), 2120–2133 (2012)
9. K.J. Yang, K.S. Hong, F. Matsuno, Robust adaptive boundary control of an axially moving string under a spatiotemporally varying tension. J. Sound Vib. **273**(4–5), 1007–1029 (2004)
10. E. Pereira, S.S. Aphale, V. Feliu, S.R. Moheimani, Integral resonant control for vibration damping and precise tip-positioning of a single-link flexible manipulator. IEEE/ASME Trans. Mechatron. **16**(2), 232–240 (2011)

11. D.S. Cho, N. Vladimir, T.M. Choi, Approximate natural vibration analysis of rectangular plates with openings using assumed mode method. Int. J. Naval Arch. Ocean Eng. **5**(3), 478–491 (2013)
12. J. Cheng, H. Xu, A. Yan, Frequency analysis of a rotating cantilever beam using assumed mode method with coupling effect. Mech. Based Des. Struct. Mach. **34**(1), 25–47 (2006)
13. A. Green, J. Sasiadek, Robot manipulator control for rigid and assumed mode flexible dynamics models, in *Aiaa Guidance, Navigation, & Control Conference & Exhibit* (2015), pp. 5435–5445
14. H. Gao, W. He, C. Zhou, C. Sun, Neural network control of a two-link flexible robotic manipulator using assumed mode method. IEEE Trans. Industr. Inf. **15**(2), 755–765 (2018)
15. D.S. Cho, B.H. Kim, J.H. Kim, N. Vladimir, T.M. Choi, Forced vibration analysis of arbitrarily constrained rectangular plates and stiffened panels using the assumed mode method. Thin-Walled Struct. **90**, 182–190 (2015)
16. D.S. Cho, B.H. Kim, J.H. Kim, N. Vladimir, T.M. Choi, Simplified dynamic analysis of stepped thickness rectangular plate structures by the assumed mode method. Proc. Inst. Mech. Eng. Part M: J. Eng. Maritime Environ. **231**(1), 177–187 (2017)
17. H. Du, M.K. Lim, K.M. Liew, A nonlinear finite element model for dynamics of flexible manipulators. Mech. Mach. Theory **31**(8), 1109–1119 (1996)
18. J.M. Martins, Z. Mohamed, M.O. Tokhi, J.S.D. Costa, M.A. Botto, Approaches for dynamic modelling of flexible manipulator systems. IEE Proc.-Control Theory Appl. **50**(4), 401–411 (2003)
19. M.O. Tokhi, Z. Mohamed, S.H.M. Amin, R. Mamat, Dynamic characterisation of a flexible manipulator system: theory and experiments. Intell. Syst. Technol. New Millennium 167–172 (2000)
20. P. Moseley, J.M. Florez, H.A. Sonar, G. Agarwal, W. Curtin, J. Paik, Modeling, design, and development of soft pneumatic actuators with finite element method. Adv. Eng. Mater. **18**(6), 978–988 (2016)
21. T. Koike, H. Wada, T. Kobayashi, Modeling of the human middle ear using the finite-element method. J. Acoust. Soc. Am. **111**(3), 1306–1317 (2002)
22. S.S. Ge, T.H. Lee, G. Zhu, A nonlinear feedback controller for a single-link flexible manipulator based on a finite element model. J. Robotic Syst. **14**(3), 165–178 (1997)
23. J. Grande, C. Lehrenfeld, A. Reusken, Analysis of a high-order trace finite element method for PDEs on level set surfaces. SIAM J. Numer. Anal. **56**(1), 228–255 (2018)
24. H.S.H. Sun, The finite segment method for recursive approach to flexible multibody dynamics, in *International Conference on Information & Computing Science* (2009), pp. 345–348
25. C.M. Chen, V. Thomee, The lumped mass finite element method for a parabolic problem. ANZIAM J. **26**(3), 329–354 (1985)
26. Q.Z. Luo, J. Tang, Q.S. Li, Finite segment method for shear lag analysis of cable-stayed bridges. J. Struct. Eng. **128**(12), 1617–1622 (2002)
27. M.B. Hamper, K.E. Zaazaa, A.A. Shabana, Modeling railroad track structures using the finite segment method. Acta Mech. **223**(8), 1707–1721 (2012)
28. S.G. Sakharwade, S. Nagpal, Study of transient characteristics of belt conveyor by lumped mass method. i-Manager's J. Future Eng. Technol. **14**(4), 33–38 (2019)
29. H. Fuxiang, K. Matuda, T. Tokai, K. Haruyuki, Dynamic analysis of midwater trawl system by a two-dimensional lumped mass method. Fish. Sci. **61**(2), 229–233 (1995)
30. B. Cockburn, G.E. Karniadakis, C.W. Shu, *The Development of Discontinuous Galerkin Methods* (Discontinuous Galerkin Methods. Springer, Berlin, Heidelberg, 2000), pp. 3–50
31. M. Marion, R. Temam, Nonlinear Galerkin methods. SIAM J. Numer. Anal. **26**(5), 1139–1157 (1989)
32. C. Johnson, J. Pitkaranta, An analysis of the discontinuous Galerkin method for a scalar hyperbolic equation. Math. Comput. **46**(173), 1–26 (1986)
33. Q. Zhan, M. Zhuang, Q. Sun, Q. Ren, Y. Mao, Y. Liu, Q. Huo, Efficient ordinary differential equation-based discontinuous Galerkin method for viscoelastic wave modeling. IEEE Trans. Geosci. Remote Sens. **55**(10), 5577–5584 (2017)

34. N. Sukumar, B. Moran, T. Black, An element-free Galerkin method for three-dimensional fracture mechanics. Comput. Mech. **20**(1–2), 170–175 (1997)
35. M.C. Shaker, A. Ghosal, Nonlinear modeling of flexible manipulators using nondimensional variables. J. Comput. Nonlinear Dyn. **1**(2), 123–134 (2006)
36. Y. Li, B. Tang, Z. Shi, Y. Lu, Experimental study for trajectory tracking of a two-link flexible manipulator. Int. J. Syst. Sci. **31**(1), 3–9 (2000)
37. R.F. Fung, H.C. Chang, Dynamic modelling of a non-linearly constrained flexible manipulator with a tip mass by Hamilton's principle. J. Sound Vib. **216**(5), 751–769 (1998)
38. L. Zhang, J. Liu, Observer-based partial differential equation boundary control for a flexible two-link manipulator in task space. Control Theory Appl. **6**(13), 2120–2133 (2012)
39. L. Zhang, J. Liu, Nonlinear PDE observer design for a flexible two-link manipulator, in *American Control Conference (ACC)* (2012), pp. 5336–5341
40. L. Zhang, J. Liu, Adaptive boundary control for flexible two-link manipulator based on partial differential equation dynamic model. Control Theory Appl. **7**(1), 43–51 (2013)
41. H.G. Tanner, K.J. Kyriakopoulos, Modeling of a mobile manipulator with the use of Kane's dynamical equations, in *IFAC Meeting on Intelligent Autonomous Vehicles* (1998), pp. 14–40
42. L. Bascetta, G. Ferretti, and B. Scaglioni, Closed form Newton–Euler dynamic model of flexible manipulators. Robotica **35**(5), 1006–1030 (2015)
43. B.M. Krstic, A. Smyshlyaev, *Boundary Control of PDEs: A Course on Backstepping Designs* (A Course on Backstepping Designs. Society for Industrial and Applied Mathematics, Boundary Control of Pdes, 2008).
44. T. Li, Z. Hou, Exponential stabilization of an axially moving string with geometrical nonlinearity by linear boundary feedback. J. Sound Vib. **296**(4–5), 861–870 (2006)
45. T. Li, Z. Hou, J. Li, *Stabilization Analysis of a Generalized Nonlinear Axially Moving String by Boundary Velocity Feedback* (Pergamon Press, Inc., 2008)
46. S. Zhang, S.S. Ge, W. He, K.S. Hong, Modeling and control of a nonuniform vibrating string under spatiotemporally varying tension and disturbance. IFAC Proc. Volumes **44**(1), 7678–7768 (2011)
47. S.S. Ge, T.H. Lee, G. Zhu, F. Hong, Variable structure control of a distributed-parameter flexible beam. J. Robotic Syst. **18**(1), 17–27 (2001)
48. G. Zhu, S.S. Ge, A quasi-tracking approach for finite-time control of a mass-beam system. Automatica **34**(7), 881–888 (1998)
49. T.H. Lee, S.S. Ge, Z.P. Wang, Adaptive robust controller design for multi-link flexible robots. Mechatronics **11**(8), 951–967 (2011)
50. O. Morgul, Dynamic boundary control of a Euler-Bernoulli beam. IEEE Trans. Autom. Control **37**(5), 639–642 (2002)
51. H. Geniele, R.V. Patel, K. Khorasani, End-point control of a flexible-link manipulator: theory and experiments. IEEE Trans. Control Syst. Technol. **5**(6), 556–570 (1997)
52. M.P. Fard, S.I. Sagatun, Exponential stabilization of a transversely vibrating beam via boundary control. J. Sound Vib. **240**(4), 613–622 (2001)
53. B. Bamieh, F. Paganini, M.A. Dahleh, Distributed control of spatially invariant systems. IEEE Trans. Autom. Control **47**(7), 1091–1107 (2002)
54. F. Wu, Distributed control for interconnected linear parameter-dependent systems. IEE Proc.-Control Theory Appl. **150**(5), 518–527 (2003)
55. S.S. Ge, T.H. Lee, J.Q. Gong, A robust distributed controller of a single-link SCARA/Cartesian smart materials robot. Mechatronics **9**(1), 65–93 (1999)
56. J.W. Wang, H.N. Wu, H.X. Li, Distributed fuzzy control design of nonlinear hyperbolic PDE systems with application to nonisothermal plug-flow reactor. IEEE Trans. Fuzzy Syst. **19**(3), 514–526 (2011)
57. J. Zhang, J. Qi, Distributed control laws for multi-agent system formation tracking modeled by a wave PDE, in *2016 Chinese Control and Decision Conference (CCDC)* (2016), pp. 2629–2634
58. J.L. Pitarch, M. Rakhshan, M.M. Mardani, M. Sadeghi, Distributed saturated control for a class of semilinear PDE systems: an SOS approach. IEEE Trans. Fuzzy Syst. **26**(2), 749–760 (2018)

59. P.D. Christofides, Output feedback control of two-time-scale hyperbolic PDE systems, in *Control Conference* (1998), pp. 1963–1968
60. S.S. Ge, T.H. Lee, G. Zhu, Genetic algorithm tuning of Lyapunov-based controllers: an application to a single-link flexible robot system. IEEE Trans. Industr. Electron. **43**(5), 567–574 (1996)
61. Z. Qu, J.X. Xu, Model-based learning controls and their comparisons using lyapunov direct method. Asian J. Control **4**(1), 99–110 (2010)
62. D. Wang, N.H. McClamroch, Position and force control for constrained manipulator motion: Lyapunov's direct method. IEEE Trans. Robot. Autom. 308–313 (1993)
63. R.E. Precup, M.L. Tomescu, S. Preitl, Fuzzy logic control system stability analysis based on Lyapunov's direct method. Int. J. Comput. Commun. Control 415–426 (2009)
64. S.M. Shahruz, L.G. Krishna, Boundary control of a non-linear string. J. Sound Vib. **195**(1), 169–174 (1996)
65. L. Yan, Y.Q. Chen, I. Podlubny, Stability of fractional-order nonlinear dynamic systems: Lyapunov direct method and generalized Mittag-Leffler stability. Comput. Math. Appl. **59**(5), 1810–1821 (2010)
66. Z.P. Jiang, Global tracking control of underactuated ships by Lyapunov's direct method. Automatica **38**(2), 301–309 (2002)
67. M.A. Pai, *Power System Stability: Analysis by the Direct Method of Lyapunov. Sole Distributors for the U.S.A. and Canada* (Elsevier, North-Holland, 1981)
68. G.A. Leonov, V.A. Boichenko, Lyapunov's direct method in the estimation of the Hausdorff dimension of attractors. Acta Applicandae Mathematica **26**(1), 1–60 (1992)
69. M. Krstic, A. Smyshlyaev, Adaptive boundary control for unstable parabolic PDEs. I. Lyapunov design. IEEE Trans. Autom. Control **53**(7), 1575–1591 (2008)
70. A. Smyshlyaev, Lyapunov adaptive boundary control for parabolic PDEs with spatially varying coefficients, in *Proceedings of the 2006 American Control Conference* (2006), pp. 41–48
71. K.J. Yang, K.S. Hong, F. Matsuno, Boundary control of a translating tensioned beam with varying speed. IEEE/ASME Trans. Mechatron. **10**(5), 594–597 (2005)
72. M.P. Fard, S.I. Sagatun, Exponential stabilization of a transversely vibrating beam by boundary control via Lyapunov's direct method. J. Sound Vib. **240**(4), 613–622 (2001)
73. C.W. Kim, K.S. Hong, H. Park, Boundary control of an axially moving string: Actuator dynamics included. J. Mech. Sci. Technol. **19**(1), 40–50 (2005)
74. A. Pazy, *Semigroups of Linear Operators and Applications to Partial Differential Equations* (Springer, New York, 1983).
75. G.L. Cain, G.H. Meyer, *Separation of Variables for Partial Differential Equations: An Eigenfunction Approach. Separation of Variables for Partial Differential Equations* (Chapman & Hall/CRC, 2006)
76. B. Chentouf, J.F. Couchouron, Nonlinear feedback stabilization of a rotating body-beam without damping. ESAIM Control Optim. Calculus Var. **4**, 515–535 (1999)
77. B. Chentouf, Dynamic boundary controls of a rotating body-beam system with time-varying angular velocity. J. Appl. Math. **2004**(2), 107–126 (2014)
78. T.D. Nguyen, O. Egeland, Observer design for a flexible robot arm with a tip load, in *Proceedings of the 2005, American Control Conference* (2005), pp. 1389–1394
79. C.Z. Xu, J. Baillieul, Stabilizability and stabilization of a rotating body-beam system with torque control. IEEE Trans. Autom. Control **38**(12), 1754–1765 (1993)
80. B.Z. Guo, F.F. Jin, Arbitrary decay rate for two connected strings with joint anti-damping by boundary output feedback. Automatica **46**(7), 1203–1209 (2010)
81. T.D. Nguyen, Boundary output feedback of second-order distributed parameter systems. Syst. Control Lett. **58**(7), 519–528 (2009)
82. M.A. Demetriou, Natural second-order observers for second-order distributed parameter systems. Syst. Control Lett. **51**(3), 225–234 (2004)
83. T. Endo, F. Matsuno, H. Kawasaki, Simple boundary cooperative control of two one-link flexible arms for grasping. IEEE Trans. Autom. Control **54**(10), 2470–2476 (2009)

84. A. Baz, Dynamic boundary control of beams using active constrained layer damping. Mech. Syst. Signal Process. **11**(6), 811–825 (1997)
85. N. Tanaka, H. Iwamoto, Active boundary control of an Euler-Bernoulli beam for generating vibration-free state. J. Sound Vib. **304**(3–5), 570–586 (2007)
86. W. He, S. Zhang, S.S. Ge, Boundary output-feedback stabilization of a Timoshenko beam using disturbance observer. IEEE Trans. Industr. Electron. **60**(11), 5186–5194 (2013)
87. C. Liu, S.S. Ge, W. He, C. Liu, Adaptive boundary control for a class of inhomogeneous Timoshenko beam equations with constraints. IET Control Theory Appl. **8**(14), 1285–1292 (2014)
88. Q.C. Nguyen, K.S. Hong, Simultaneous control of longitudinal and transverse vibrations of an axially moving string with velocity tracking. J. Sound Vib. **331**(13), 3006–3019 (2012)
89. O. Morgul, A dynamic control law for the wave equation. Automatica **30**(11), 1785–1792 (1994)
90. Z. Qu, An iterative learning algorithm for boundary control of a stretched moving string. Automatica **38**(5), 821–827 (2002)
91. F. Cao, J. Liu, Vibration control for a rigid-flexible manipulator with full state constraints via Barrier Lyapunov Function. J. Sound Vib. **406**, 237–252 (2017)
92. K.P. Tee, S.S. Ge, E.H. Tay, Barrier Lyapunov functions for the control of output constrained nonlinear systems. Automatica **45**(4), 918–927 (2009)
93. Z. Liu, J. Liu, W. He, Boundary control of an Euler-Bernoulli beam with input and output restrictions. Nonlinear Dyn. **92**(2), 531–541 (2018)
94. A.K. Kostarigka, Z. Doulgeri, G.A. Rovithakis, Prescribed performance tracking for flexible joint robots with unknown dynamics and, variable elasticity, in *IEEE International Conference on Robotics & Automation* (2012), pp. 1137–1147
95. W. He, S.S. Ge, Vibration control of a flexible beam with output constraint. IEEE Trans. Industr. Electron. **62**(8), 5023–5030 (2015)
96. Y. Jiang, Q. Hu, G. Ma, Adaptive backstepping fault-tolerant control for flexible spacecraft with unknown bounded disturbances and actuator failures. ISA Trans. **49**(1), 57–69 (2010)
97. B. Xiao, Q. Hu, Y. Zhang, Adaptive sliding mode fault tolerant attitude tracking control for flexible spacecraft under actuator saturation. IEEE Trans. Control Syst. Technol. **20**(6), 1605–1612 (2012)
98. F. Cao, J. Liu, Adaptive actuator fault compensation control for a rigid-flexible manipulator with ODEs-PDEs model. Int. J. Syst. Sci. **49**(8), 1748–1759 (2018)

Chapter 2
Preliminaries

2.1 Finite Difference Method

When the analytical solution of the system is hard to obtain, the finite difference method is an effective approach to get the approximate solution and realize the simulation of continuous PDE systems [1]. Consequently, the finite difference method is generally used in the simulation of PDE control systems and its effectiveness has been confirmed by [2–6]. Referring to aforementioned researches, the finite difference method is applied in the simulation in this chapter. To guarantee the validity of simulations, the time and position steps can be chosen as small as possible to make the approximate solution closer to the exact solution of the PDE system.

Define the sampling time step as Δt and the space step as Δx, which satisfy the condition $\Delta t \leq \frac{1}{2}\Delta x$ referred by [7]. There are three kinds of discrete methods to express $y(x, t)$ as follows:

(1) Backward difference: $\frac{\partial y}{\partial t}\big|_{t=j} = \frac{y(i,j)-y(i,j-1)}{\Delta t}$ and $\frac{\partial y}{\partial x}\big|_{x=i} = \frac{y(i,j)-y(i-1,j)}{\Delta x}$;
(2) Forward difference: $\frac{\partial y}{\partial t}\big|_{t=j} = \frac{y(i,j+1)-y(i,j)}{\Delta t}$ and $\frac{\partial y}{\partial x}\big|_{x=i} = \frac{y(i+1,j)-y(i,j)}{\Delta x}$;
(3) Central difference: $\frac{\partial y}{\partial t}\big|_{t=j} = \frac{y(i,j+1)-y(i,j-1)}{2\Delta t}$ and $\frac{\partial y}{\partial x}\big|_{x=i} = \frac{y(i+1,j)-y(i-1,j)}{2\Delta x}$.

More details of the finite difference method can be found in [8].

In the practical simulation, smaller sampling step will enhance the precision of simulation. For example, a PDE model of a flexible manipulator is given as follows [9]

$$\rho\big(x\ddot{\theta}(t) + \ddot{y}(x, t)\big) = -EIy_{xxxx}(x, t) \tag{2.1}$$

$$0 = I_{\mathrm{h}}\ddot{\theta}(t) - EIy_{xx}(0, t) \tag{2.2}$$

$$0 = m\big(L\ddot{\theta}(t) + \ddot{y}(L, t)\big) - EIy_{xxx}(L, t) \tag{2.3}$$

X. Xing and J. Liu, *Vibration Control Methods of Mechanical Distributed Parameter Systems*, Springer Tracts in Mechanical Engineering,
https://doi.org/10.1007/978-981-16-1532-0_2

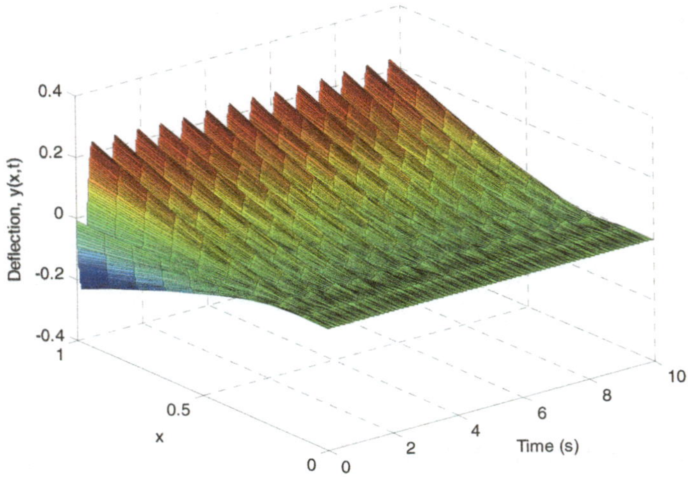

Fig. 2.1 The distributed elastic deflection of the flexible manipulator in case 1

$$y(0,t) = 0,\ y_x(0,t) = 0,\ y_{xx}(L,t) = 0. \tag{2.4}$$

Let $EI = 3\ \text{Nm}^2$, $\rho = 0.2$ kg/m, $m = 0.1$ kg, $L = 1$ m, $I_h = 0.1\ \text{kgm}^2$, $y(x,0) = 10^{-3}x$, and $\dot{y}(x,0) = 0$. Then the states of the open-loop system in following three cases are given:

Case 1. $\Delta t = 5 \times 10^{-4}$ s and $\Delta x = 0.1$ m;

Case 2. $\Delta t = 1.25 \times 10^{-3}$ s and $\Delta x = 0.1$ m;

Case 3. $\Delta t = 5 \times 10^{-4}$ s and $\Delta x = 0.2$ m.

From Figs. 2.1, 2.2 and 2.3, it can be observed that Fig. 2.1 with the simulation results with smallest time and space step has the highest precision and it is smoother than Figs. 2.2 and 2.3.

The simulation program of the PDE model in Eqs. (2.1)–(2.4) is given in the Appendix (2.1) as chapter2.m.

2.2 Hamilton Principle

Hamilton principle is adopted in this chapter to establish the PDE model of the overhead crane bridge system. It is noteworthy that Hamilton principle is widely applied in PDE modeling and its effectiveness has already been proved [10, 11]. By accurately calculating the kinetic energy, potential energy of the system and virtual work, the PDE models of distributed parameter systems can be finally attained.

Hamilton principle [10, 11] is represented by

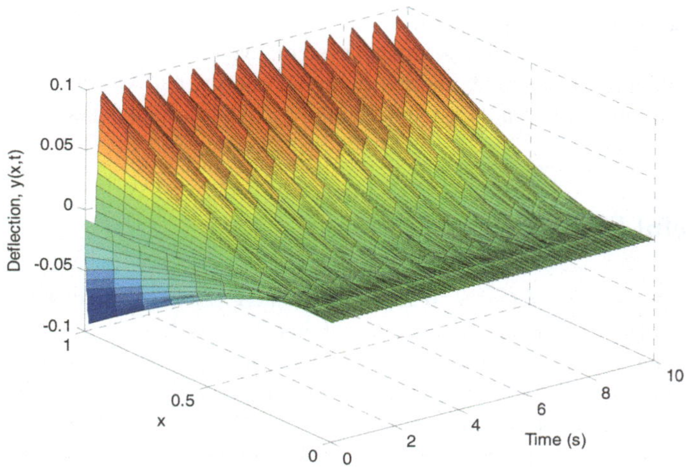

Fig. 2.2 The distributed elastic deflection of the flexible manipulator in case 2

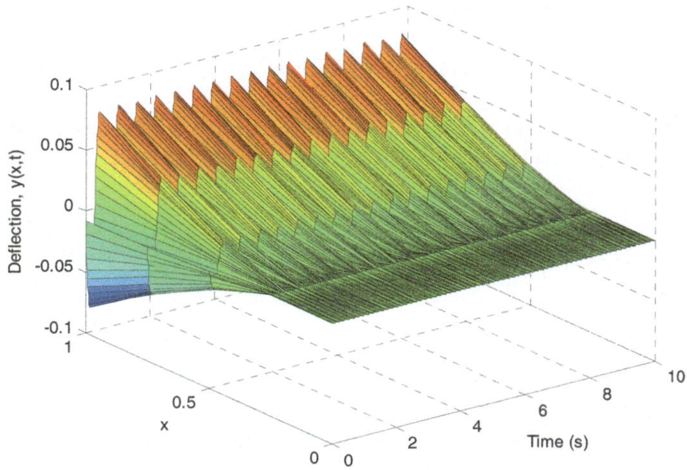

Fig. 2.3 The distributed elastic deflection of the flexible manipulator in case 3

$$\int_{t_1}^{t_2} \delta\left(E_{\mathrm{k}} - E_{\mathrm{p}} + W\right)dt = 0 \tag{2.5}$$

where t_1 and t_2 are two time instants, $t_1 < t < t_2$ is the operating interval, δ denotes the variational operator, E_{k} and E_{p} are the kinetic and potential energies of the system, respectively, and W is the work done by the nonconservative force acting on the system.

It can be seen from Eq. (2.5) that the application of Hamilton principle does not involve the selection of coordinate system, which simplifies the process of modeling. Besides, the kinetic energy, potential energy, and virtual work done by the nonconservative force are easy to obtained by dynamics analysis.

2.3 Useful Definitions and Lemmas

Definition 2.1 Matrix $\boldsymbol{\Gamma}(z) > 0$ represents that $\boldsymbol{\Gamma}(z)$ is positive definite in the defined area of z. The set of singular values of matrix $(*)$ is denoted by $\sigma(*)$. $\sigma_{\max}(*)$ denotes the maximum singular value of matrix $(*)$ and $\sigma_{\min}(*)$ denotes the minimum singular value of matrix $(*)$. $\boldsymbol{I}$ denotes the identity matrix, and $n \times n$-dimensional identity matrices and zero matrices are respectively defined by $\boldsymbol{I}_n$ and $\boldsymbol{0}_n$. Besides, R, R^+, R^n and $R^{m\times n}$ are defined as the sets of all real numbers, positive real numbers, n-dimensional column vectors, and $m \times n$ dimensional matrices, respectively.

Lemma 2.1 [12] *If $\theta(z,t) \in \mathcal{L}^2(\Omega)$ is an absolutely continuous scalar function, where $\theta_z(z,t)$ for $z \in [0,l]$ is square integrable and $\theta(0,t) = 0$, we have*

$$\|\theta(z,t)\|_2^2 \leq 4l^2\pi^{-2}\|\theta_z(z,t)\|_2^2. \tag{2.6}$$

Lemma 2.2 [13] *Define $\nu(z,t)$ as a continuous and first order derivative function. Assume that $\nu(0,t) = 0$ holds for $\forall t \in [0,\infty)$, then we have*

$$\nu^2 \leq l\int_0^l \nu_z^2 dz, \forall z \in [0,l]. \tag{2.7}$$

Definition 2.2 ([14]) A BLF is a scalar function $V(x)$, defined with respect to the system $\dot{x} = f(x)$ on an open region $\mathcal{D}$ containing the origin, that is continuous, positive definite, has continuous first-order partial derivatives at every point of $\mathcal{D}$, has the property $V(x) \to \infty$ as x approaches the boundary of $\mathcal{D}$, and satisfies $V(x) \leq b$ for $\forall t \geq 0$ along the solution of $\dot{x} = f(x)$ for $x(0) \in \mathcal{D}$ and some positive constant b.

A typical symmetric BLF is presented in Fig. 2.4.

Lemma 2.3 ([14]) *For any positive constants k_1, k_2, let $\mathrm{Z}_1 := \{z_1 \in R : -k_1 < z_1 < k_2\} \subset R$ and $\mathcal{N} := R^l \times \mathrm{Z}_1 \subset R^{l+1}$ be open sets. Consider the system $\dot{\eta} = h(t,\eta)$ where $\eta := [w, z_1]^T \in \mathcal{N}$, and $h : R^+ \times \mathcal{N} \to R^{l+1}$ is piecewise continuous in t and locally Lipschitz in z, uniformly in t, on $R^+ \times \mathcal{N}$. Suppose that there exist functions $U : R^l \to R^+$ and $V_1 : \mathrm{Z}_1 \to R^+$, continuously differentiable and positive definite in their respective domains, such that $V_1(z_1) \to \infty$ as $z_1 \to -k_1$ or $z_1 \to k_2$ and $\gamma_1(\|w\|) \leq U(w) \leq \gamma_2(\|w\|)$, where γ_1 and γ_2 are class K_∞ functions. Let $V(\eta) := V_1(z_1) + U(w)$, and $z_1(0)$ belong*

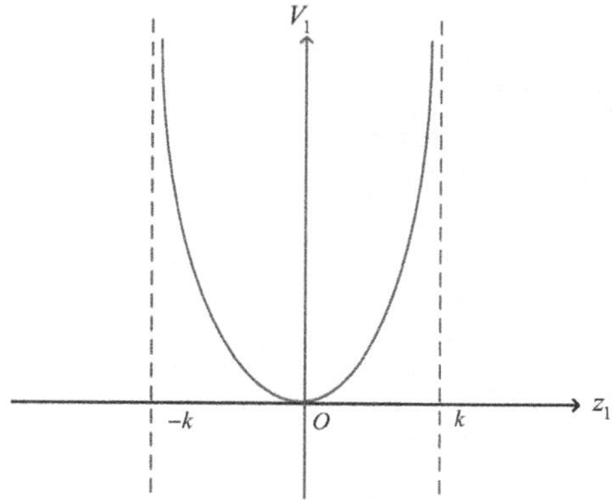

Fig. 2.4 A typical symmetric BLF

to the set $z_1 \in (-k_1, k_2)$. *If* $\dot{V} = \frac{\partial V}{\partial \eta} h \le 0$ *holds, then* z_1 *remains in the open set* $z_1 \in (-k_1, k_2)$ *for* $\forall t \in [0, \infty)$.

Lemma 2.4 [15] *Let* $\psi_1(s, t)$, $\psi_2(s, t) \in R$ *with* $s \in [0, L]$ *and* $t \in [0, \infty)$, *an inequality holds:*

$$\psi_1(s, t)\psi_2(s, t) \le \frac{1}{\gamma}\psi_1^2(s, t) + \gamma\psi_2^2(s, t) \tag{2.8}$$

where $\gamma > 0$.

Lemma 2.5 *Let* $\bar{f}(s, t) \in R$ *be a function whose definition domain is* $s \in [0, L]$ *and* $t \in [0, \infty)$, *and has the property* $\bar{f}(L, t) = 0$ *for* $\forall t \in [0, \infty)$, *then we have:*

$$\bar{f}^2(s, t) \le L \int_0^L \bar{f}_s^2(s, t)\mathrm{d}s, \forall s \in [0, L]. \tag{2.9}$$

Proof An inner product is defined as

$$\left(\bar{f}_1(s, t), \bar{f}_2(s, t)\right) = \int_0^L \bar{f}_1(s, t)\bar{f}_2(s, t)\mathrm{d}s. \tag{2.10}$$

Using Cauchy-Schwarz inequality, we have

$$\left(\int_0^L \bar{f}_1(s, t)\bar{f}_2(s, t)\mathrm{d}s\right)^2 \le \int_0^L \bar{f}_1^2(s, t)\mathrm{d}s \int_0^L \bar{f}_2^2(s, t)\mathrm{d}s. \tag{2.11}$$

Let $\bar{f}_2(s, t) = \bar{f}_s(s, t)$ and

$$\bar{f}_1(s,t)=H(s-h)=\begin{cases}0, & s<h\\ 1, & s\geq h\end{cases}, \quad \text{for } \forall h\in[0,L]. \tag{2.12}$$

Then it can be obtained that

$$\bar{f}^2(h,t)=\left(\bar{f}(L,t)-\bar{f}(h,t)\right)^2\leq(L-h)\int_0^L\bar{f}_s^2(s,t)\mathrm{d}s\leq L\int_0^L\bar{f}_s^2(s,t)\mathrm{d}s. \tag{2.13}$$

Thus we have $\bar{f}^2(s,t)\leq L\int_0^L\bar{f}_s^2(s,t)\mathrm{d}s$, $\forall s\in[0,L]$.

Lemma 2.6 [16] *For any $\vartheta(s,t)$ continuously differentiable on $[L_1,L_2]$, we have following inequalities*

$$\int_{L_1}^{L_2}\vartheta(s,t)\mathrm{d}s\leq 2(L_2-L_1)\vartheta^2(L_2,t)+4(L_2-L_1)^2\int_{L_1}^{L_2}\vartheta_s(s,t)\mathrm{d}s \tag{2.14}$$

$$\int_{L_1}^{L_2}\vartheta(s,t)\mathrm{d}s\leq 2(L_2-L_1)\vartheta^2(L_1,t)+4(L_2-L_1)^2\int_{L_1}^{L_2}\vartheta_s(s,t)\mathrm{d}s. \tag{2.15}$$

Proof Utilizing integration by part, we have

$$2\int_{L_1}^{L_2}(s-L_1)\vartheta(s,t)\vartheta_s(s,t)\mathrm{d}s=(L_2-L_1)\vartheta^2(L_2,t)-\int_{L_1}^{L_2}\vartheta^2(s,t)\mathrm{d}s. \tag{2.16}$$

Then we get

$$\begin{aligned}\int_{L_1}^{L_2}\vartheta^2(s,t)\mathrm{d}s&=(L_2-L_1)\vartheta^2(L_2,t)-2\int_{L_1}^{L_2}(s-L_1)\vartheta(s,t)\vartheta_s(s,t)\mathrm{d}s\\&\leq(L_2-L_1)\vartheta^2(L_2,t)+\frac{1}{2}\int_{L_1}^{L_2}\vartheta^2(s,t)\mathrm{d}s+2(L_2-L_1)^2\int_{L_1}^{L_2}\vartheta_s^2(s,t)\mathrm{d}s.\end{aligned} \tag{2.17}$$

From inequality (2.17), it can be attained

$$\int_{L_1}^{L_2} \vartheta^2(s,t)\mathrm{ds} \le 2(L_2 - L_1)\vartheta^2(L_2,t) + 4(L_2 - L_1)^2 \int_{L_1}^{L_2} \vartheta_s^2(s,t)\mathrm{ds}. \tag{2.18}$$

Inequality (2.15) can be achieved in a similar way.

Definition 2.3 [17] A function $\mathcal{N}(s)$ can be called the Nussbaum function when it satisfies bilateral market characteristics given as

$$\lim_{k\to\pm\infty} \sup \frac{1}{k}\int_0^k \mathcal{N}(s)\mathrm{d}s = \infty \tag{2.19}$$

$$\lim_{k\to\pm\infty} \inf \frac{1}{k}\int_0^k \mathcal{N}(s)\mathrm{d}s = -\infty. \tag{2.20}$$

Lemma 2.7 [18] *If $V(t)$ and $k(\cdot)$ are smooth functions with $V(t) \ge 0$ for $\forall t \in [0, t_f)$, $\mathcal{N}(\cdot)$ is an even smooth Nussbaum-type function, and μ is a nonzero constant, then $V(t)$, $k(\cdot)$ and $\int_0^t (\mathcal{N}(k(\tau)) + 1)\dot{k}(\tau)\mathrm{d}\tau$ are bounded on $\forall t \in [0, t_f)$ if the following inequality holds*

$$V(t) \le \int_0^t (\mu\mathcal{N}(k(\tau)) + 1)\dot{k}(\tau)\mathrm{d}\tau + \upsilon \tag{2.21}$$

where υ is a suitable constant.

Appendix 2.1: Simulation Program: chapter2.m

```
close all;
clear all;

nx=10;
nt=20000;%Case 1

nx=10;
nt=8000;%Case 2

nx=5;
nt=8000;%Case 3

tmax=10;L=1;

dx=L/(nx-1);
T=tmax/(nt-1);
t=linspace(0,nt*T,nt);
x=linspace(0,L,nx);

EI=3;rho=0.2;m=0.1;Ih=0.1;
F_1=0;

y=zeros(nx,nt);

for j=1:nt
      th(j)=0;
      tol(j)=0;
      F(j)=0;
end

for i=1:nx
y(i,1)=0.001*(i-1)*dx;
end

for j=3:nt
if j==10000
    tol(j-1)=0;
end
```

```
y(1,:)=0;
y(2,:)=0;

yxx0=(y(3,j-1)-2*y(2,j-1)+y(1,j-1))/dx^2;
th(j)=0;

dth(j)=(th(j)-th(j-1))/T;
ddth(j)=(th(j)-2*th(j-1)+th(j-2))/T^2;

for i=3:nx-2
    yxxxx=(y(i+2,j-1)-4*y(i+1,j-1)+6*y(i,j-1)-4*y(i-1,j-1)+y(i-2,j-1))/dx^4;
    y(i,j)=T^2*(-i*dx*ddth(j)-(EI*yxxxx)/rho)+2*y(i,j-1)-y(i,j-2);
    dy(i,j-1)=(y(i,j-1)-y(i,j-2))/T;
end

yxxxx(nx-1,j-1)=(-2*y(nx,j-1)+5*y(nx-1,j-1)-4*y(nx-2,j-1)+y(nx-3,j-1))/dx^4;
y(nx-1,j)=T^2*(-(nx-1)*dx*ddth(j)-EI*yxxxx(nx-1,j-1)/rho)+2*y(nx-1,j-1)-y(nx-1,j-2);
dy(nx-1,j)=(y(nx-1,j)-y(nx-1,j-1))/T;

yxxxL(j-1)=(-y(nx,j-1)+2*y(nx-1,j-1)-y(nx-2,j-1))/dx^3;
y(nx,j)=T^2*(-L*ddth(j-1)+(EI*yxxxL(j-1)+F_1)/m)+2*y(nx,j-1)-y(nx,j-2);

dy(nx,j)=(y(nx,j)-y(nx,j-1))/T;

F(j-1)=0;
F_1=F(j-1);
end

tshort=linspace(0,tmax,nt/10);
yshort=zeros(nx,nt/10);
dyshort=zeros(nx,nt/10);
for j=1:nt/10
        for i=1:nx
              yshort(i,j)=y(i,j*10);
              dyshort(i,j)=dy(i,j*10);
        end
end

figure(1);
surf(tshort,x,yshort);
xlabel('Time (s)'); ylabel('x');zlabel('Deflection, y(x,t)');
```

References

1. N.S. Abhyankar, E.K. Hall, S.V. Hanagud, Chaotic vibrations of beams: numerical solution of partial differential equations. J. Appl. Mech. **60**(1), 167–174 (1993)
2. W. He, C. Yang, T. Meng, C. Sun, Distributed control of a class of flexible mechanical systems with global constraint. Int. J. Control **89**(1), 128–139 (2016)
3. Z. Zhao, X. He, Z. Ren, G. Wen, Boundary adaptive robust control of a flexible riser system with input nonlinearities. IEEE Trans. Syst. Man Cybern. Syst. (2018). https://doi.org/10.1109/tsmc.2018.2882734
4. S. Zhang, W. He, D. Huang, Active vibration control for a flexible string system with input backlash. IET Control Theory Appl. **10**(7), 800–805 (2016)
5. Z. Liu, J. Liu, W. He, Robust adaptive fault tolerant control for a linear cascaded ODE-beam systems. Automatica **98**, 42–50 (2018)
6. Q.C. Nguyen, K.-S. Hong, Asymptotic stabilization of a nonlinear axially moving string by adaptive boundary control. J. Sound Vib. **329**(22), 4588–4603 (2010)
7. N.S. Abhyankar, E.K. Hall, S.V. Hanagud, Chaotic vibrations of beams: numerical solution of partial differential equations. J. Appl. Mech. **60**, 167–174 (1993)
8. J. Liu, W. He, *Distributed Parameter Modeling and Boundary Control of Flexible Manipulators* (Springer Singapore, Singapore, 2018)
9. Z. Liu, J. Liu, Boundary control of a flexible robotic manipulator with output constraints. Asian J. Control **19**(1), 332–345 (2017)
10. H. Goldstein, *Classical Mechanics* (Addison-Wesley, Massachusetts, 1951)
11. L. Meirovitch, *Analytical Methods in Vibration* (Macmillan, New York, 1967)
12. H.N. Wu, J.W. Wang, Static output feedback control via PDE boundary and ODE measurements in linear cascaded ODE–beam systems. Automatica **50**(11), 2787–2798 (2014)
13. W. He, S.S. Ge, B.V.E. How, Y.S. Choo, *Dynamics and Control of Mechanical Systems in Offshore Engineering* (Springer, London, 2014)
14. K.P. Tee, S.S. Ge, E.H. Tay, Barrier Lyapunov functions for the control of output-constrained nonlinear systems. Automatica **45**, 918–927 (2009)
15. C.D. Rahn, *Mechatronic Control of Distributed Noise and Vibration: A Lyapunov Approach* (Heidelberg, New York, Berlin, 2001)
16. W. He, S.S. Ge, Dynamic modeling and vibration control of a flexible satellite. IEEE Trans. Aerosp. Electron. Syst. **51**(2), 1422–1431 (2015)
17. R.D. Nussbaum, Some remark on the conjecture in parameter adaptive control. Syst. Control Lett. **3**(4), 243–246 (1983)
18. C. Wen, J. Zhou, Z. Liu, H. Su, Robust adaptive control of uncertain nonlinear systems in the presence of input saturation and external disturbance. IEEE Trans. Autom. Control **56**(7), 1672–1678 (2011)

Chapter 3
LMI-Based Robust Adaptive Neural Network Control for Euler-Bernoulli Beam with Unknown Parametric and Modeling Uncertainties

3.1 Introduction

During the past decades, due to the extensive existence of elastic beam structures in practical projects, the control of Euler-Bernoulli beams has received a great deal of attention. However, the flexible system is very sensitive to uncertainties due to its structure and thus even small modeling uncertainties may also induce the vibration. The excessive vibration always impacts the performance of control systems, so the vibration attenuation is a tough task in the control area.

The Euler-Bernoulli beam with uncertain parameters and modeling uncertainties is chosen as the control object in this chapter whose elastic vibration is expected to be eliminated by the proposed robust adaptive neural network control strategy. In view of the deficiency in the application of LMI approach to distributed parameter systems, the LMI-based boundary control strategy is developed with reserving all the system modal information, which complicates the analysis. Since the control design conditions are given in the form of a set of LMIs in this chapter, the feasibility of the designed controller is easy to be demonstrated. Besides, feasible regions of control gains can be systematically and efficiently achieved by involving LMIs.

Notation (Applicable for this chapter, Chaps. 4 and 5) Define $\mathcal{L}^2(\Omega) = \left\{ f(z) \middle| \sqrt{\int_0^L f^2(z)\mathrm{d}z} < \infty, z \in \Omega \right\}$, $\Omega = (0, L)$, and the space $H_0^k(\Omega) = \left\{ f \middle| f, f', \ldots, f^{(k)} \in \mathcal{L}^2(\Omega), f(0) = f'(0) = 0 \right\}$. For clarity, the notations are introduced throughout as $(*)_z = \frac{\partial(*)}{\partial z}$, $(*)_{zz} = \frac{\partial^2(*)}{\partial z^2}$, $(*)_{zzz} = \frac{\partial^3(*)}{\partial z^3}$, $(*)_{zzzz} = \frac{\partial^4(*)}{\partial z^4}$, $(*)_t = \frac{\partial(*)}{\partial t}$, $(*)_{tt} = \frac{\partial^2(*)}{\partial t^2}$.

X. Xing and J. Liu, *Vibration Control Methods of Mechanical Distributed Parameter Systems*, Springer Tracts in Mechanical Engineering,
https://doi.org/10.1007/978-981-16-1532-0_3

3.2 Problem Formulation

The configuration of the Euler-Bernoulli beam is illustrated in Fig. 3.1.

The physical meanings of the system are described as follows. L is the length of the beam; ρ is beam's uniform density; EI is beam's flexural rigidity; $w(z,t)$ is the vibratory deformation at z for time t; $\boldsymbol{\omega}$ is an independent variable defined as $\boldsymbol{\omega} = \left[\, w(L,t)\ w_t(L,t)\, \right]^{\mathrm{T}}$, which is also regarded as the input signal of the designed RBF neural network; $d(\boldsymbol{\omega})$ is the state-dependent boundary modeling uncertainty acting at the tip of the beam, which is caused by external environmental changes; $u(t)$ is the boundary control force.

The governing equation of the beam can be written as [1]

$$\rho w_{tt}(z,t) + EI w_{zzzz}(z,t) = 0 \tag{3.1}$$

for $z \in (0, L)$, and the boundary conditions can be written as

$$EI w_{zzz}(L,t) + u(t) + d(\boldsymbol{\omega}) = 0 \tag{3.2}$$

$$w_{zz}(L,t) = 0 \tag{3.3}$$

$$w(0,t) = w_z(0,t) = 0. \tag{3.4}$$

Remark 3.1 The beam structure discussed in this chapter is common in both theory and engineering. For instance, control systems (flexible manipulator, flexible marine installation system, flexible spacecraft) mentioned by [2–4] are all extensions of the model described by Eqs. (3.1)–(3.4). The boundary controller acts on the end of the beam which can be implemented by a micro-trailing edge effector [5].

Assumption 3.1 All the parameters of the Euler-Bernoulli beam are uncertain and bounded. The ranges of the system parameters are given as: $0 < \bar{\rho}_{\min} \le \rho \le \bar{\rho}_{\max}$ and $0 < \overline{EI}_{\min} \le EI \le \overline{EI}_{\max}$, where $\bar{\rho}_{\min}$, $\bar{\rho}_{\max}$, $\overline{EI}_{\min}$ and $\overline{EI}_{\max}$ are all positive constants.

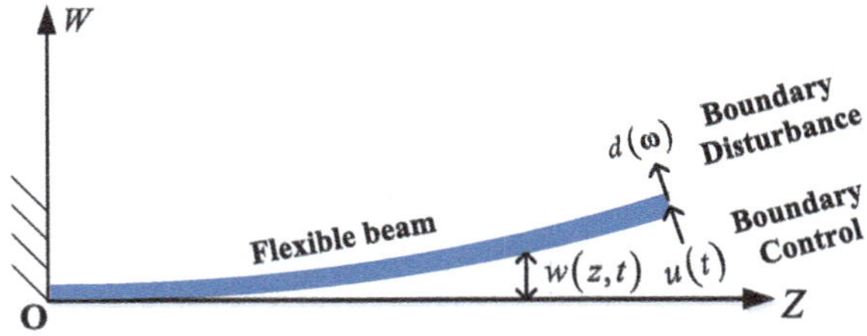

Fig. 3.1 Configuration of the Euler-Bernoulli beam

Assumption 3.2 Due to the finite energy, the uncertainty $d(\boldsymbol{\omega})$ in Eq. (3.2) is bounded.

3.3 Control Design and Stability Analysis

The control scheme aims at designing the LMI-based robust neural network control to restrict the elastic vibration of the beam with unknown parameters and modeling uncertainty affecting the system.

3.3.1 RBF Neural Network Design

The RBF neural network is proposed in this part to approximate the uncertainty and make preparations for the control design.

We first design

$$d(\boldsymbol{\omega}) = \boldsymbol{Q}^{*\mathrm{T}}\boldsymbol{h}(\boldsymbol{\omega}) + \varepsilon \tag{3.5}$$

where $\boldsymbol{Q}^* \in \Re^n$ is the ideal weight vector; n is the network node number of the hidden layer and $\boldsymbol{h}(\boldsymbol{\omega}) = \begin{bmatrix} h_1 \ h_2 \cdots h_n \end{bmatrix}^{\mathrm{T}} \in R^n$, $i = 1, 2, \ldots n$, is a vector of basis function; ε is the bounded approximation error meeting $|\varepsilon| \leq \mu$ where $\mu > 0$.

Then the RBF neural network strategy can be expressed as

$$h_i = \exp\left(\frac{\|\omega - \boldsymbol{f}_i\|^2}{2g_i^2}\right) \tag{3.6}$$

in which $\boldsymbol{f}_i \in R^2, g_i \in R, j = 1, 2, \cdots n$, are the parameters of the Gaussian function, and h_i is the output signal of the Gaussian function.

The adaptive law is then proposed as

$$\dot{\hat{\boldsymbol{Q}}} = \gamma \boldsymbol{x}^{\mathrm{T}}(t)\tilde{\boldsymbol{p}}\boldsymbol{h}(\boldsymbol{\omega}) - \eta\gamma\hat{\boldsymbol{Q}} \tag{3.7}$$

$$\hat{d}(\boldsymbol{\omega}) = \hat{\boldsymbol{Q}}^{\mathrm{T}}\boldsymbol{h}(\boldsymbol{\omega}) \tag{3.8}$$

where $\hat{\boldsymbol{Q}} \in R^n$ is the time related vector, $\boldsymbol{x}(t) = \begin{bmatrix} w_t(L,t) \ w_z(L,t) \end{bmatrix}^{\mathrm{T}}$, $\tilde{\boldsymbol{p}} = \begin{bmatrix} p_{11} \ p_{12}L \end{bmatrix}^{\mathrm{T}}$, and γ and η are positive constants.

Remark 3.3 The first term of Eq. (3.7) is constructed to cancel the undesirable term $2\boldsymbol{x}^{\mathrm{T}}(t)\tilde{\boldsymbol{p}}\tilde{\boldsymbol{Q}}^{\mathrm{T}}\boldsymbol{h}(\boldsymbol{\omega})$ occurring in $\dot{V}_1(t)$ in inequality (3.33) given later and the second term of Eq. (3.7) is designed to adjust the control effect.

Define $\tilde{\boldsymbol{Q}} = \boldsymbol{Q}^* - \hat{\boldsymbol{Q}}$ and $\tilde{d}(\boldsymbol{\omega}) = d(\boldsymbol{\omega}) - \hat{d}(\boldsymbol{\omega})$. Thus we have

$$\tilde{d}(\boldsymbol{\omega}) = \boldsymbol{Q}^{*\mathrm{T}}\boldsymbol{h}(\boldsymbol{\omega}) + \varepsilon - \hat{\boldsymbol{Q}}^{\mathrm{T}}\boldsymbol{h}(\boldsymbol{\omega}) = \tilde{\boldsymbol{Q}}^{\mathrm{T}}\boldsymbol{h}(\boldsymbol{\omega}) + \varepsilon. \tag{3.9}$$

The above designed RBF network algorithm lays the foundation for the LMI-based control design in following subsection.

3.3.2 Controller Design and Stability Analysis

To realize the control objective, the boundary control $u(t)$ is designed as

$$u(t) = \boldsymbol{K}\boldsymbol{x}(t) - \hat{d}(\boldsymbol{\omega}) - \mu \operatorname{sgn}\left(\boldsymbol{x}^{\mathrm{T}}(t)\tilde{\boldsymbol{p}}\right) \tag{3.10}$$

where $\boldsymbol{K} = \begin{bmatrix} K_1 & K_2 \end{bmatrix} \in R^{1\times 2}$.

Remark 3.4 Notice that states $w_z(L,t)$, $w(L,t)$, and $w_t(L,t)$ are required to be measured in Eqs. (3.7), (3.8), and (3.10), and can be achieved by inclinometer, laser displacement sensor, and backward difference algorithm, respectively [6].

Theorem 3.1 *Define*

$$\boldsymbol{\Lambda}(z) = \begin{bmatrix} p_{11}\bar{\rho}_{\min} & p_{12}\bar{\rho}_{\max}z \\ p_{12}\bar{\rho}_{\max}z & p_{22} \end{bmatrix} \tag{3.11}$$

$$\boldsymbol{\Theta} = \begin{bmatrix} p_{12}L\bar{\rho}_{\max} + 2K_1p_{11} & p_{22} + K_2p_{11} + K_1p_{12}L & \boldsymbol{0}_n^{\mathrm{T}} \\ p_{22} + K_2p_{11} + K_1p_{12}L & 2K_2p_{12}L & \boldsymbol{0}_n^{\mathrm{T}} \\ \boldsymbol{0}_n & \boldsymbol{0}_n & -\eta\boldsymbol{I}_n \end{bmatrix} \tag{3.12}$$

$$\boldsymbol{\Xi}(z) = \begin{bmatrix} -p_{12}\bar{\rho}_{\min} & -p_{22} & 0 \\ -p_{22} & -\frac{3}{2}p_{12}\overline{EI}_{\min} & 0 \\ 0 & 0 & -\frac{3}{8}\pi^2L^{-2}p_{12}\overline{EI}_{\min} \end{bmatrix}. \tag{3.13}$$

If there exist $p_{11}, p_{12}, p_{22} \in R^+$ *satisfying following LMIs*

$$\boldsymbol{\Lambda}(z) > 0 \tag{3.14}$$

$$\boldsymbol{\Theta} < 0 \tag{3.15}$$

$$\boldsymbol{\Xi}(z) < 0 \tag{3.16}$$

then the boundary control law proposed by Eq. (3.10) *can ensure that the state of the closed-loop system is uniformly bounded and exponentially converges to a small region of the origin* Ψ*, which is defined by* $\Psi := \left\{ w(z,t) \in \Re \left| \lim_{t\to\infty} |w(z,t)| \le D, z \in [0, L] \right. \right\}$ *for* $t \to \infty$*, where* $D = \sqrt{\frac{4L^3\kappa}{\pi^2 p_1}}$*, in which* $\kappa = \frac{\eta}{\lambda}\boldsymbol{Q}^{*\mathrm{T}}\boldsymbol{Q}^* > 0$, $\lambda = p_3 p_2^{-1} > 0$, $p_2 = \max\left\{ \max_{z\in[0,L]} \{\sigma(\boldsymbol{P}(z))\}, p_{11}EI, \frac{1}{\gamma} \right\} > 0$, $\boldsymbol{P}(z) = \begin{bmatrix} p_{11}\rho & p_{12}\rho z \\ p_{12}\rho z & p_{22} \end{bmatrix}$*, and* $p_3 = \min\left\{ \min_{z\in[0,L]} \{\sigma(-\boldsymbol{\Xi}(z))\}, \min_{z\in[0,L]} \{\sigma(-\boldsymbol{\Theta})\} \right\} > 0$.

Proof The Lyapunov candidate function is designed to be

$$V(t) = V_1(t) + V_2(t) + V_3(t) \tag{3.17}$$

where

$$V_1(t) = \int_0^L \boldsymbol{\xi}^{\mathrm{T}}(z,t)\boldsymbol{P}(z)\boldsymbol{\xi}(z,t)\mathrm{d}z \tag{3.18}$$

$$V_2(t) = \int_0^L p_{11}EIw_{zz}^2(z,t)\mathrm{d}z \tag{3.19}$$

$$V_3(t) = \frac{1}{\gamma}\tilde{\boldsymbol{Q}}^{\mathrm{T}}\tilde{\boldsymbol{Q}} \tag{3.20}$$

where $\boldsymbol{\xi}(z,t) = \left[w_t(z,t) \; w_z(z,t) \right]^{\mathrm{T}}$ and $p_{11}, p_{12}, p_{22} \in R^+$.

Differentiating $V_1(t)$, $V_2(t)$ and $V_3(t)$ in Eqs. (3.18)–(3.20) with respect to time respectively leads to

$$\begin{aligned}\dot{V}_1(t) = {} & 2\int_0^L p_{11}\rho w_t(z,t)w_{tt}(z,t)\mathrm{d}z + 2\int_0^L p_{22}w_z(z,t)w_{zt}(z,t)\mathrm{d}z \\ & + 2\int_0^L p_{12}\rho z w_t(z,t)w_{zt}(z,t)\mathrm{d}z + 2\int_0^L p_{12}\rho z w_z(z,t)w_{tt}(z,t)\mathrm{d}z\end{aligned} \tag{3.21}$$

$$\dot{V}_2(t) = 2\int_0^L p_{11}EIw_{zz}(z,t)w_{zzt}(z,t)\mathrm{d}z \tag{3.22}$$

$$\dot{V}_3(t) = -2\tilde{\boldsymbol{Q}}^{\mathrm{T}}\boldsymbol{x}^{\mathrm{T}}(t)\tilde{\boldsymbol{p}}\boldsymbol{h}(\boldsymbol{\omega}) + 2\eta\tilde{\boldsymbol{Q}}^{\mathrm{T}}\hat{\boldsymbol{Q}}. \tag{3.23}$$

Following steps aim at calculating items $\int_0^L \rho w_t(z,t)w_{tt}(z,t)\mathrm{d}z$, $\int_0^L w_z(z,t)w_{zt}(z,t)\mathrm{d}z$, $\int_0^L \rho z w_t(z,t)w_{zt}(z,t)\mathrm{d}z$, and $\int_0^L \rho z w_z(z,t)w_{tt}(z,t)\mathrm{d}z$ in

Eq. (3.21) to achieve $\dot{V}_1(t)$ for further obtaining $\dot{V}(t)$ by arrangement in form of LMIs.

First, substituting Eqs. (3.1)–(3.4) into the first term of Eq. (3.21) and using integration by parts yield

$$\begin{aligned}
&\int_0^L \rho w_t(z,t)w_{tt}(z,t)\mathrm{d}z\\
&= -\int_0^L w_t(z,t)EIw_{zzzz}(z,t)\mathrm{d}z\\
&= -EIw_t(z,t)w_{zzz}(z,t)\big|_0^L + EI\int_0^L w_{zt}(z,t)w_{zzz}(z,t)\mathrm{d}z\\
&= -EIw_t(z,t)w_{zzz}(z,t)\big|_0^L + EIw_{zt}(z,t)w_{zz}(z,t)\big|_0^L - \int_0^L EIw_{zzt}(z,t)w_{zz}(z,t)\mathrm{d}z\\
&= -EIw_t(L,t)w_{zzz}(L,t) - \int_0^L EIw_{zzt}(z,t)w_{zz}(z,t)\mathrm{d}z\\
&= w_t(L,t)\Big(\boldsymbol{K}\boldsymbol{x}(t) - \mu\mathrm{sgn}\big(\boldsymbol{x}^{\mathrm{T}}(t)\tilde{\boldsymbol{p}}\big) + \tilde{d}(\boldsymbol{\omega})\Big) - \int_0^L EIw_{zzt}(z,t)w_{zz}(z,t)\mathrm{d}z. \qquad (3.24)
\end{aligned}$$

Similarly, through integration by parts, the second and third terms of Eq. (3.21) can be achieved as

$$\int_0^L w_z(z,t)w_{zt}(z,t)\mathrm{d}z = w_t(L,t)w_z(L,t) - \int_0^L w_t(z,t)w_{zz}(z,t)\mathrm{d}z \qquad (3.25)$$

and

$$\int_0^L \rho z w_t(z,t)w_{zt}(z,t)\mathrm{d}z = \frac{1}{2}\rho L w_t^2(L,t) - \frac{1}{2}\int_0^L \rho w_t^2(z,t)\mathrm{d}z. \qquad (3.26)$$

Next, substituting Eq. (3.1) into the last term of Eq. (3.21) and applying twice integration by parts, the following equation is deduced using Eq. (3.10)

$$
\begin{aligned}
&\int_0^L \rho z w_z(z,t) w_{tt}(z,t)\mathrm{d}z \\
&= -\int_0^L EIz w_z(z,t) w_{zzzz}(z,t)\mathrm{d}z \\
&= -EIz w_z(z,t) w_{zzz}(z,t)\big|_0^L + \int_0^L EIz w_{zz}(z,t) w_{zzz}(z,t)\mathrm{d}z + \int_0^L EI w_z(z,t) w_{zzz}(z,t)\mathrm{d}z \\
&= -EIL w_z(L,t) w_{zzz}(L,t) + \int_0^L EIz w_{zz}(z,t) w_{zzz}(z,t)\mathrm{d}z \\
&\quad + EI w_z(z,t) w_{zz}(z,t)\big|_0^L - \int_0^L EI w_{zz}^2(z,t)\mathrm{d}z \\
&= L w_z(L,t)\Big(\boldsymbol{K}\boldsymbol{x}(t) - \mu \mathrm{sgn}\big(\boldsymbol{x}^{\mathrm{T}}(t)\tilde{\boldsymbol{p}}\big) + \tilde{d}(\boldsymbol{\omega})\Big) \\
&\quad - \int_0^L EI w_{zz}^2(z,t)\mathrm{d}z + \int_0^L EIz w_{zz}(z,t) w_{zzz}(z,t)\mathrm{d}z.
\end{aligned} \tag{3.27}
$$

Due to

$$
\begin{aligned}
&\int_0^L EIz w_{zz}(z,t) w_{zzz}(z,t)\mathrm{d}z \\
&= EIz w_{zz}^2(z,t)\big|_0^L - \int_0^L EIz w_{zzz}(z,t) w_{zz}(z,t)\mathrm{d}z - \int_0^L EI w_{zz}^2(z,t)\mathrm{d}z \\
&= -\int_0^L EI w_{zz}^2(z,t)\mathrm{d}z - \int_0^L EIz w_{zzz}(z,t) w_{zz}(z,t)\mathrm{d}z,
\end{aligned} \tag{3.28}
$$

one gets

$$
\int_0^L EIz w_{zz}(z,t) w_{zzz}(z,t)\mathrm{d}z = -\frac{1}{2}\int_0^L EI w_{zz}^2(z,t)\mathrm{d}z. \tag{3.29}
$$

Substituting Eq. (3.29) into Eq. (3.27), Eq. (3.27) can be rewritten as

$$\int_0^L \rho z w_z(z,t) w_{tt}(z,t)\mathrm{d}z$$
$$= L w_z(L,t)\left(\boldsymbol{K}\boldsymbol{x}(t) - \mu \operatorname{sgn}\left(\boldsymbol{x}^{\mathrm{T}}(t)\tilde{\boldsymbol{p}}\right) + \tilde{d}(\omega)\right) - \frac{3}{2}\int_0^L EI w_{zz}^2(z,t)\mathrm{d}z. \tag{3.30}$$

Then substituting four terms in Eqs. (3.30) and (3.24)–(3.26) into Eq. (3.21), the time derivative of $V_1(t)$ can be received as

$$\begin{aligned}\dot{V}_1(t) = &-2p_{11}\int_0^L EI w_{zzt}(z,t) w_{zz}(z,t)\mathrm{d}z - 2p_{22}\int_0^L w_t(z,t) w_{zz}(z,t)\mathrm{d}z \\ &- 3p_{12}\int_0^L EI w_{zz}^2(z,t)\mathrm{d}z - p_{12}\int_0^L \rho w_t^2(z,t)\mathrm{d}z + p_{12}\rho L w_t^2(L,t) \\ &+ 2p_{11} w_t(L,t)\left(\boldsymbol{K}\boldsymbol{x}(t) - \mu \operatorname{sgn}\left(\boldsymbol{x}^{\mathrm{T}}(t)\tilde{\boldsymbol{p}}\right) + \tilde{d}(\omega)\right) \\ &+ 2p_{12} L w_z(L,t)\left(\boldsymbol{K}\boldsymbol{x}(t) - \mu \operatorname{sgn}\left(\boldsymbol{x}^{\mathrm{T}}(t)\tilde{\boldsymbol{p}}\right) + \tilde{d}(\omega)\right) \\ &+ 2p_{22} w_t(L,t) w_z(L,t).\end{aligned} \tag{3.31}$$

For the convenience of constructing LMIs, $\dot{V}_1(t)$ is restated using Lemma 2.1 and Eq. (3.31) as

$$\begin{aligned}\dot{V}_1(t) \le &-2p_{11}\int_0^L EI w_{zzt}(z,t) w_{zz}(z,t)\mathrm{d}z - 2p_{22}\int_0^L w_t(z,t) w_{zz}(z,t)\mathrm{d}z \\ &- \frac{3}{2}p_{12}\int_0^L EI w_{zz}^2(z,t)\mathrm{d}z - \frac{3}{8}L^{-2}\pi^2 p_{12}\int_0^L EI w_z^2(z,t)\mathrm{d}z - p_{12}\int_0^L \rho w_t^2(z,t)\mathrm{d}z \\ &+ 2(p_{11} w_t(L,t) + p_{12} L w_z(L,t))\left(\boldsymbol{K}\boldsymbol{x}(t) - \mu \operatorname{sgn}\left(\boldsymbol{x}^{\mathrm{T}}(t)\tilde{\boldsymbol{p}}\right) + \tilde{d}(\omega)\right) \\ &+ 2p_{22} w_t(L,t) w_z(L,t) + p_{12}\rho L w_t^2(L,t) \\ \le &-2p_{11}\int_0^L EI w_{zzt}(z,t) w_{zz}(z,t)\mathrm{d}z - 2p_{22}\int_0^L w_t(z,t) w_{zz}(z,t)\mathrm{d}z \\ &- \frac{3}{2}p_{12}\int_0^L \bar{E}\bar{I}_{\min} w_{zz}^2(z,t)\mathrm{d}z - \frac{3}{8}L^{-2}\pi^2 p_{12}\int_0^L \bar{E}\bar{I}_{\min} w_z^2(z,t)\mathrm{d}z \\ &- p_{12}\int_0^L \bar{\rho}_{\min} w_t^2(z,t)\mathrm{d}z + 2\boldsymbol{x}^{\mathrm{T}}(t)\tilde{\boldsymbol{p}}\left(\boldsymbol{K}\boldsymbol{x}(t) - \mu \operatorname{sgn}\left(\boldsymbol{x}^{\mathrm{T}}(t)\tilde{\boldsymbol{p}}\right) + \tilde{d}(\omega)\right) \\ &+ 2p_{22} w_t(L,t) w_z(L,t) + p_{12}\bar{\rho}_{\max} L w_t^2(L,t).\end{aligned} \tag{3.32}$$

By observing inequality (3.32), it is obvious that matrixes $\boldsymbol{\Xi}(z)$ and $\boldsymbol{\Theta}_2$ can be used to express $\dot{V}_1(t)$ as

$$
\begin{aligned}
\dot{V}_1(t) \leq & \int_0^L \boldsymbol{\zeta}^{\mathrm{T}}(z,t)\boldsymbol{\Xi}(z)\boldsymbol{\zeta}(z,t)\mathrm{d}z + \boldsymbol{x}^{\mathrm{T}}(t)\boldsymbol{\Theta}_1\boldsymbol{x}(t) - 2p_{11}\int_0^L EIw_{zzt}(z,t)w_{zz}(z,t)\mathrm{d}z \\
& + 2\boldsymbol{x}^{\mathrm{T}}(t)\tilde{\boldsymbol{p}}\left(\boldsymbol{K}\boldsymbol{x}(t) - \mu\operatorname{sgn}\left(\boldsymbol{x}^{\mathrm{T}}(t)\tilde{\boldsymbol{p}}\right) + \tilde{d}(\boldsymbol{\omega})\right) \\
= & \int_0^L \boldsymbol{\zeta}^{\mathrm{T}}(z,t)\boldsymbol{\Xi}(z)\boldsymbol{\zeta}(z,t)\mathrm{d}z + \boldsymbol{x}^{\mathrm{T}}(t)\boldsymbol{\Theta}_1\boldsymbol{x}(t) + \boldsymbol{x}^{\mathrm{T}}(t)\left((\tilde{\boldsymbol{p}}\boldsymbol{K}) + (\tilde{\boldsymbol{p}}\boldsymbol{K})^{\mathrm{T}}\right)\boldsymbol{x}(t) \\
& + 2\boldsymbol{x}^{\mathrm{T}}(t)\tilde{\boldsymbol{p}}\left(\tilde{d}(\boldsymbol{\omega}) - \mu\operatorname{sgn}\left(\boldsymbol{x}^{\mathrm{T}}(t)\tilde{\boldsymbol{p}}\right)\right) - 2p_{11}\int_0^L EIw_{zzt}(z,t)w_{zz}(z,t)\mathrm{d}z \\
\leq & \int_0^L \boldsymbol{\zeta}^{\mathrm{T}}(z,t)\boldsymbol{\Xi}(z)\boldsymbol{\zeta}(z,t)\mathrm{d}z + \boldsymbol{x}^{\mathrm{T}}(t)\boldsymbol{\Theta}_2\boldsymbol{x}(t) + 2\boldsymbol{x}^{\mathrm{T}}(t)\tilde{\boldsymbol{p}}\tilde{\boldsymbol{Q}}^{\mathrm{T}}\boldsymbol{h}(\boldsymbol{\omega}) \\
& - 2p_{11}\int_0^L EIw_{zzt}(z,t)w_{zz}(z,t)\mathrm{d}z
\end{aligned}
\tag{3.33}
$$

where

$$
\boldsymbol{\Theta}_1 = \begin{bmatrix} p_{12}L\bar{\rho}_{\max} & p_{22} \\ p_{22} & 0 \end{bmatrix} \tag{3.34}
$$

$$
\boldsymbol{\Theta}_2 = \begin{bmatrix} p_{12}L\bar{\rho}_{\max} + 2K_1p_{11} & p_{22} + K_2p_{11} + K_1p_{12}L \\ p_{22} + K_2p_{11} + K_1p_{12}L & 2K_2p_{12}L \end{bmatrix} \tag{3.35}
$$

$$
\varsigma(z,t) = [w_t(z,t)\ w_{zz}(z,t)\ w_z(z,t)]^{\mathrm{T}}. \tag{3.36}
$$

Then it is obvious that the last two terms in Eq. (3.33) can be canceled out by $\dot{V}_3(t)$ in Eq. (3.23) and $\dot{V}_2(t)$ in Eq. (3.22), respectively. Consequently, we combine Eqs. (3.22), (3.23) and inequality (3.33) to express $\dot{V}(t)$ in the following form of LMIs

$$
\begin{aligned}
\dot{V}(t) \leq & \int_0^L \boldsymbol{\zeta}^{\mathrm{T}}(z,t)\boldsymbol{\Xi}(z)\boldsymbol{\zeta}(z,t)\mathrm{d}z + \boldsymbol{x}^{\mathrm{T}}(t)\boldsymbol{\Theta}_2\boldsymbol{x}(t) + 2\eta\tilde{\boldsymbol{Q}}^{\mathrm{T}}\left(\boldsymbol{Q}^* - \tilde{\boldsymbol{Q}}\right) \\
\leq & \int_0^L \boldsymbol{\zeta}^{\mathrm{T}}(z,t)\boldsymbol{\Xi}(z)\boldsymbol{\zeta}(z,t)\mathrm{d}z + \boldsymbol{x}^{\mathrm{T}}(t)\boldsymbol{\Theta}_2\boldsymbol{x}(t) - \eta\tilde{\boldsymbol{Q}}^{\mathrm{T}}\tilde{\boldsymbol{Q}} + \eta\boldsymbol{Q}^{*\mathrm{T}}\boldsymbol{Q}^* \\
= & \int_0^L \boldsymbol{\zeta}^{\mathrm{T}}(z,t)\boldsymbol{\Xi}(z)\boldsymbol{\zeta}(z,t)\mathrm{d}z + \bar{\boldsymbol{x}}^{\mathrm{T}}(t)\boldsymbol{\Theta}\bar{\boldsymbol{x}}(t) + \eta\boldsymbol{Q}^{*\mathrm{T}}\boldsymbol{Q}^*
\end{aligned}
\tag{3.37}
$$

where $\bar{\boldsymbol{x}}(t) = \left[\boldsymbol{x}^{\mathrm{T}}(t)\,\tilde{\boldsymbol{Q}}^{\mathrm{T}}\right]$ and $\boldsymbol{\Theta} = \begin{bmatrix} \boldsymbol{\Theta}_2 & 0 \\ 0 & -\eta \boldsymbol{I}_n \end{bmatrix}$.

If LMI (3.14) is satisfied, considering the definition of $\boldsymbol{P}(z)$, we can obtain

$$\boldsymbol{P}(z) > 0 \tag{3.38}$$

and thus

$$V(t) \geq p_1 \int_0^L \left(\boldsymbol{\xi}^{\mathrm{T}}(z,t)\boldsymbol{\xi}(z,t) + w_{zz}^2(z,t)\right)\mathrm{d}z > 0 \tag{3.39}$$

$$V(t) \leq p_2\left(\int_0^L \left(\boldsymbol{\xi}^{\mathrm{T}}(z,t)\boldsymbol{\xi}(z,t) + w_{zz}^2(z,t)\right)\mathrm{d}z + \bar{\boldsymbol{x}}^{\mathrm{T}}(t)\bar{\boldsymbol{x}}(t)\right) \tag{3.40}$$

in which $p_1 = \min\left\{\min\limits_{z\in[0,L]}\{\sigma(\boldsymbol{P}(z))\}, p_{11}EI\right\} > 0$.

If LMIs (3.15) and (3.16) are satisfied, applying inequalities (3.37) and (3.40) yields

$$\begin{aligned}
\dot{V}(t) &\leq -p_3\left(\int_0^L \boldsymbol{\zeta}^{\mathrm{T}}(z,t)\boldsymbol{\zeta}(z,t)\mathrm{d}z + \bar{\boldsymbol{x}}^{\mathrm{T}}(t)\bar{\boldsymbol{x}}(t)\right) + \eta \boldsymbol{Q}^{*\mathrm{T}}\boldsymbol{Q}^* \\
&= -p_3\bar{\boldsymbol{x}}^{\mathrm{T}}(t)\bar{\boldsymbol{x}}(t) - p_3\int_0^L w_t^2(z,t)\mathrm{d}z - p_3\int_0^L w_z^2(z,t)\mathrm{d}z \\
&- p_3\int_0^L w_{zz}^2(z,t)\mathrm{d}z + \eta \boldsymbol{Q}^{*\mathrm{T}}\boldsymbol{Q}^* \\
&= -p_3\left(\bar{\boldsymbol{x}}^{\mathrm{T}}(t)\bar{\boldsymbol{x}}(t) + \int_0^L \left(\boldsymbol{\xi}^{\mathrm{T}}(z,t)\boldsymbol{\xi}(z,t) + w_{zz}^2(z,t)\right)\mathrm{d}z\right) + \eta \boldsymbol{Q}^{*\mathrm{T}}\boldsymbol{Q}^* \\
&\leq -p_3 p_2^{-1} V(t) + \eta \boldsymbol{Q}^{*\mathrm{T}}\boldsymbol{Q}^*.
\end{aligned} \tag{3.41}$$

Then inequality (3.41) can be calculated as

$$V(t) \leq V(0)e^{-\lambda t} + \kappa. \tag{3.42}$$

From inequality (3.39), we can get

$$V(t) > p_1 \int_0^L w_{zz}^2(z,t)\mathrm{d}z. \tag{3.43}$$

Using Lemmas 2.1 and 2.2, we have

$$0.25L^{-3}\pi^2 w^2(z,t) \le \int_0^L w_{zz}^2(z,t)\mathrm{d}z \le \frac{1}{p_1}\big(V(0)e^{-\lambda t} + \kappa\big). \tag{3.44}$$

Hence, we have

$$|w(z,t)| \le \sqrt{\frac{4L^3}{\pi^2 p_1}\big(V(0)e^{-\lambda t} + \kappa\big)} \tag{3.45}$$

which implies

$$\lim_{t\to\infty} |w(z,t)| \le D. \tag{3.46}$$

From inequalities (3.45) and (3.46), the elastic deformation of the beam is bounded if initial states of the system are bounded. In addition, the deflection $w(z,t)$ of the beam exponentially converges to a region of zero for $t \to \infty$.

To conclude, the proposed boundary robust adaptive controller based on the RBF neural network algorithm can stabilize the system when there exist $p_{11}, p_{12}, p_{22} \in R^+$ and the LMIs (3.14)–(3.16) are satisfied.

Remark 3.5 According to Eqs. (3.11)–(3.13), the feasible solution of the system is related to the uncertain parameters. The broader the scope of uncertain parameters, the harder it is to find feasible solutions. It is noteworthy that the control performance is related to system actual parameters EI and ρ, and can be improved by increasing γ or decreasing η due to the region of attraction D in inequality (3.46). When the size of the uncertain parameters changes, we can properly select γ and η to obtain desired control effect and adjust other parameters $p_{11}, p_{12}, p_{22}, n, K_1$, and K_2 to get a set of a feasible solution of the LMIs of inequalities (3.14)–(3.16).

Remark 3.6 In this chapter, we can give the two-step procedure to find the feasible solutions of LMIs in inequalities (3.14)–(3.16): First, solve the LMIs (3.14) and (3.16) for positive scalars p_{11}, p_{12}, p_{22}. And then solving the LMI (3.15) gives the control gains K_1 and K_2 by specifying values for n and η. It is worth mentioning that the used two-step procedure is simple and can be easily implemented [7]. Consequently, the solution of the proposed LMIs comes with minor calculation and can be easily obtained in an effective way.

3.4 Simulations

Numerical simulations are provided in this section for verifying the validity of the control law in Eqs. (3.7), (3.8), and (3.10). The parameters of the beam are listed as $EI = 2\ \text{Nm}^2$, $\rho = 5\ \text{kgm}^{-1}$, and $L = 2$ m.

If system parameters are uncertain, their ranges are given as $\overline{EI}_{\min} = 1.7\ \text{Nm}^2$, $\overline{EI}_{\max} = 2.5\ \text{Nm}^2$, $\bar{\rho}_{\min} = 4.7\ \text{kgm}^{-1}$, and $\bar{\rho}_{\max} = 5.3\ \text{kgm}^{-1}$. In order to verify the robustness of the designed control more convincingly, the interference is considered in following two cases:

$$\textbf{\textit{Case}}\ \mathbf{1.}\ d(\omega) = 0.1(w(L,t)+1)(w_t(L,t)+1) \tag{3.47}$$

$$\textbf{\textit{Case}}\ \mathbf{2.}\ d(\omega) = 0.3\cos(w(L,t))\cos(w_t(L,t)). \tag{3.48}$$

By observing the state-related uncertainty defined in cases 1 and 2, it is obvious that the uncertainty finally converges to a nonzero constant with beam deflection and vibration velocity (0.1 in case 1 and 0.3 in case 2) when the closed-loop system is stabilized with the proposed control. With such uncertainty in cases 1 and 2, the control performance can be decisively verified and simulation results can be exactly reflected with constant sample rate.

Then to further demonstrate the validity of the LMI-based control method, two sets of feasible solutions to LMIs (3.14)–(3.16) are obtained by using LMI toolbox of MATLAB as follows:

Feasible Solutions 1 Control law in Eq. (3.10) with

$$p_{11} = 0.1264, p_{12} = 0.0120, p_{22} = 0.0329 \tag{3.49}$$

$$\boldsymbol{K} = [-2.8253, -0.5942]. \tag{3.50}$$

Feasible Solutions 2 Control law in Eq. (3.10) with

$$p_{11} = 0.8217, p_{12} = 0.0957, p_{22} = 0.2716 \tag{3.51}$$

$$\boldsymbol{K} = [-1.2059, -0.3047]. \tag{3.52}$$

Parameters of the RBF neural network are designed as $\gamma = 200$, $n = 5$, $\eta = 0.1$, $\boldsymbol{f}_1 = \begin{bmatrix}-2 & -2\end{bmatrix}^{\mathrm{T}}, \boldsymbol{f}_2 = \begin{bmatrix}-1 & -1\end{bmatrix}^{\mathrm{T}}, \boldsymbol{f}_3 = \begin{bmatrix}0 & 0\end{bmatrix}^{\mathrm{T}}, \boldsymbol{f}_4 = \begin{bmatrix}1 & 1\end{bmatrix}^{\mathrm{T}}, \boldsymbol{f}_5 = \begin{bmatrix}2 & 2\end{bmatrix}^{\mathrm{T}}$, $i = 1, 2, \ldots, 5$. Besides, $g_i = 2$ and $g_i = 20$ are chosen for feasible solutions 1 and 2 respectively.

The finite difference method is used to simulate the system performance. The discrete time and position steps are chosen as $\Delta t = 1 \times 10^{-3}$ s and $\Delta x = 0.1$ m.

Without control input, the dynamic response of $w(z, t)$ with uncertainty in cases 1 and 2 is shown in Fig. 3.2 and the dynamic response of the end of the beam is delineated by Fig. 3.3. It is obvious that the vibration of the beam is intense with perturbation and the open-loop system is unstable within a given time.

According to Fig. 3.4, we can see the elastic deformation $w(z, t)$ of the beam under the proposed boundary robust adaptive neural network control input with feasible solutions 1 in case 1 and feasible solutions 2 in case 2. Figure 3.5 underlines the boundary deflection of the beam with control in Eqs. (3.7), (3.8), and (3.10).

By comparing Figs. 3.2 and 3.3 with Figs. 3.4 and 3.5, it is clear that the state values of the closed-loop system are much smaller than the ones of the open-loop system. We also notice that the vibration of the disturbed beam converges to zero after 20 s even with persistent interference using feasible solutions 1 in case 1 and

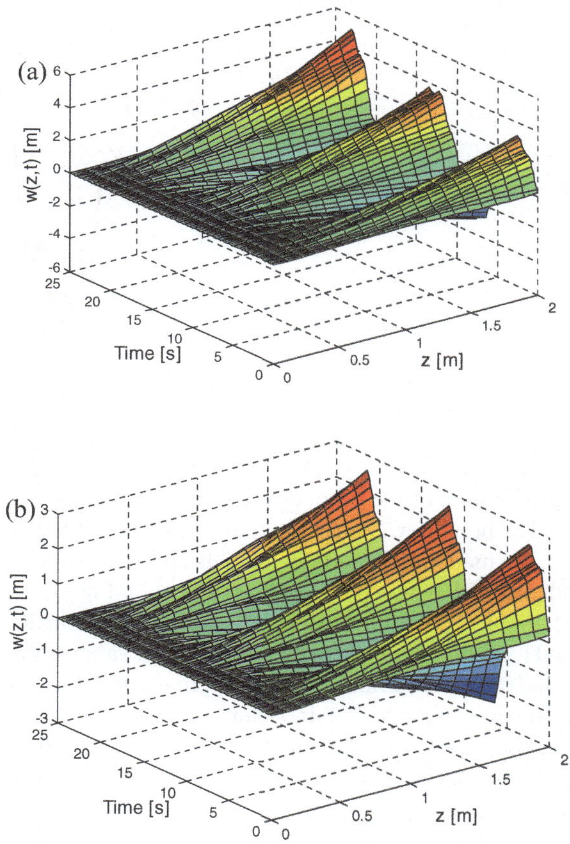

Fig. 3.2 $w(z, t)$ in the open-loop system. **a** In case 1. **b** In case 2

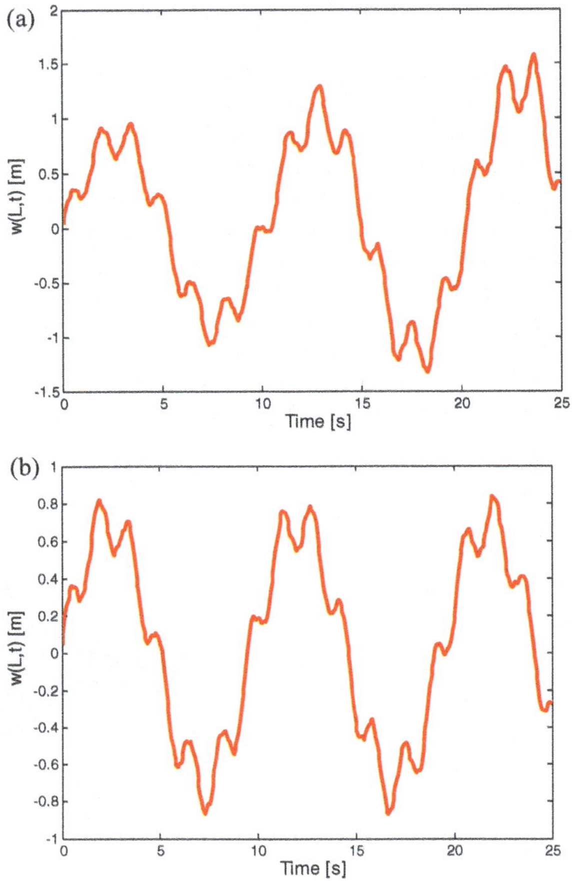

Fig. 3.3 $w(L, t)$ in the open-loop system. **a** In case 1. **b** In case 2

feasible solutions 2 in case 2, which indicates that the proposed control can eliminate the vibration of the beam quickly despite modeling and parameter uncertainties.

The boundary control input $u(t)$ is shown in Fig. 3.6.

The modeling uncertainty and its approximation value generated by the designed adaptive RBF neural network are shown in Fig. 3.7.

Figure 3.7 illustrates that the output signal of the RBF neural network in Eqs. (3.7) and (3.8) finally trends to the true constant value of the uncertainty, that is, 0.1 in case 1 and 0.3 in case 2, and $\hat{d}(\boldsymbol{\omega})$ can approximate $d(\boldsymbol{\omega})$ with time growing. Therefore, the estimation of the uncertainty can compensate the additive persistent uncertainty with bounded approximation and has certain effect on the overall control performance to guarantee the stability of the system. The effectiveness of the neural network is thus verified.

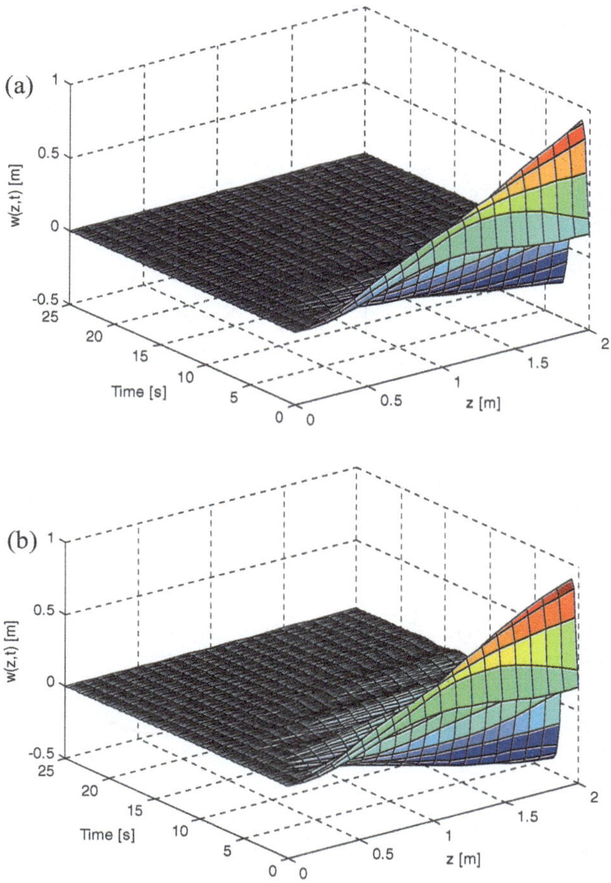

Fig. 3.4 Dynamic response of $w(z, t)$. **a** With feasible solutions 1 in case 1. **b** With feasible solutions 2 in case 2

To summarize, simulation results in Figs. 3.2, 3.3, 3.4, 3.5, 3.6 and 3.7 are consistent with theoretical analysis and demonstrate the correctness of the control strategy in Eqs. (3.7), (3.8), and (3.10).

The simulation programs for feasible solutions 1 and case 1 are given in the Appendixes 3.1, 3.2 and 3.3, where chapter3_LMI1.m in Appendix 3.1 is used to solve LMIs (3.14) and (3.16) to obtain p_{12}, p_{22}, chapter3_LMI2.m in Appendix 3.2 is used to solve LMIs (3.15), and the program chapter3.m in Appendix 3.3 is the main program.

Fig. 3.5 Dynamic response of $w(L, t)$. **a** With feasible solutions 1 in case 1. **b** With feasible solutions 2 in case 2

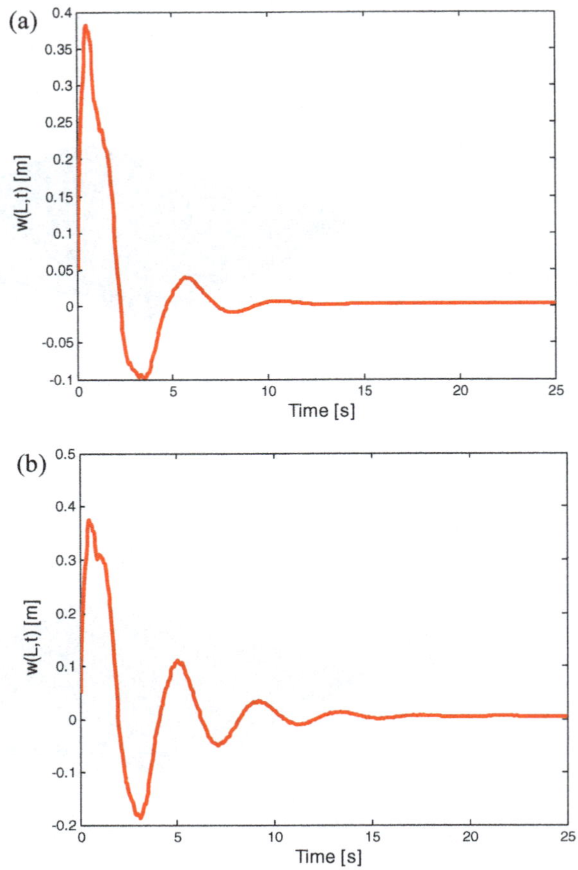

Fig. 3.6 Control signal. **a** With feasible solutions 1 in case 1. **b** With feasible solutions in case 2

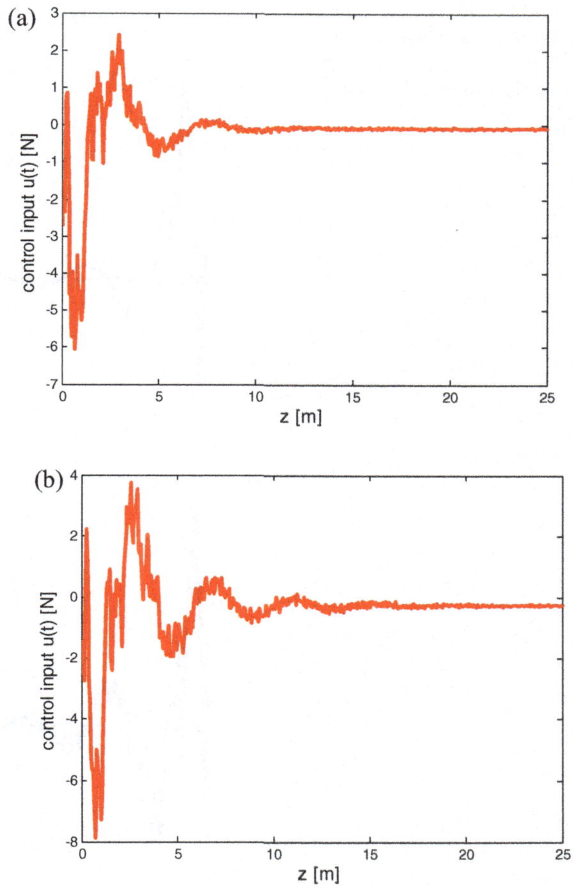

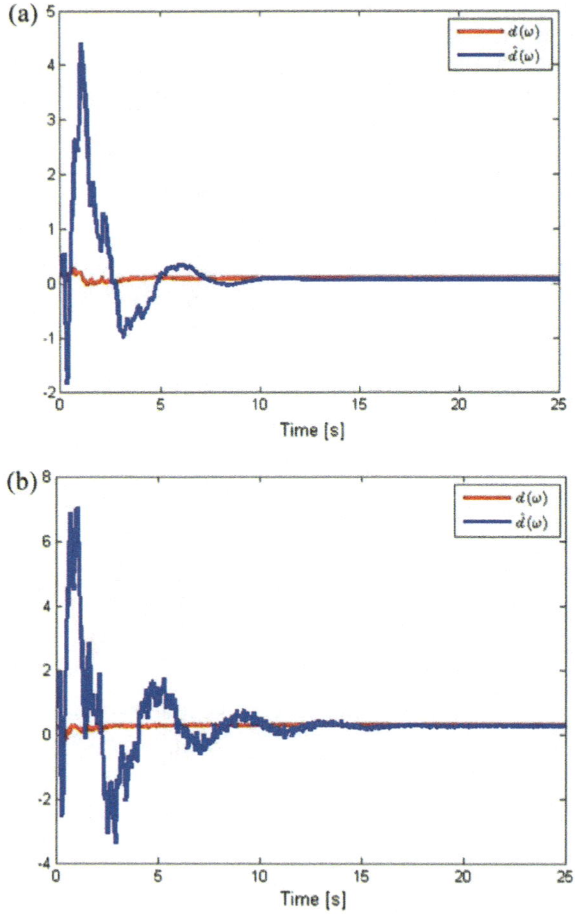

Fig. 3.7 Uncertainty and its estimation. **a** With feasible solutions 1 in case 1. **b** With feasible solutions 2 in case 2

3.5 Conclusions

A robust adaptive neural network control has been proposed to compensate the modeling uncertainty and restrict the vibration of the Euler-Bernoulli beam. Based on the designed RBF neural network and PDE model, LMI techniques are introduced to design boundary control laws for the beam with uncertain parameters. By resolving a set of LMIs, control parameters can be effectively determined. The system stability has been analyzed by Lyapunov's direct method. With the proposed LMI-based control, the vibration suppression is realized even under uncertainties. The effectiveness of the proposed control scheme has been finally proved by numerical simulations containing the proposed RBF adaptive law in (3.7) and (3.8) and control law in (3.10).

Appendix 3.1: Simulation Program: chapter3_LMI1.m

```
clear all;
close all;

L=2;

mmin=4.7;
m=5;
mmax=5.3;

EImin=1.7;
EI=2;
EImax=2.5;

z=2;
z1=0;

p11=0.1264;
p12=sdpvar(1,1);
p22=sdpvar(1,1);

P=sdpvar(2,2,'symmetric');
T=sdpvar(3,3,'symmetric');
T0=sdpvar(4,4,'symmetric');

L2=(p12>0);

L3=(p22>0);

P=[p11*mmin p12*mmax*L;p12*mmax*L p22];
L4=(P>0);

T=[-p12*mmin -p22 0;-p22 -1.5*p12*EImin 0;0 0 -3*p12*EImin*0.125*(pi^2)*(L^(-2))];
L5=(T<0);

L=L2+L3+L4+L5;

solvesdp(L);

M=[p11 p12 p22];

p12=double(M(1,2))
p22=double(M(1,3))
```

Appendix 3.2: Simulation Program: chapter3_LMI2.m

```
clear all;
close all;

L=2;
mmin=4.7;
m=5;
mmax=5.3;
deta=1;

EImin=1.7;
EI=2;
EImax=2.5;

p11=0.1264;
p12=0.0120;
p22=0.0329;

K1=sdpvar(1,1);
K2=sdpvar(1,1);
n=0.1;

N=[p12*mmax*L+2*K1*p11 p22+K2*p11+K1*p12*L 0;p22+K2*p11+K1*p12*L 2*K2*p12*L
0;0 0 -n];
L6=(N<0);

solvesdp(L6);

M=[K1 K2];

K1=double(M(1,1))
K2=double(M(1,2))
```

Appendix 3.3: Simulation Program: chapter3.m

```
close all;
clear all;
N=20;
n=50000;

t=25;L=2;
h=L/N;
T=t/n;

EI=2;rho=5;
n1=0.1;

K1=-2.8253;
K2=-0.5942;
K3=0;
p11=0.1264;
p12=0.0120;
p22=0.0329;

pp=[p11;p12*L];
b=(4*T^2*EI)/(rho*h^4);
c=-(T^2*EI)/(rho*h^4);
a=2-(6*T^2*EI)/(rho*h^4);
c1ij=1.5*[-2 -1 0 1 2; -2 -1 0 1 2];

b1j=20;
q=200;
for j=1:n+1

      u(j)=0;
      dL(j)=0;
end

z=zeros(n+1,N+4);
h1=zeros(5,1);
W=zeros(n+1,5);
for j=1:n+1
   z(j,3)=0;
```

```
   z(j,4)=0;
end

for i=5:N+2
   z(1,i)=0.04*sin(i*0.01*h);

   z(2,i)=z(1,i)+T;
end

z(2,1)=0;
z(2,2)=0;
dL(1)=0;

zLt(2)=(z(2,N+2)-z(1,N+2))/T;
zL(2)=(z(2,N+2)-z(2,N+1))/h;

z(2,N+3)=2*z(2,N+2)-z(2,N+1);
z(2,N+4)=z(2,N)-4*z(2,N+1)+4*z(2,N+2)-(2*h^3/EI)*(u(1)+dL(1));

u(2)=K1*zLt(2)+K2*zL(2);

for j=3:n+1
dzt(j,1)=(z(j,3)-z(j-1,3))/T;
dzt(j,2)=(z(j,4)-z(j-1,4))/T;
dz(j,1)=(z(j,3)-z(j,2))/h;
dz(j,2)=(z(j,4)-z(j,3))/h;
end

for j=3:n+1
      for i=5:N+2
            z(j,i)=c*(z(j-1,i-2)+z(j-1,i+2))+b*(z(j-1,i-1)+z(j-1,i+1))+a*z(j-1,i)-z(j-2,i);
            dzt(j,i-2)=(z(j,i)-z(j-1,i))/T;
            dz(j,i-2)=(z(j,i)-z(j,i-1))/h;
      end
            z(j,1)=0;
            z(j,2)=0;

            zLt(j)=(z(j,N+2)-z(j-1,N+2))/T;
```

```
            zL(j)=(z(j,N+2)-z(j,N+1))/h;

            z(j,N+3)=2*z(j,N+2)-z(j,N+1);
            z(j,N+4)=z(j,N)-4*z(j,N+1)+4*z(j,N+2)-(2*h^3/EI)*(u(j-1)+dL(j-1));

dL(j)=0.1*(zL(j)+1)*(zLt(j)+1);% Case 1

x1i=[zL(j);zLt(j)];

for k=1:1:5
    h1(k)=exp(-norm(x1i-c1ij(:,k))^2/(2*b1j^2));
end
    W(j,:)=W(j-1,:)+T*(q*x1i'*pp*h1.'-n1*q*W(j-1,:));
    dp(j)= W(j,:)*h1;

            u(j)=K1*zLt(j)+K2*zL(j)-dp(j);%with control

            %u(j)=0;%without control
end

zshort=zeros(n/400,N);
dzshort=zeros(n/400,N);
dztshort=zeros(n/400,N);
for i=1:N
for j=1:n/400
            zshort(j,i)=z(j*400,i+2);
            dztshort(j,i)=dzt(j*400,i);
            dzshort(j,i)=dz(j*400,i);

        end
end
for j=1:n/100
            ushort(j)=u(j*100);
            z1(j)=z(j*100,12);
            z2(j)=dzt(j*100,10);
            z3(j)=dz(j*100,10);
            dL_1(j)=dL(j*100);
            dp_1(j)=dp(j*100);
```

```
end
dp_1(1)=dp(1);
tshort1=linspace(0,t,n/400);
tshort3=linspace(0,t,n/100);
y=linspace(0,L,N);
[y1,tshort2]=meshgrid(y,tshort1);
tt=linspace(0,n*T,n);
tt1=linspace(0,(n+1)*T,n+1);

figure(1);

surf(y1,tshort1,zshort);
xlabel('z [m]'); ylabel('Time [s]');zlabel('w(z,t) [m]');
ylim([0,25]);

figure(2);
plot(tshort3,ushort,'r','linewidth',2.5);
xlabel('z [m]');ylabel('control input u(t) [N]');

figure(3);
plot(tshort3,z1,'r','linewidth',2.5);
xlabel('Time [s]');ylabel('w(L,t) [m]');

figure(4);
plot(tshort3,dL_1,'r',tshort3,dp_1,'b','linewidth',2.5);
xlabel('Time [s]');
h0=legend('${d}\left( \mathbf{\omega } \right)$','$\hat{d}\left( \mathbf{\omega } \right)$');
set(h0,'interpreter','latex','fontsize',10)
```

References

1. O. Morgul, Dynamic boundary control of a Euler-Bernoulli beam. IEEE Trans. Autom. Control **37**(5), 639–642 (1992)
2. W. He, Y. Ouyang, J. Hong, Vibration control of a flexible robotic manipulator in the presence of input deadzone. IEEE Trans. Industr. Inf. **13**(1), 48–59 (2017)
3. S.S. Ge, W. He, B.V.E. How, Y.S. Choo, Boundary control of a coupled nonlinear flexible marine riser. IEEE Trans. Control Syst. Technol. **18**(5), 1080–1091 (2010)
4. Y. Liu, Y. Fu, W. He, Q. Hui, Modeling and observer-based vibration control of a flexible spacecraft with external disturbances. IEEE Trans. Indus. Electron (2019). https://doi.org/10.1109/tie.2018.2884172

5. S. Bieniawski, I.M. Kroo, Flutter suppression using micro-trailing edge effectors. AIAA paper 1941 (2003)
6. S. Zhang, W. He, D. Huang, Active vibration control for a flexible string system with input backlash. IET Control Theory Appl. **10**(7), 800–805 (2016)
7. H.N. Wu, J.W. Wang, Static output feedback control via PDE boundary and ODE measurements in linear cascaded ODE-beam systems. Automatica **50**, 2787–2798 (2014)

Chapter 4
LMI-Based Vibration Control Design for Euler–Bernoulli Beam with Uncertain Parameters and Distributed Disturbance

4.1 Introduction

During the past decades, due to the extensive application of the flexible beam in practical engineering, control of the Euler–Bernoulli beam has received a great deal of attention. However, the excessive vibration may impact the performance of control systems, so the vibration suppression has been a challenging task in the field of control design.

A vibration control problem of the Euler–Bernoulli beam with uncertain parameters and external distributed disturbances is investigated in this chapter. A distributed disturbance observer is proposed to eliminate the influence of the external disturbances on the beam. Based on the observer, the boundary and distributed controllers are designed by LMI and Lyapunov's direct method to exponentially stabilize the Euler–Bernoulli beam. The existence and uniqueness of the solution of the closed-loop system is verified by semigroup theory. Finally, numerical simulations are provided to validate the effectiveness of the proposed control.

4.2 Problem Formulation

For simplification, we consider that the Euler–Bernoulli beam without rotating angle and payload on the free end moves in the horizontal direction. Moreover, the tension of the beam and the viscous force are neglected for the sake of simplicity. The configuration of the beam without payload is shown in Fig. 4.1.

In Fig. 4.1, $d(z, t)$ is the spatially disturbance, which is caused by the interaction with external environment, $F(z, t)$ is the distributed control input generated by the piezoelectric actuator [1, 2]. Other physical parameters of the system are defined as ones given in Chap. 3.

The governing equation of the Euler–Bernoulli beam can be written as [3]

X. Xing and J. Liu, *Vibration Control Methods of Mechanical Distributed Parameter Systems*, Springer Tracts in Mechanical Engineering,
https://doi.org/10.1007/978-981-16-1532-0_4

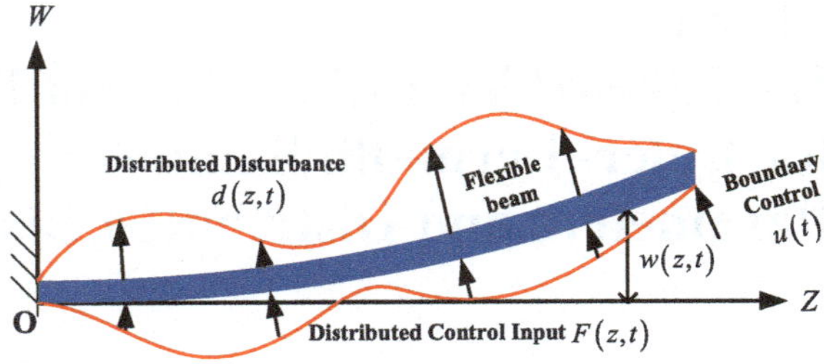

Fig. 4.1 Configuration of an Euler–Bernoulli beam without payload

$$\rho w_{tt}(z,t) + EIw_{zzzz}(z,t) = F(z,t) + d(z,t) \tag{4.1}$$

and the boundary conditions can be written as

$$EIw_{zzz}(L,t) + u(t) = 0 \tag{4.2}$$

$$w_{zz}(L,t) = 0 \tag{4.3}$$

$$w(0,t) = w_z(0,t) = 0. \tag{4.4}$$

Remark 4.1 In the following control design, the distributed control $F(z,t)$ will contain an observer feedback item to compensate for the uncertainty of the external disturbance. If the effect of the external disturbances can be ignored, that is, if $d(z,t)$ can be considered as $d(z,t)=0$ for $(z,t) \in [0,L] \times [0,\infty)$, then it takes only boundary control $u(t)$ to achieve vibration elimination.

Assumption 4.1 In this chapter, assume that all the parameters of the beam are uncertain and bounded. The ranges of the system parameters are given as: $0 < \overline{\rho}_{\min} \leq \rho \leq \overline{\rho}_{\max}$ and $0 < \overline{EI}_{\min} \leq EI \leq \overline{EI}_{\max}$, where $\overline{\rho}_{\min}$, $\overline{\rho}_{\max}$, $\overline{EI}_{\min}$ and $\overline{EI}_{\max}$ are all positive constants.

Assumption 4.2 Due to the limited disturbance energy, assume that the uncertain external disturbance is bounded. Define the upper and the lower bounds of $|d(z,t)|$ as D_1 and D_2, respectively, where $D_1 > 0$ and $D_2 > 0$. Then, it satisfies $D_1 < |d(z,t)| < D_2$.

Assumption 4.3 [4] We assume that $\dot{d}(z,t) = 0$ for $(z,t) \in [0,L] \times [0,\infty)$. In fact, since a disturbance observer is used in the control design in this chapter and the disturbance varies slowly relative to the observer dynamics, it is reasonable to consider the derivative of the disturbances as zero.

4.3 Control Design and Analysis

The control objective of this chapter is to design a controller to suppress the elastic vibration of the Euler–Bernoulli beam with uncertain parameters and eliminate the external disturbance affecting to the system based on the proposed disturbance observer.

4.3.1 *Observer Design and Analysis*

An exponentially stable disturbance observer is proposed to estimate the external distributed disturbance in this part.

Define the estimation error of $d(z, t)$ as follows

$$\tilde{d}(z, t) = d(z, t) - \hat{d}(z, t) \tag{4.5}$$

where $\hat{d}(z, t)$ is the estimation of $d(z, t)$.

In order to correct the error between $d(z, t)$ and $\hat{d}(z, t)$, we propose that

$$\begin{aligned}\dot{\hat{d}}(z, t) &= k\tilde{d}(z, t) = k\Big(d(z, t) - \hat{d}(z, t)\Big) \\ &= k\Big(\rho w_{tt}(z, t) + EIw_{zzzz}(z, t) - F(z, t) - \hat{d}(z, t)\Big)\end{aligned} \tag{4.6}$$

where k is a positive constant.

Then the disturbance observer is designed as

$$\dot{\eta}(z, t) = -k(-EIw_{zzzz}(z, t) + F(z, t)) - k\hat{d}(z, t) \tag{4.7}$$

$$\hat{d}(z, t) = \eta(z, t) + k\rho w_t(z, t) \tag{4.8}$$

where $\eta(z, t)$ is an auxiliary variable.

Theorem 4.1 *The proposed observer in Eqs.* (4.7) *and* (4.8) *is exponentially stable which means* $\hat{d}(z, t) \to d(z, t)$ for $t \to \infty$.

Proof According to the designed observer, a Lyapunov function is defined as.

$$V_{ob}(t) = \frac{1}{2}\int_0^L \tilde{d}^2(z, t)\mathrm{d}z. \tag{4.9}$$

After differentiating Eq. (4.9) and substituting Eqs. (4.7) and (4.8) into $\dot{V}_{ob}(t)$, we get

$$
\begin{aligned}
\dot{V}_{ob}(t) &= \int_0^L \tilde{d}(z,t)\dot{\tilde{d}}(z,t)\mathrm{d}z \\
&= -\int_0^L \tilde{d}(z,t)\dot{\hat{d}}(z,t)\mathrm{d}z = -\int_0^L \tilde{d}(z,t)(\dot{\eta}(z,t) + k\rho w_{tt}(z,t))\mathrm{d}z \\
&= -\int_0^L \tilde{d}(z,t)\Big(-k(-EIw_{zzzz}(z,t) + F(z,t) - \rho w_{tt}(z,t)) - k\hat{d}(z,t)\Big)\mathrm{d}z \\
&= -\int_0^L \tilde{d}(z,t)\Big(kd(z,t) - k\hat{d}(z,t)\Big)\mathrm{d}z \\
&= -k\int_0^L \tilde{d}^2(z,t)\mathrm{d}z \\
&= -2kV_{ob}(t).
\end{aligned}
\tag{4.10}
$$

By integrating Eq. (4.10) from 0 to t, the solution of Eq. (4.10) is

$$
V_{ob}(t) = V_{ob}(0)\exp(-2kt). \tag{4.11}
$$

From Eq. (4.11), $V_{ob}(t) \to 0$ for $t \to \infty$ can be obtained. Hence we can deduce from Eq. (4.9) that $\tilde{d}(z,t) \to 0$ for $t \to \infty$, that is, $\hat{d}(z,t) \to d(z,t)$ for $t \to \infty$.

To summarize, we can validate that the observer is exponentially stable and the distributed disturbance of the Euler–Bernoulli beam can be estimated effectively by the proposed observer.

4.3.2 Controller Design and Stability Analysis

To realize the control objective, the boundary control $u(t)$ and the distributed control $F(z,t)$ are designed as

$$
u(t) = \boldsymbol{K}\boldsymbol{x}(t) \tag{4.12}
$$

$$
F(z,t) = K_3 w_t(z,t) - \hat{d}(z,t) \tag{4.13}
$$

where $\boldsymbol{x}(t) = \begin{bmatrix} w_t(L,t) & w_z(L,t) \end{bmatrix}^{\mathrm{T}}$ and $\boldsymbol{K} = \begin{bmatrix} K_1 & K_2 \end{bmatrix} \in R^{1\times 2}$.

Substituting $u(t)$ and $F(z,t)$ into Eqs. (4.1)–(4.4), we have the closed-loop system as follows

$$
\rho w_{tt}(z,t) + EIw_{zzzz}(z,t) = K_3 w_t(z,t) + \tilde{d}(z,t) \tag{4.14}
$$

$$
-EIw_{zzz}(L,t) = \boldsymbol{K}\boldsymbol{x}(t) \tag{4.15}
$$

$$w_{zz}(L,t) = 0 \tag{4.16}$$

$$w(0,t) = w_z(0,t) = 0. \tag{4.17}$$

Theorem 4.2 *Defining* $\boldsymbol{\Lambda}(z) = \begin{bmatrix} p_{11}\overline{\rho}_{\min} & p_{12}\overline{\rho}_{\max}z \\ p_{12}\overline{\rho}_{\max}z & p_{22} \end{bmatrix}$,

$$\boldsymbol{\Theta} = \begin{bmatrix} p_{12}L\overline{\rho}_{\max} + 2K_1p_{11} & p_{22} + K_2p_{11} + K_1p_{12}L \\ p_{22} + K_2p_{11} + K_1p_{12}L & 2K_2p_{12}L \end{bmatrix},$$

$$\boldsymbol{\Xi}(z) = \begin{bmatrix} -p_{12}\overline{\rho}_{\min} + 2p_{11}K_3 & -p_{22} & p_{12}K_3z & p_{11} \\ -p_{22} & -\frac{3}{2}p_{12}\overline{EI}_{\min} & 0 & 0 \\ p_{12}K_3z & 0 & -\frac{3}{8}\pi^2L^{-2}p_{12}\overline{EI}_{\min} & p_{12}z \\ p_{11} & 0 & p_{12}z & -k \end{bmatrix}$$, if there exist $p_{11}, p_{12}, p_{22} \in \Re^+$ and the following LMIs are satisfied

$$\boldsymbol{\Lambda}(z) > 0 \tag{4.18}$$

$$\boldsymbol{\Theta} < 0 \tag{4.19}$$

$$\boldsymbol{\Xi}(z) < 0 \tag{4.20}$$

the boundary and distributed control laws proposed in Eqs. (4.12) and (4.13) can ensure that the closed-loop system is exponentially stable in the following sense: $w(z,t)$, $w_z(z,t)$ and $w_t(z,t)$ converge to zero, $\forall z \in [0, L]$, for $t \to \infty$.

Proof A Lyapunov candidate function is designed to be

$$\begin{aligned} V(t) &= V_1(t) + V_2(t) + V_{ob}(t) \\ &= \int_0^L \boldsymbol{\xi}^{\mathrm{T}}(z,t)\boldsymbol{P}(z)\boldsymbol{\xi}(z,t)\mathrm{d}z + \int_0^L p_{11}EIw_{zz}^2(z,t)\mathrm{d}z + \frac{1}{2}\int_0^L \tilde{d}^2(z,t)\mathrm{d}z \end{aligned} \tag{4.21}$$

where $V_1(t)$ and $V_2(t)$ are defined as follows

$$V_1(t) = \int_0^L \boldsymbol{\xi}^{\mathrm{T}}(z,t)\boldsymbol{P}(z)\boldsymbol{\xi}(z,t)\mathrm{d}z \tag{4.22}$$

$$V_2(t) = \int_0^L p_{11}EIw_{zz}^2(z,t)\mathrm{d}z \tag{4.23}$$

where

$\boldsymbol{P}(z)=\begin{bmatrix} p_{11}\rho & p_{12}\rho z \\ p_{12}\rho z & p_{22} \end{bmatrix}$, $\boldsymbol{\xi}(z,t)=\begin{bmatrix} w_t(z,t) & w_z(z,t) \end{bmatrix}^{\mathrm{T}}$, $p_{11}, p_{12}, p_{22} \in R^{+}$.
Differentiating $V_1(t)$ and $V_2(t)$ with respect to time respectively yields

$$\begin{aligned}\dot{V}_1(t) =& 2\int_0^L p_{11}\rho w_t(z,t)w_{tt}(z,t)\mathrm{d}z + 2\int_0^L p_{22}w_z(z,t)w_{zt}(z,t)\mathrm{d}z \\ &+ 2\int_0^L p_{12}\rho z w_t(z,t)w_{zt}(z,t)\mathrm{d}z + 2\int_0^L p_{12}\rho z w_z(z,t)w_{tt}(z,t)\mathrm{d}z\end{aligned} \tag{4.24}$$

$$\dot{V}_2(t) = 2\int_0^L p_{11}EIw_{zz}(z,t)w_{zzt}(z,t)\mathrm{d}z. \tag{4.25}$$

Substituting Eqs. (4.14)–(4.16) into each term of $\dot{V}_1(t)$ and using integration by parts, we can obtain

$$\begin{aligned}\int_0^L \rho w_t(z,t)w_{tt}(z,t)\mathrm{d}z =& -\int_0^L w_t(z,t)EIw_{zzzz}(z,t)\mathrm{d}z + \int_0^L w_t(z,t)\tilde{d}(z,t)\mathrm{d}z \\ &+ \int_0^L K_3 w_t^2(z,t)\mathrm{d}z \\ =& -EIw_t(z,t)w_{zzz}(z,t)\big|_0^L + EI\int_0^L w_{zt}(z,t)w_{zzz}(z,t)\mathrm{d}z \\ &+ \int_0^L w_t(z,t)\tilde{d}(z,t)\mathrm{d}z + \int_0^L K_3 w_t^2(z,t)\mathrm{d}z \\ =& -EIw_t(z,t)w_{zzz}(z,t)\big|_0^L + EIw_{zt}(z,t)w_{zz}(z,t)\big|_0^L \\ &- \int_0^L EIw_{zzt}(z,t)w_{zz}(z,t)\mathrm{d}z + \int_0^L w_t(z,t)\tilde{d}(z,t)\mathrm{d}z \\ &+ \int_0^L K_3 w_t^2(z,t)\mathrm{d}z \\ =& -EIw_t(L,t)w_{zzz}(L,t) - \int_0^L EIw_{zzt}(z,t)w_{zz}(z,t)\mathrm{d}z \\ &+ \int_0^L w_t(z,t)\tilde{d}(z,t)\mathrm{d}z + \int_0^L K_3 w_t^2(z,t)\mathrm{d}z \\ =& w_t(L,t)\boldsymbol{K}\boldsymbol{x}(t) - \int_0^L EIw_{zzt}(z,t)w_{zz}(z,t)\mathrm{d}z \\ &+ \int_0^L w_t(z,t)\tilde{d}(z,t)\mathrm{d}z + \int_0^L K_3 w_t^2(z,t)\mathrm{d}z\end{aligned} \tag{4.26}$$

$$\int_0^L w_z(z,t)w_{zt}(z,t)\mathrm{d}z = w_t(L,t)w_z(L,t) - \int_0^L w_t(z,t)w_{zz}(z,t)\mathrm{d}z \tag{4.27}$$

$$\int_0^L \rho z w_t(z,t)w_{zt}(z,t)\mathrm{d}z = \frac{1}{2}\rho L w_t^2(L,t) - \frac{1}{2}\int_0^L \rho w_t^2(z,t)\mathrm{d}z \tag{4.28}$$

$$\begin{aligned}
\int_0^L \rho z w_z(z,t)w_{tt}(z,t)\mathrm{d}z =& -\int_0^L EIzw_z(z,t)w_{zzzz}(z,t)\mathrm{d}z + \int_0^L zw_z(z,t)\tilde{d}(z,t)\mathrm{d}z \\
&+ \int_0^L K_3 z w_z(z,t)w_t(z,t)\mathrm{d}z \\
=& -EIzw_z(z,t)w_{zzz}(z,t)\Big|_0^L + \int_0^L EIzw_{zz}(z,t)w_{zzz}(z,t)\mathrm{d}z \\
&+ \int_0^L EIw_z(z,t)w_{zzz}(z,t)\mathrm{d}z + \int_0^L zw_z(z,t)\tilde{d}(z,t)\mathrm{d}z \\
&+ \int_0^L K_3 z w_z(z,t)w_t(z,t)\mathrm{d}z \\
=& -EILw_z(L,t)w_{zzz}(L,t) + \int_0^L EIzw_{zz}(z,t)w_{zzz}(z,t)\mathrm{d}z \\
&+ EIw_z(z,t)w_{zz}(z,t)\Big|_0^L - \int_0^L EIw_{zz}^2(z,t)\mathrm{d}z \\
&+ \int_0^L zw_z(z,t)\tilde{d}(z,t)\mathrm{d}z + \int_0^L K_3 z w_z(z,t)w_t(z,t)\mathrm{d}z \\
=& Lw_z(L,t)\boldsymbol{K}\boldsymbol{x}(t) - \int_0^L EIw_{zz}^2(z,t)\mathrm{d}z \\
&+ \int_0^L EIzw_{zz}(z,t)w_{zzz}(z,t)\mathrm{d}z + \int_0^L zw_z(z,t)\tilde{d}(z,t)\mathrm{d}z \\
&+ \int_0^L K_3 z w_z(z,t)w_t(z,t)\mathrm{d}z.
\end{aligned} \tag{4.29}$$

Since the following equation holds

$$\begin{aligned}
\int_0^L EIzw_{zz}(z,t)w_{zzz}(z,t)\mathrm{d}z =& EIzw_{zz}^2(z,t)\Big|_0^L - \int_0^L EIzw_{zzz}(z,t)w_{zz}(z,t)\mathrm{d}z \\
&- \int_0^L EIw_{zz}^2(z,t)\mathrm{d}z \\
=& -\int_0^L EIw_{zz}^2(z,t)\mathrm{d}z - \int_0^L EIzw_{zzz}(z,t)w_{zz}(z,t)\mathrm{d}z,
\end{aligned} \tag{4.30}$$

we can get

$$\int_0^L EIzw_{zz}(z,t)w_{zzz}(z,t)\mathrm{d}z = -\frac{1}{2}\int_0^L EIw_{zz}^2(z,t)\mathrm{d}z. \tag{4.31}$$

Substituting Eq. (4.31) into Eq. (4.29), we have

$$\int_0^L \rho z w_z(z,t)w_{tt}(z,t)\mathrm{d}z = Lw_z(L,t)\boldsymbol{K}\boldsymbol{x}(t) - \frac{3}{2}\int_0^L EIw_{zz}^2(z,t)\mathrm{d}z + \int_0^L zw_z(z,t)\tilde{d}(z,t)\mathrm{d}z + \int_0^L K_3 zw_z(z,t)w_t(z,t)\mathrm{d}z. \tag{4.32}$$

Substituting Eqs. (4.32) and (4.26)-(4.28) into Eq. (4.24) and using Lemma 2.1, we can get

$$\begin{aligned}
\dot{V}_1(t) &\le -2p_{11}\int_0^L EIw_{zzt}(z,t)w_{zz}(z,t)\mathrm{d}z + 2p_{11}\int_0^L w_t(z,t)\tilde{d}(z,t)\mathrm{d}z \\
&+2p_{11}\int_0^L K_3 w_t^2(z,t)\mathrm{d}z - 2p_{22}\int_0^L w_t(z,t)w_{zz}(z,t)\mathrm{d}z \\
&-\frac{3}{2}p_{12}\int_0^L \overline{EI}_{\min}w_{zz}^2(z,t)\mathrm{d}z - \frac{3}{8}L^{-2}\pi^2 p_{12}\int_0^L \overline{EI}_{\min}w_z^2(z,t)\mathrm{d}z \\
&-p_{12}\int_0^L \overline{\rho}_{\min}w_t^2(z,t)\mathrm{d}z + 2p_{12}\int_0^L zw_z(z,t)\tilde{d}(z,t)\mathrm{d}z \\
&+2p_{12}\int_0^L K_3 zw_z(z,t)w_t(z,t)\mathrm{d}z + 2p_{11}w_t(L,t)\boldsymbol{K}\boldsymbol{x}(t) + 2p_{12}Lw_z(L,t)\boldsymbol{K}\boldsymbol{x}(t) \\
&+2p_{22}w_t(L,t)w_z(L,t) + p_{12}\overline{\rho}_{\max}Lw_t^2(L,t) \\
&\le \int_0^L \boldsymbol{\zeta}^{\mathrm{T}}(z,t)\boldsymbol{\Xi}_1(z)\boldsymbol{\zeta}(z,t)\mathrm{d}z + \boldsymbol{x}^{\mathrm{T}}(t)\boldsymbol{\Theta}_1\boldsymbol{x}(t) + 2\boldsymbol{x}^{\mathrm{T}}(t)\tilde{\boldsymbol{p}}\boldsymbol{K}\boldsymbol{x}(t) \\
&-2p_{11}\int_0^L EIw_{zzt}(z,t)w_{zz}(z,t)\mathrm{d}z \\
&\le \int_0^L \boldsymbol{\zeta}^{\mathrm{T}}(z,t)\boldsymbol{\Xi}_1(z)\boldsymbol{\zeta}(z,t)\mathrm{d}z + \boldsymbol{x}^{\mathrm{T}}(t)\boldsymbol{\Theta}_1\boldsymbol{x}(t) + \boldsymbol{x}^{\mathrm{T}}(t)\Big((\tilde{\boldsymbol{p}}\boldsymbol{K}) + (\tilde{\boldsymbol{p}}\boldsymbol{K})^{\mathrm{T}}\Big)\boldsymbol{x}(t) \\
&-2p_{11}\int_0^L EIw_{zzt}(z,t)w_{zz}(z,t)\mathrm{d}z \\
&= \int_0^L \boldsymbol{\zeta}^{\mathrm{T}}(z,t)\boldsymbol{\Xi}_1(z)\boldsymbol{\zeta}(z,t)\mathrm{d}z + \boldsymbol{x}^{\mathrm{T}}(t)\Theta\mathbf{x}(t) - 2p_{11}\int_0^L EIw_{zzt}(z,t)w_{zz}(z,t)\mathrm{d}z
\end{aligned} \tag{4.33}$$

where

$$\boldsymbol{\zeta}(z,t) = \left[w_t(z,t)\ w_{zz}(z,t)\ w_z(z,t)\ \tilde{d}(z,t)\right]^{\mathrm{T}}$$

$$\boldsymbol{\Xi}_1(z) = \begin{bmatrix} -p_{12}\overline{\rho}_{\min} + 2p_{11}K_3 & -p_{22} & p_{12}K_3 z & p_{11} \\ -p_{22} & -\frac{3}{2}p_{12}\overline{EI}_{\min} & 0 & 0 \\ p_{12}K_3 z & 0 & -\frac{3}{8}\pi^2 L^{-2}p_{12}\overline{EI}_{\min} & p_{12}z \\ p_{11} & 0 & p_{12}z & 0 \end{bmatrix},$$

$\tilde{\boldsymbol{p}} = \left[\, p_{11}\ p_{12}L \,\right]^{\mathrm{T}}$, and $\boldsymbol{\Theta}_1 = \begin{bmatrix} p_{12}L\overline{\rho}_{\max} & p_{22} \\ p_{22} & 0 \end{bmatrix}$.

Substituting Eqs. (4.10), (4.25) and inequality (4.33) into $\dot{V}(t)$, we have

$$\begin{aligned}\dot{V}(t) &= \dot{V}_1(t) + \dot{V}_2(t) + \dot{V}_{ob}(t) \\ &\le \int_0^L \boldsymbol{\zeta}^{\mathrm{T}}(z,t)\boldsymbol{\Xi}_1(z)\boldsymbol{\zeta}(z,t)\mathrm{d}z + \boldsymbol{x}^{\mathrm{T}}(t)\Theta\mathbf{x}(t) - k\int_0^L \tilde{d}^2(z,t)\mathrm{d}z \\ &= \int_0^L \boldsymbol{\zeta}^{\mathrm{T}}(z,t)\boldsymbol{\Xi}(z)\boldsymbol{\zeta}(z,t)\mathrm{d}z + \boldsymbol{x}^{\mathrm{T}}(t)\Theta\mathbf{x}(t). \end{aligned} \tag{4.34}$$

If LMI (4.18) is satisfied, considering the definition of $\boldsymbol{P}(z)$, we can obtain

$$\boldsymbol{P}(z) > 0 \tag{4.35}$$

and

$$V(t) \ge p_1 \int_0^L \Big(\boldsymbol{\xi}^{\mathrm{T}}(z,t)\boldsymbol{\xi}(z,t) + w_{zz}^2(z,t) + \tilde{d}^2(z,t)\Big)\mathrm{d}z > 0 \tag{4.36}$$

$$V(t) \le p_2 \int_0^L \Big(\boldsymbol{\xi}^{\mathrm{T}}(z,t)\boldsymbol{\xi}(z,t) + w_{zz}^2(z,t) + \tilde{d}^2(z,t) + \boldsymbol{x}^{\mathrm{T}}(t)\boldsymbol{x}(t)\Big)\mathrm{d}z \tag{4.37}$$

where $p_1 = \min\left\{ \min\limits_{z\in[0,L]} (\sigma(\boldsymbol{P}(z))),\ p_{11}EI,\ \frac{1}{2} \right\} > 0$ and $p_2 = \max\left\{ \max\limits_{z\in[0,L]} (\sigma(\boldsymbol{P}(z))),\ p_{11}EI,\ \frac{1}{2} \right\} > 0$.

Considering Lemma 2.1, we have

$$-\int_0^L w_{zz}^2(z,t)\mathrm{d}z \le -0.25L^{-2}\pi^2 \int_0^L w_z^2(z,t)\mathrm{d}z. \tag{4.38}$$

If inequalities (4.19) and (4.20) are satisfied, applying inequalities (4.37) and (4.38) to inequality (4.34) yields

$$\begin{aligned}\dot{V}(t) &\le \int_0^L \boldsymbol{\zeta}^{\mathrm{T}}(z,t)\boldsymbol{\Xi}(z)\boldsymbol{\zeta}(z,t)\mathrm{d}z + \boldsymbol{x}^{\mathrm{T}}(t)\Theta\mathbf{x}(t) \\ &\le -p_3\Big(\int_0^L \boldsymbol{\zeta}^{\mathrm{T}}(z,t)\boldsymbol{\zeta}(z,t)\mathrm{d}z + \boldsymbol{x}^{\mathrm{T}}(t)\boldsymbol{x}(t)\Big) \\ &= -p_3\boldsymbol{x}^{\mathrm{T}}(t)\boldsymbol{x}(t) - p_3\int_0^L w_t^2(z,t)\mathrm{d}z - p_3\int_0^L w_z^2(z,t)\mathrm{d}z \\ &\quad - p_3\int_0^L w_{zz}^2(z,t)\mathrm{d}z - p_3\int_0^L \tilde{d}^2(z,t)\mathrm{d}z \end{aligned}$$

$$
\begin{aligned}
&= -p_3\left(\boldsymbol{x}^{\mathrm{T}}(t)\boldsymbol{x}(t) + \int_0^L \left(\boldsymbol{\xi}^{\mathrm{T}}(z,t)\boldsymbol{\xi}(z,t) + w_{zz}^2(z,t) + \tilde{d}^2(z,t)\right)\mathrm{d}z\right) \\
&\leq -p_3 p_2^{-1} V(t)
\end{aligned} \tag{4.39}
$$

where $p_3 = \min\left\{\min\limits_{z\in[0,L]}(\sigma(-\boldsymbol{\Xi}(z))), \sigma(-\boldsymbol{\Theta})\right\} > 0$.

By integrating inequality (4.39), we have that

$$
V(t) \leq V(0)\exp\left(-p_3 p_2^{-1} t\right). \tag{4.40}
$$

According inequality (4.40), we can further obtain that $V(t) \to 0$ for $t \to \infty$ which implies $w_t(z,t) \to 0$ for $t \to \infty$.

From inequality (4.36), we can get

$$
V(t) \geq p_1 \int_0^L \left(\boldsymbol{\xi}^{\mathrm{T}}(z,t)\boldsymbol{\xi}(z,t) + w_{zz}^2(z,t) + \tilde{d}^2(z,t)\right)\mathrm{d}z > p_1 \int_0^L w_{zz}^2(z,t)\mathrm{d}z. \tag{4.41}
$$

Using Lemmas 2.1, 2.2, inequalities (4.40) and (4.41), we have

$$
\frac{1}{L} w_z^2(z,t) \leq \int_0^L w_{zz}^2(z,t)\mathrm{d}z \leq \frac{1}{p_1} V(t) \leq \frac{1}{p_1} V(0)\exp\left(-p_3 p_2^{-1} t\right) \tag{4.42}
$$

$$
\begin{aligned}
0.25L^{-3}\pi^2 w^2(z,t) &\leq 0.25L^{-2}\pi^2 \int_0^L w_z^2(z,t)\mathrm{d}z \\
&\leq \int_0^L w_{zz}^2(z,t)\mathrm{d}z \leq \frac{1}{p_1} V(t) \leq \frac{1}{p_1} V(0)\exp\left(-p_3 p_2^{-1} t\right).
\end{aligned} \tag{4.43}
$$

Hence, we have

$$
|w_z(z,t)| \leq \sqrt{\frac{L}{p_1} V(0)\exp\left(-p_3 p_2^{-1} t\right)} \tag{4.44}
$$

$$
|w(z,t)| \leq \sqrt{\frac{4L^3}{\pi^2 p_1} V(0)\exp\left(-p_3 p_2^{-1} t\right)}. \tag{4.45}
$$

From inequalities (4.44) and (4.45), we can obtain that $w(z,t)$ as well as $w_z(z,t)$ exponentially converges to zero for $t \to \infty$.

Then the existence and uniqueness of solutions of the system will be demonstrated as follows.

Let

$$\boldsymbol{q} = \left[w(z,t)\ w_t(z,t)\ \tilde{d}(z,t) \right] = \left[q_1\ q_2\ q_3 \right]. \tag{4.46}$$

The closed-loop system can be compactly expressed in the following form

$$\dot{\boldsymbol{q}} = \boldsymbol{A}\boldsymbol{q},\ \boldsymbol{q}(0) \in H \tag{4.47}$$

where

$$\boldsymbol{A}\boldsymbol{q} = \begin{bmatrix} q_2 \\ -\frac{EI}{\rho} q_{1,zzzz} + \frac{K_3}{\rho} q_2 + \frac{1}{\rho} q_3 \\ -k q_3 \end{bmatrix},\ \forall \boldsymbol{q} \in \mathcal{D}(\boldsymbol{A}). \tag{4.48}$$

Define the spaces mentioned above as

$$H = H_0^2(\Omega) \times \mathcal{L}^2(\Omega) \times \Re,\ \ \Omega \in [0, L]$$

$$\mathcal{D}(\boldsymbol{A}) = \left\{ \boldsymbol{q} \in H_0^4(\Omega) \times H_0^2(\Omega) \times R \middle| q_{1,zz}|_L = 0,\ EI q_{1,zzz}|_L = -K_1 q_2|_L - K_2 q_{1,z}|_L \right\}$$

where

$$H_0^k(\Omega) = \left\{ f \middle| f, f', \ldots, f^{(k)} \in \mathcal{L}^2(\Omega),\ f(0) = f'(0) = 0 \right\}.$$

An inner-product is then defined in H as follows

$$\langle \boldsymbol{q}, \boldsymbol{t} \rangle_H = \int_\Omega \rho q_2 t_2 \mathrm{d}z + \int_\Omega EI q_{1,zz} t_{1,zz} \mathrm{d}z + \int_\Omega q_3 t_3 \mathrm{d}z \tag{4.49}$$

where $\boldsymbol{q} = \left[q_1\ q_2\ q_3 \right] \in H$ and $\boldsymbol{t} = \left[t_1\ t_2\ t_3 \right] \in H$.

Define $(H, \langle \cdot, \cdot \rangle_H)$ as a Hilbert space [5] and the energy of Eq. (4.47) can be expressed as

$$\varepsilon(t) = \frac{1}{2} \langle \boldsymbol{q}, \boldsymbol{q} \rangle_H = \frac{1}{2} \|\boldsymbol{q}\|_H^2 = \frac{1}{2} \int_0^L \rho w_t^2(z,t) \mathrm{d}z + \frac{1}{2} \int_0^L EI w_{zz}^2(z,t) \mathrm{d}z$$
$$+ \frac{1}{2} \int_0^L \tilde{d}^2(z,t) \mathrm{d}z. \tag{4.50}$$

If inequality (4.18) is satisfied, $\boldsymbol{P}(z) > 0$ holds for $\forall z \in [0, L]$. Applying Eqs. (4.21) and (4.50), we can get the inequality as follows

$$p_4 \frac{1}{2} \|\boldsymbol{q}\|_H^2 \leq V(t) \leq p_5 \left(\frac{1}{2} \|\boldsymbol{q}\|_H^2 + \frac{1}{2} \int_0^L \rho w_z^2(z,t) \mathrm{d}z \right) \tag{4.51}$$

where $p_4 = \min\left\{2\rho^{-1} \min_{z\in[0,L]} (\sigma(\boldsymbol{P}(z)), 2p_{11}, 1\right\} > 0$ and $p_5 = \max\left\{2\rho^{-1} \max_{z\in[0,L]} (\sigma(\boldsymbol{P}(z)), 2p_{11}, 1\right\} > 0$.

According to Lemma 2.1, we obtain

$$\int_0^L \rho w_z^2(z,t)\mathrm{d}z \le 4L^2\pi^{-2} \int_0^L \rho w_{zz}^2(z,t)\mathrm{d}z. \tag{4.52}$$

Hence, we have

$$\begin{aligned}
&\frac{1}{2}\|\boldsymbol{q}\|_H^2 + \frac{1}{2}\int_0^L \rho w_z^2(z,t)\mathrm{d}z \\
&= \frac{1}{2}\int_0^L \rho w_t^2(z,t)\mathrm{d}z + \frac{1}{2}\int_0^L EIw_{zz}^2(z,t)\mathrm{d}z + \frac{1}{2}\int_0^L \rho w_z^2(z,t)\mathrm{d}z + \frac{1}{2}\int_0^L \tilde{d}^2(z,t)\mathrm{d}z \\
&\le \frac{1}{2}\int_0^L \rho w_t^2(z,t)\mathrm{d}z + \frac{1}{2}\int_0^L EIw_{zz}^2(z,t)\mathrm{d}z \\
&+ 4L^2\pi^{-2}\frac{\rho}{EI}\frac{1}{2}\int_0^L EIw_{zz}^2(z,t)\mathrm{d}z + \frac{1}{2}\int_0^L \tilde{d}^2(z,t)\mathrm{d}z \\
&= \frac{1}{2}\int_0^L \rho w_t^2(z,t)\mathrm{d}z + \left(1 + 4L^2\pi^{-2}\frac{\rho}{EI}\right)\frac{1}{2}\int_0^L EIw_{zz}^2(z,t)\mathrm{d}z + \frac{1}{2}\int_0^L \tilde{d}^2(z,t)\mathrm{d}z.
\end{aligned} \tag{4.53}$$

From inequalities (4.52) and (4.53), we can further obtain

$$\begin{aligned}
p_4\frac{1}{2}\|\boldsymbol{q}\|_H^2 &\le V(t) \\
&\le p_5\left(\frac{1}{2}\|\boldsymbol{q}\|_H^2 + \frac{1}{2}\int_0^L \rho w_z^2(z,t)\mathrm{d}z\right) \\
&\le p_5\left(\frac{1}{2}\int_0^L \rho w_t^2(z,t)\mathrm{d}z + \left(1 + 4L^2\pi^{-2}\frac{\rho}{EI}\right)\frac{1}{2}\int_0^L EIw_{zz}^2(z,t)\mathrm{d}z + \frac{1}{2}\int_0^L \tilde{d}^2(z,t)\mathrm{d}z\right) \\
&\le p_5\left(1 + 4L^2\pi^{-2}\frac{\rho}{EI}\right)\left(\frac{1}{2}\int_0^L \rho w_t^2(z,t)\mathrm{d}z + \frac{1}{2}\int_0^L EIw_{zz}^2(z,t)\mathrm{d}z + \frac{1}{2}\int_0^L \tilde{d}^2(z,t)\mathrm{d}z\right) \\
&= p_5\left(1 + 4L^2\pi^{-2}\frac{\rho}{EI}\right)\frac{1}{2}\|\boldsymbol{q}\|_H^2.
\end{aligned} \tag{4.54}$$

Based on inequality (4.54), we can obtain that $V(t)$ is equivalent to $\frac{1}{2}\|\boldsymbol{q}\|_H^2$ in H. If inequalities (4.19) and (4.20) are satisfied, we can deduce from inequality (4.34) that $\dot{V}(t) \le 0$. Therefore, $\boldsymbol{A}$ is a dissipative operator.

Let $\boldsymbol{g} = \begin{bmatrix} g_1 & g_2 & g_3 \end{bmatrix} \in H$ be given. Consider the following equation

$$\boldsymbol{A}\boldsymbol{q} = \boldsymbol{g} \tag{4.55}$$

where

$$g_1 = q_2 \tag{4.56}$$

$$g_2 = -\frac{EI}{\rho} q_{1,zzzz} + \frac{K_3}{\rho} q_2 + \frac{1}{\rho} q_3 \tag{4.57}$$

$$g_3 = -kq_3. \tag{4.58}$$

The unique solution of Eq. (4.51) can be obtained from Eqs. (4.56)–(4.58) as follows

$$q_1(z) = -\frac{\rho}{EI} \int_0^z \int_0^{\xi_1} \int_0^{\xi_2} \int_0^{\xi_3} \left(g_2(\xi_4) - \frac{K_3}{\rho} g_1(\xi_4) + \frac{1}{\rho k} g_3(\xi_4) \right) \mathrm{d}\xi_4 \mathrm{d}\xi_3 \mathrm{d}\xi_2 \mathrm{d}\xi_1 + \sum_{j=0}^{3} \sigma_j z^j \tag{4.59}$$

$$q_2 = g_1 \tag{4.60}$$

$$q_3 = -\frac{1}{k} g_3 \tag{4.61}$$

where $\sigma_0, \sigma_1, \sigma_2, \sigma_3$ are uniquely determined by boundary conditions Eqs. (4.2)–(4.4).

Hence the Eq. (4.55) has a unique solution $\boldsymbol{q} \in \mathcal{D}(\boldsymbol{A})$, which means that $\boldsymbol{A}^{-1}$ exists and maps H into $H_0^4(\Omega) \times H_0^2(\Omega) \times R$. Furthermore, since $\boldsymbol{A}^{-1}$ maps every bounded set of H into bounded sets of $H_0^4(\Omega) \times H_0^2(\Omega) \times R$ and the embedding of the latter space into H is compact [6], it follows that $\boldsymbol{A}^{-1}$ is a compact operator.

Given $\lambda > 0$ and $\boldsymbol{g} \in H$, let

$$(\lambda \boldsymbol{I} - \boldsymbol{A})\boldsymbol{q} = \boldsymbol{A}\left(\lambda \boldsymbol{A}^{-1} - \boldsymbol{I}\right)\boldsymbol{q} = \boldsymbol{g} \tag{4.62}$$

where $\boldsymbol{I}$ is a unit operator.

It proves that for $0 < \lambda < \left\|\boldsymbol{A}^{-1}\right\|^{-1}$ in Eq. (4.62) has a unique solution $\boldsymbol{q} \in \mathcal{D}(\boldsymbol{A})$ by contraction mapping theorem [7]. Hence the operator $\lambda \boldsymbol{I} - \boldsymbol{A}$ is onto H for $0 < \lambda < \left\|\boldsymbol{A}^{-1}\right\|^{-1}$. Moreover, the operator $\lambda \boldsymbol{I} - \boldsymbol{A}$ is onto H for all $\lambda > 0$ [8]. Therefore, we can have that $\mathcal{D}(\boldsymbol{A})$ is dense in H [7]. Based on the Lumer-Phillips theorem [9], the operator $\boldsymbol{A}$ generates a C_0-semigroup of contraction $T(t)$ on H.

Hence there exists a unique solution of Eq. (4.47) for all $\boldsymbol{q}(0) \in H$ [10] and the closed-loop system exists a unique solution.

The results of this section are presented as: boundary and distributed controllers are designed based on the proposed disturbance observer. When there exist $p_{11}, p_{12}, p_{22} \in R^+$ and the LMIs (4.18)–(4.20) are satisfied, the existence of the unique solution and the exponential stability of the closed-loop system can be verified by the operator semigroup theory and Lyapunov's direct method respectively.

4.4 Simulations

Numerical simulations are provided in this section to verify the validness of the proposed control law. The parameters of the Euler–Bernoulli beam are listed below

$$L = 1\,m$$

$$EI = 2\,Nm^2$$

$$\rho = 5\,kgm^{-1}.$$

If the parameters of the beam are uncertain, the ranges of parameters are given as

$$\overline{EI}_{\min} = 1.7\,Nm^2$$

$$\overline{EI}_{\max} = 2.5\,Nm^2$$

$$\overline{\rho}_{\min} = 4.7\,kgm^{-1}$$

$$\overline{\rho}_{\max} = 5.3\,kgm^{-1}.$$

By using LMI toolbox of MATLAB, the feasible solutions to LMIs (4.18)–(4.20) can be obtained as

$$p_{11} = 1.2875,\ p_{12} = 0.4028,\ \ p_{22} = 0.9834,\ \ k = 5$$

$$\boldsymbol{K} = [-1.3238, -0.5693]$$

$$K_3 = -0.4165.$$

In this chapter, the discrete time step and time step of the finite difference method are chosen as $\Delta t = 1.5\times10^{-3}$s and $\Delta x = 0.1$m. The external distributed disturbance is given as $d(z,t) = -0.1\sin(0.01\pi(z+t))$. The simulation results are shown in Figs. 4.2, 4.3, 4.4, 4.5, 4.6, 4.7, 4.8, 4.9, 4.10, 4.11, 4.12 and 4.13 with equivalent interval sampling.

Without control input, the dynamic responses of $w(z,t)$, $w_z(z,t)$ and $w_t(z,t)$ are respectively shown in Figs. 4.2, 4.3 and 4.4 and the dynamic responses of $w(L,t)$, $w_z(L,t)$ and $w_t(L,t)$ are shown in Fig. 4.5. It is clear that the vibration of the beam is large and the open-loop system is unstable within a given time.

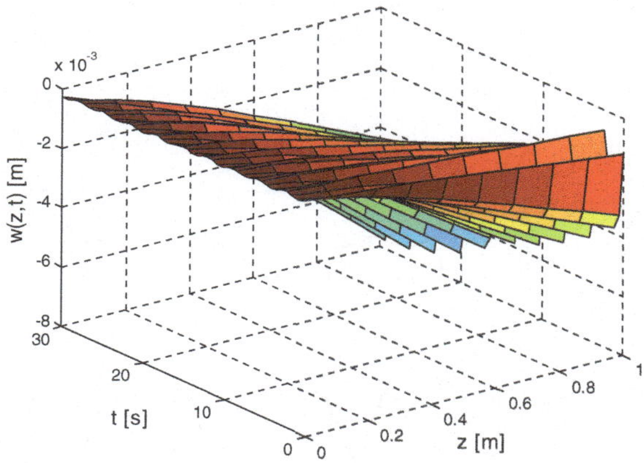

Fig. 4.2 $w(z, t)$ in the open-loop system

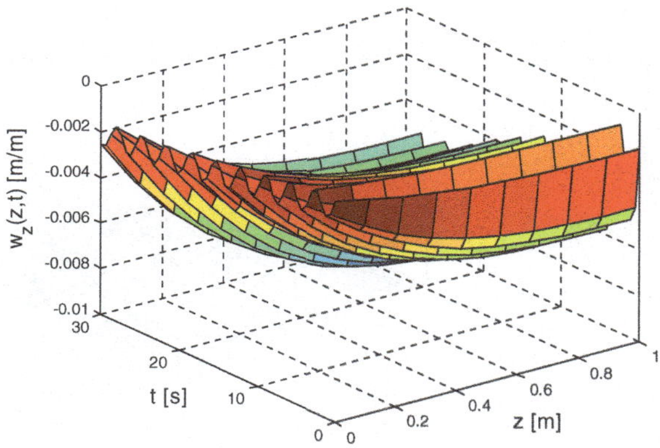

Fig. 4.3 $w_z(z, t)$ in the open-loop system

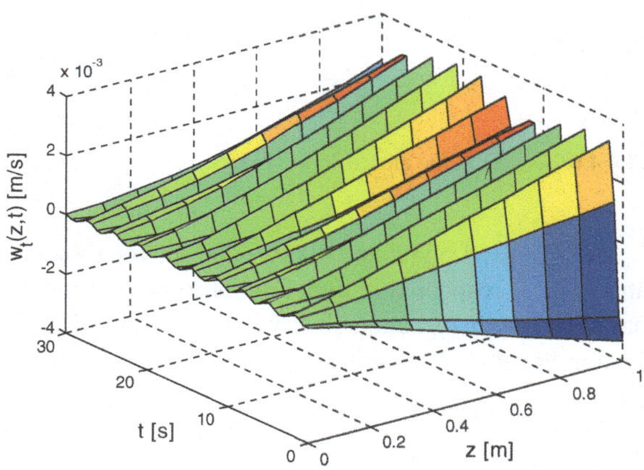

Fig. 4.4 $w_t(z, t)$ in the open-loop system

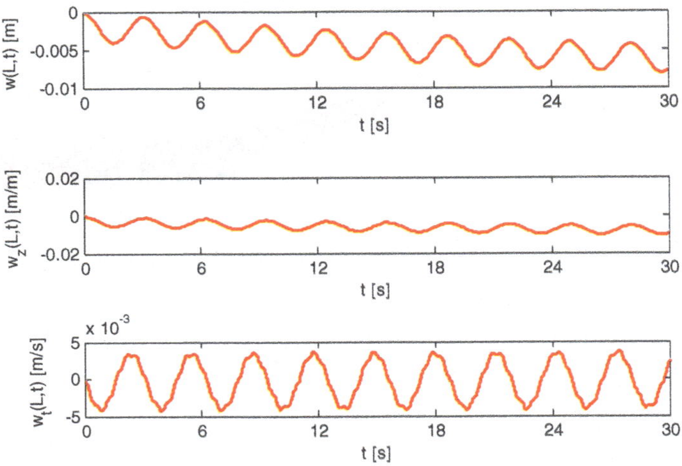

Fig. 4.5 $w(L,t)$, $w_z(L,t)$ and $w_t(L,t)$ in the open-loop system

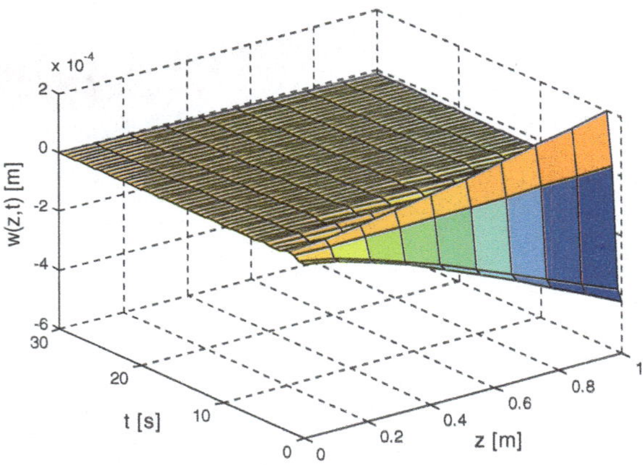

Fig. 4.6 $w(z,t)$ in the closed-loop system

From Figs. 4.6, 4.7 and 4.8, we can see the dynamic responses of $w(z,t)$, $w_z(z,t)$ and $w_t(z,t)$ respectively under the proposed boundary and distributed control inputs based on the disturbance observer. $w(L,t)$, $w_z(L,t)$ and $w_t(L,t)$ in the closed-loop system are shown in Fig. 4.9.

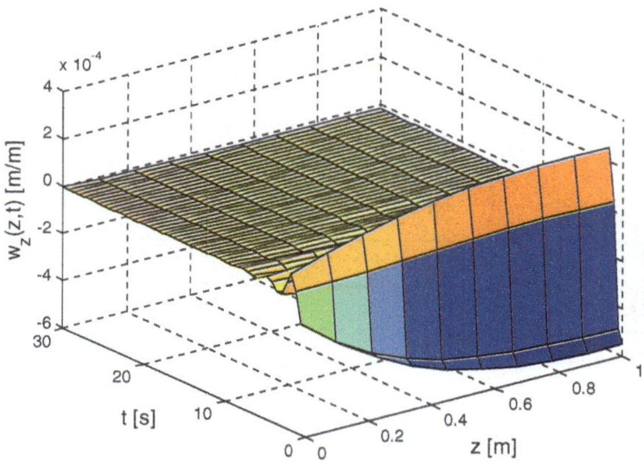

Fig. 4.7 $w_z(z, t)$ in the closed-loop system

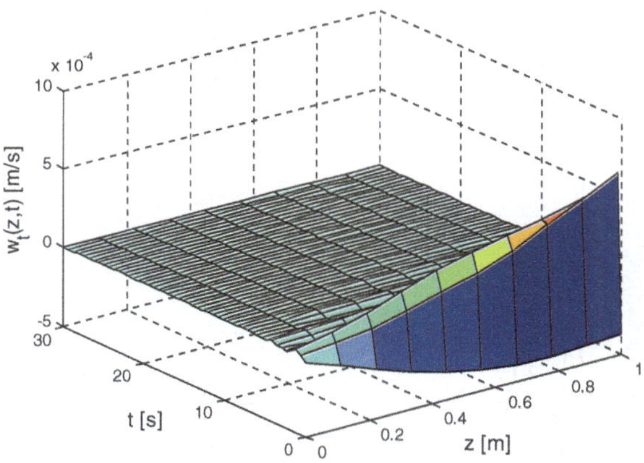

Fig. 4.8 $w_t(z, t)$ in the closed-loop system

By comparing Figs. 4.2, 4.3, 4.4 and 4.5 with Figs. 4.6, 4.7, 4.8 and 4.9, it is clear that the values of $w(z, t)$, $w_z(z, t)$ and $w_t(z, t)$ in the closed-loop system are much smaller than the values in the open-loop system. It is apparent that values of $w(z, t)$, $w_z(z, t)$ and $w_t(z, t)$ converge to zero after several seconds, which means that based on the observer, the proposed control can suppress the vibration quickly and make the system exponentially stable.

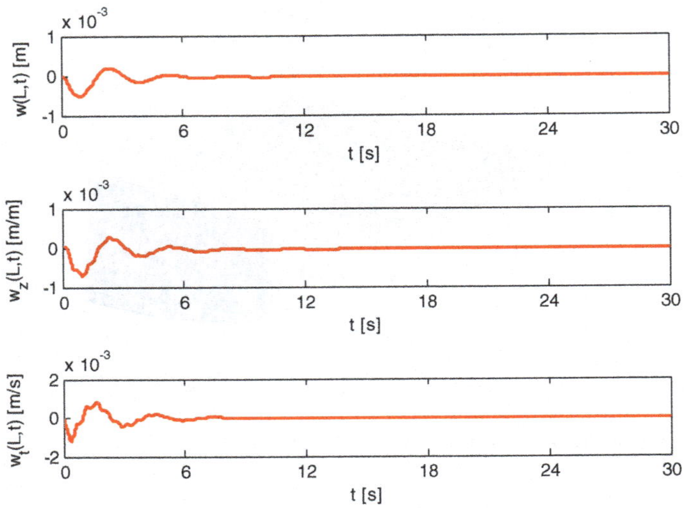

Fig. 4.9 $w(L,t)$, $w_z(L,t)$ and $w_t(L,t)$ in the closed-loop system

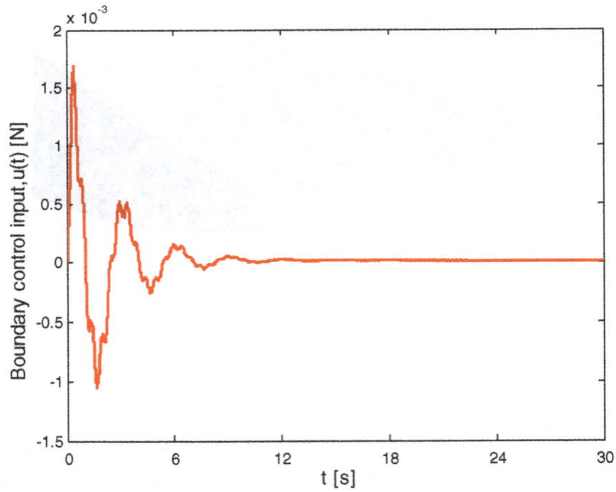

Fig. 4.10 Boundary control input $u(t)$

The boundary control law $u(t)$ and distributed control $F(z,t)$ are shown in Figs. 4.10 and 4.11 respectively.

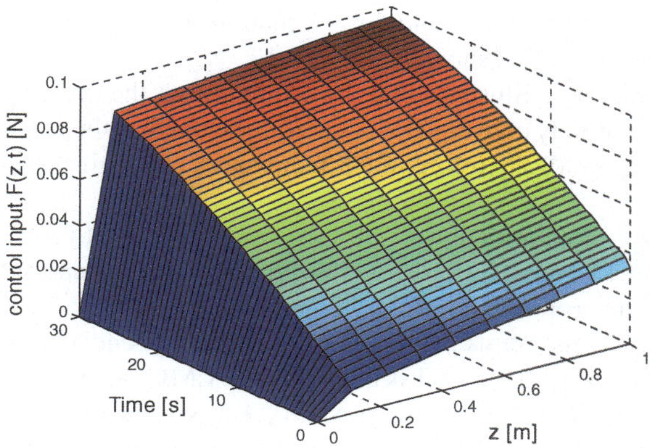

Fig. 4.11 Distributed control input $F(z, t)$

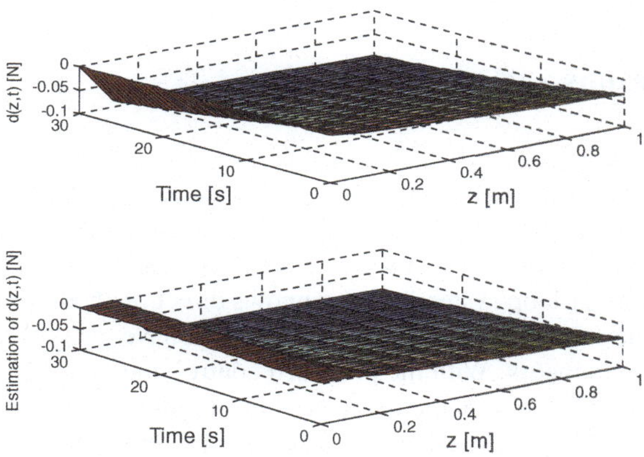

Fig. 4.12 External distributed disturbance $d(z, t)$ and its estimation $\hat{d}(z, t)$

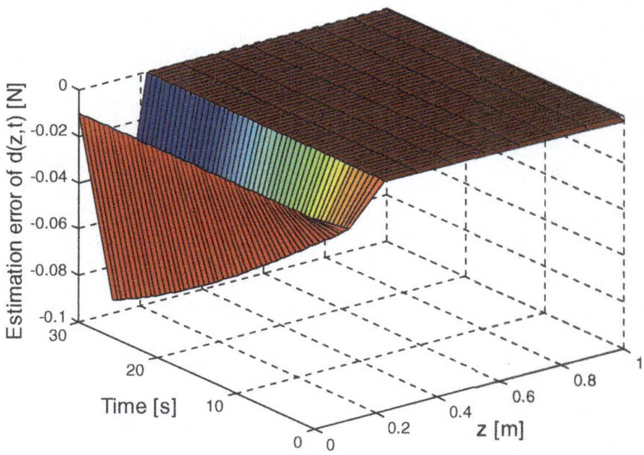

Fig. 4.13 Estimation error of $d(z, t)$

The distributed disturbance and its estimation are shown in Fig. 4.12. The estimation error of $d(z,t)$ is showed in Fig. 4.13.

Figures 4.12 and 4.13 illustrate that the estimation of the distributed disturbance can be fast and accurately converge to its true values, that is, when $t \to \infty, \hat{d}(z,t) \to d(z,t)$ and $\tilde{d}(z,t) \to 0$. Therefore, the effectiveness of the disturbance observer is verified.

To summarize, the simulation results in Figs. 4.2, 4.3, 4.4, 4.5, 4.6, 4.7, 4.8, 4.9, 4.10, 4.11, 4.12 and 4.13 agree well with the stability analysis and prove the effectiveness of the proposed control scheme.

The simulation programs are given in the Appendixes 4.1–4.3, where chapter4_LMI1.m in Appendix 4.1 is used to solve LMIs (4.18) and (4.20) to obtain p_{11}, p_{12}, p_{22}, chapter4_LMI2.m in Appendix 4.2 is used to solve LMI (4.19) to obtain $\boldsymbol{K}$, and chapter4.m in Appendix 4.3 is the main program containing proposed control laws in Eqs. (4.6)–(4.8) and observers in Eqs. (4.12) and (4.13).

4.5 Conclusions

In this chapter, an exponentially stable disturbance observer has been proposed to compensate the external distributed disturbance of the system. Based on the observer and original PDE model, LMI techniques have been introduced to design control law for the Euler–Bernoulli beam with uncertain parameters and analyze the exponential stability of the closed-loop system by Lyapunov's direct method. The existence and uniqueness of the solution of the closed-loop system have also been demonstrated by using semigroup theory. With the proposed control, the vibration suppression is realized even under the systematic parameter uncertainty and external disturbance. The effectiveness of the proposed disturbance observer and control scheme has been proved by numerical simulations.

Appendix 4.1: Simulation Program: chapter4_LMI1.m

```
clear all;
close all;

L=1;

mmin=4.7;
m=5;
mmax=5.3;

EImin=1.7;
EI=2;
EImax=2.5;

z=1;
z1=0;
k=5;
k3=-0.4165;

p11=sdpvar(1,1);
p12=sdpvar(1,1);
p22=sdpvar(1,1);

P=sdpvar(2,2,'symmetric');
T=sdpvar(4,4,'symmetric');
T0=sdpvar(4,4,'symmetric');

L1=(p11>0);

L2=(p12>0);

L3=(p22>0);

P=[p11*mmin p12*mmax*L;p12*mmax*L p22];
L4=(P>0);

T=[-p12*mmin+k3*2*p11  -p22  p12*k3*z  p11;-p22  -1.5*p12*EImin  0  0;p12*k3*z  0  -
    3*p12*EImin*0.125*(pi^2)*(L^(-2)) p12*z;p11 0 p12*z -k];
L5=(T<0);

T0=[-p12*mmin+k3*2*p11  -p22  p12*k3*z1  p11;-p22  -1.5*p12*EImin  0  0;p12*k3*z1  0  -
    3*p12*EImin*0.125*(pi^2)*(L^(-2)) p12*z1;p11 0 p12*z1 -k];
L6=(T0<0);

L=L1+L2+L3+L4+L5+L6;

solvesdp(L);

M=[p11 p12 p22 k3];

p11=double(M(1,1));
p12=double(M(1,2));
p22=double(M(1,3));
```

Appendix 4.2: Simulation Program: chapter4_LMI2.m

```
clear all;
close all;

L=1;
mmin=4.7;
m=5;
mmax=5.3;

EImin=1.7;
EI=2;
EImax=2.5;

p11=1.287507600188427;
p12=0.402778567018183;
p22=0.983443562737251;
k3=-0.4165;


k=5;

K1=sdpvar(1,1);
K2=sdpvar(1,1);

N=[p12*mmax*L+2*K1*p11 p22+K2*p11+K1*p12*L;p22+K2*p11+K1*p12*L 2*K2*p12*L];
L6=(N<0);

solvesdp(L6);

M=[K1 K2];

K1=double(M(1,1));
K2=double(M(1,2));
```

Appendix 4.3: Simulation Program: chapter4.m

```
close all;
clear all;
N=10;
n=20000;

t=30;L=1;
h=L/N;
T=t/n;

EI=2;rho=5;

K1=-1.323848214370248;
K2=-0.569348386014983;
K3=-0.4165;

k=5;

b=(4*T^2*EI)/(rho*h^4);
c=-(T^2*EI)/(rho*h^4);
a=2-(6*T^2*EI)/(rho*h^4);
```

```
for j=1:n+1
      u(j)=0;

      for i=2:N+1

      d0(j,i)=-0.1*sin(0.01*pi*j*T+0.01*pi*i);

      end

      d0(j,N+2)=0;
end

z=zeros(n+1,N+4);

for j=1:n+1
      z(j,3)=0;
end

for j=1:n+1
      for i=1:N+2
    u0(j,i)=0;
    ds(j,i)=0;
    dp(j,i)=0;
      end
end

for i=5:N+2
   z(1,i)=0;
   dz1=0;
   z(2,i)=0;
end

z(2,1)=0;
z(2,2)=0;

zLt(2)=(z(2,N+2)-z(1,N+2))/T;
zL(2)=(z(2,N+2)-z(2,N+1))/h;

z(2,N+3)=2*z(2,N+2)-z(2,N+1);
z(2,N+4)=z(2,N)-4*z(2,N+1)+4*z(2,N+2)-(2*h^3/EI)*(u(1));

u(2)=K1*zLt(2)+K2*zL(2);
```

```
%u(2)=0;

for j=3:n+1
dzt(j,1)=(z(j,3)-z(j-1,3))/T;
dzt(j,2)=(z(j,4)-z(j-1,4))/T;
dz(j,1)=(z(j,3)-z(j,2))/h;
dz(j,2)=(z(j,4)-z(j,3))/h;
u0(j,1)=-K3*dzt(j,1)-dp(j,1);
u0(j,2)=-K3*dzt(j,2)-dp(j,2);
end

for j=3:n+1
    for i=3:N+2

        z(j,i)=c*(z(j-1,i-2)+z(j-1,i+2))+b*(z(j-1,i-1)+z(j-1,i+1))+a*z(j-1,i)-z(j-
2,i)+((T^2)/rho)*(d0(j-1,i)+u0(j-1,i));
        dzt(j,i-2)=(z(j,i)-z(j-1,i))/T;
        dz(j,i-2)=(z(j,i)-z(j,i-1))/h;

        ds(j,i)=ds(j-1,i)+T*(-k*(-EI*(z(j-1,i+2)-4*z(j-1,i+1)+6*z(j-1,i)-4*z(j-1,i-1)+z(j-1,i-
2))/(h^4)+u0(j-1,i))-k*dp(j-1,i));
        dp(j,i)=ds(j,i)+k*rho*(z(j,i)-z(j-1,i))/T;
        u0(j,i)=-K3*dzt(j,i-2)-dp(j,i);%with distributed control
        %u0(j,i)=0;%without distributed control

    end

        z(j,1)=0;
        z(j,2)=0;

        zLt(j)=(z(j,N+2)-z(j-1,N+2))/T;
        zL(j)=(z(j,N+2)-z(j,N+1))/h;

        z(j,N+3)=2*z(j,N+2)-z(j,N+1);
        z(j,N+4)=z(j,N)-4*z(j,N+1)+4*z(j,N+2)-(2*h^3/EI)*(u(j-1));

        u(j)=K1*zLt(j)+K2*zL(j);%with boundary control
        %u(j)=0;%without boundary control

end

zshort=zeros(n/400,N);
dzshort=zeros(n/400,N);
dztshort=zeros(n/400,N);
```

```
for i=1:N
for j=1:n/400
            zshort(j,i)=z(j*400,i+2);
            dztshort(j,i)=dzt(j*400,i);
            dzshort(j,i)=dz(j*400,i);
            d0short(j,i)=d0(j*400,i);
            dpshort(j,i)=dp(j*400,i);
            u0short(j,i)=u0(j*400,i+1);
      end
end
for j=1:n
            ushort(j)=u(j);
            z1(j)=z(j,12);
            z2(j)=dzt(j,10);
            z3(j)=dz(j,10);
      end
tshort=linspace(0,t,n/400);
y=linspace(0,L,N);
[y1,tshort1]=meshgrid(y,tshort);

figure(1);

surf(y1,tshort1,zshort);
xlabel('z [m]'); ylabel('t [s]');zlabel('w(z,t) [m]');

figure(2);
surf(y1,tshort1,dztshort);
xlabel('z [m]'); ylabel('t [s]');zlabel('w_t(z,t) [m/s]');

figure(3);
surf(y1,tshort1,dzshort);
xlabel('z [m]'); ylabel('t [s]');zlabel('w_z(z,t) [m/m]');

figure(4);
plot(ushort,'r','linewidth',2);
```

```
xlabel('t [s]');ylabel('control input,u(t) [N]');
set(gca,'xticklabel',[0 3 6 9 12 15 18 21 24 27 30]);

figure(5);
subplot(3,1,1);
plot(z1,'r','linewidth',2);
xlabel('t [s]');ylabel('w(L,t) [m]');
set(gca,'xticklabel',[0 3 6 9 12 15 18 21 24 27 30]);
subplot(3,1,2);

plot(z3,'r','linewidth',2);
xlabel('t [s]');ylabel('w_z(L,t) [m/m]');
set(gca,'xticklabel',[0 3 6 9 12 15 18 21 24 27 30]);
subplot(3,1,3);
plot(z2,'r','linewidth',2);
xlabel('t [s]');ylabel('w_t(L,t) [m/s]');
set(gca,'xticklabel',[0 3 6 9 12 15 18 21 24 27 30]);

figure(6);
subplot(211);
surf(y1,tshort1,d0short);
xlabel('z [m]'); ylabel('Time [s]');zlabel('d(z,t) [N]');
subplot(212);
surf(y1,tshort1,dpshort);
xlabel('z [m]'); ylabel('Time [s]');zlabel('Estimation of d(z,t) [N]');

figure(7);
surf(y1,tshort1,d0short-dpshort);
xlabel('z [m]'); ylabel('Time [s]');zlabel('Estimation error of d(z,t) [N]');

figure(8);
surf(y1,tshort1,u0short);
xlabel('z [m]'); ylabel('Time [s]');zlabel('control input,F(z,t) [N]');
```

References

1. S.S. Ge, T.H. Lee, J.Q. Gong, A robust distributed controller of a single-link SCARA/Cartesian smart materials robot. Mechatronics **9**(1), 65–93 (1999)
2. S.S. Ge, T.H. Lee, J.Q. Gong, Dynamic modeling of a smart materials robot. AIAA J. **36**(8), 1466–1478 (1988)
3. G. Montseny, J. Audounet, D. Matignon, Fractional integrodifferential boundary control of the Euler-Bernoulli beam, in *Proceedings of the 36th Conference on Decision and Control* (San Diego, Califomia, USA, 1997), pp. 4973–4978
4. W.H. Chen, D.J. Balance, P.J. Gawthrop, J.J. Gribble, J. O'Reilly, A nonlinear disturbance observer for two link robotic manipulators., in *38th IEEE Conference on Decision and Control* (Phoenix, Arizona, USA, 1999)
5. B.Z. Guo, Riesz basis property and exponential stability of controlled Euler-Bernoulli beam equations with variable coefficients. SIAM J. Control Optim. **40**(6), 1905–1923 (2002)
6. H. Tanabe, *Equations of Evolution* (Pitman, London, UK, 1979).
7. T.D. Nguyen, O. Egeland, Observer design for a flexible robot arm with a tip load, in *American Control Conference* (Portland, OR, 2005), pp. 1389–1394
8. A. Pazy, *Semigroups of Linear Operators and Applications to Partial Differential Equations* (Springer, New York, NY, 1983).
9. Z. Liu, S. Zheng, *Semigroups Associated with Dissipative Systems* (Chapman & Hall/CRC, Boca Raton, 1999).
10. C.D. Rahn, *Mechatronic Control of Distributed Noise and Vibration* (Springer, New York, NY, 2001).

Chapter 5
LMI-Based Boundary and Distributed Control Design for a Flexible String Subject to Disturbance

5.1 Introduction

With the progress of technology, the flexible systems are widely used in various fields. Meanwhile, the flexible string systems are increasingly pervasive in numerous engineering areas such as offshore mooring lines, drill strings and conveyor belts because of their light weight, low energy consumption and fast motion. Notably, there are external disturbances which widely exist in practical processes. If the external disturbances of the flexible string system is not considered in the control design, the control performance will be greatly affected. In order to avoid this problem, disturbance observers are introduced to compensate the unknown external disturbances and improve the control performance.

In this chapter, boundary and distributed controllers are proposed to reduce the vibration of the flexible string with external disturbances. Boundary and distributed disturbance observers, which are proved exponentially stable, are designed to compensate unknown disturbance. Based on the proposed disturbance observers, the boundary and distributed controllers are developed by LMI and the asymptotic stability of the system with controllers is demonstrated by Lyapunov's direct method. Extensive numerical simulations are presented to validate the performance of the proposed control laws.

5.2 Problem Formulation

We consider that a flexible string without payload on the free end moves in the horizontal direction. The configuration of the flexible string is shown in Fig. 5.1. WOZ represents the inertial coordinate system attached to the string; L is the length of the string; ρ is the uniform density of the string; T is the string's tension; γ is the damping coefficient of the string; $w(z, t)$ denotes the vibratory deflection at

X. Xing and J. Liu, *Vibration Control Methods of Mechanical Distributed Parameter Systems*, Springer Tracts in Mechanical Engineering,
https://doi.org/10.1007/978-981-16-1532-0_5

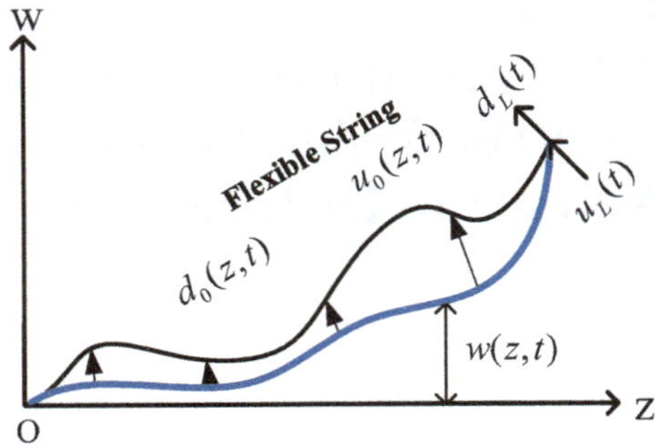

Fig. 5.1 Configuration of the flexible string

z; $d_L(t)$ is the external boundary disturbance and $d_0(z,t)$ is the external distributed disturbance. $u_L(t)$ is the boundary control force and $u_0(z,t)$ is the distributed control force.

Remark 5.1 The following notations are introduced in this chapter for clarity: $(*)_z = \frac{\partial(*)}{\partial z}$, $(*)_{zz} = \frac{\partial^2(*)}{\partial z^2}$, $(*)_t = \frac{\partial(*)}{\partial t}$, $(*)_{tt} = \frac{\partial^2(*)}{\partial t^2}$.

The governing equation of the flexible string can be expressed as [1]

$$\rho w_{tt}(z,t) = T w_{zz}(z,t) - \gamma w_t(z,t) + u_0(z,t) + d_0(z,t) \tag{5.1}$$

and the boundary conditions are

$$T w_z(L,t) = u_L(t) + d_L(t) \tag{5.2}$$

$$w(0,t) = 0. \tag{5.3}$$

Assumption 5.1 All the parameters of the flexible string are known and the uncertain boundary and distributed disturbances are bounded.

Assumption 5.2 [2] It is reasonable to assume that $\dot{d}_L(t) = 0$ and $\dot{d}_0(z,t) = 0$ in this chapter since the disturbances vary slowly relative to the observer dynamics.

Assumption 5.3 The unknown disturbances $d_L(t)$ and $d_0(z,t)$ are all bounded which means $|d_L(t)| \le D_1$ and $|d_0(z,t)| \le D_2$, where D_1 and D_2 are two positive constants.

5.3 Observer Design

Exponentially stable boundary and distributed disturbance observers are proposed to estimate the external boundary and distributed disturbances in this section.

Define the estimation error of $d_L(t)$ and $d_0(z,t)$ as follows

$$\tilde{d}_L(t) = d_L(t) - \hat{d}_L(t)$$

$$\tilde{d}_0(z,t) = d_0(z,t) - \hat{d}_0(z,t)$$

where $\hat{d}_L(t)$ is the estimation of the boundary disturbance and $\hat{d}_0(z,t)$ is the estimation of the distributed disturbance.

In order to correct the error between the observed values of observers and the real ones, we propose

$$\dot{\hat{d}}_L(t) = k_1\tilde{d}_L(t) \tag{5.4}$$

$$\dot{\hat{d}}_0(z,t) = k_2\tilde{d}_0(z,t) \tag{5.5}$$

where k_1 and k_2 are two positive constants.

Substituting Eq. (5.2) into Eq. (5.4) and substituting Eq. (5.1) into Eq. (5.5), the boundary disturbance observer is designed as

$$\dot{\hat{d}}_L(t) = -k_1(-Tw_z(L,t) + u_L(t)) - k_1\hat{d}_L(t) \tag{5.6}$$

and the distributed disturbance observer is designed as

$$\dot{\eta}(z,t) = -k_2(Tw_{zz}(z,t) - \gamma w_t(z,t) + u_0(z,t)) - k_2\hat{d}_0(z,t) \tag{5.7}$$

$$\hat{d}_0(z,t) = \eta(z,t) + k_2\rho w_t(z,t) \tag{5.8}$$

where $\eta(z,t)$ is an intermediate variable.

Theorem 5.1 *The proposed boundary and distributed observers in* Eqs. (5.6)–(5.8) *are exponentially stable which means* $d_L(t) \to 0$ *and* $d(z,t) \to 0$ *when* $t \to \infty$.

Proof According to the designed observers, two Lyapunov functions of observers are defined as

$$V_{ob1}(t) = \frac{1}{2}\tilde{d}_L^2(t) \tag{5.9}$$

$$V_{ob2}(t) = \frac{1}{2}\rho\int_0^L \tilde{d}_0^2(z,t)\mathrm{d}z. \tag{5.10}$$

After differentiating Eq. (5.9) and substituting Eq. (5.6) into $\dot{V}_{ob1}$, we get

$$\dot{V}_{ob1}(t) = \tilde{d}_L(t)\dot{\tilde{d}}_L(t)$$

$$
\begin{aligned}
&= -\tilde{d}_L(t)\Big(k_1(Tw_z(L,t) - u_L(t)) - k_1\hat{d}_L(t)\Big) \\
&= -k_1\tilde{d}_L^2(t) \\
&= -2k_1 V_{ob1}(t).
\end{aligned} \tag{5.11}
$$

By integrating Eq. (5.11) from 0 to t, the solution of Eq. (5.11) is

$$
V_{ob1}(t) = V_{ob1}(0)\exp(-2k_1 t). \tag{5.12}
$$

From Eq. (5.12), $V_{ob1}(t) \to 0$ for $t \to \infty$ can be obtained. Hence, we can deduce from Eq. (5.9) that $\tilde{d}_L(t) \to 0$ for $t \to \infty$.

Similarly, substituting Eqs. (5.7) and (5.8) into $\dot{V}_{ob2}$, we can obtain

$$
\begin{aligned}
\dot{V}_{ob2}(t) &= \rho\int_0^L \tilde{d}_0(z,t)\dot{\tilde{d}}_0(z,t)\mathrm{d}z \\
&= \rho\int_0^L \tilde{d}_0(z,t)(-\dot{\eta}(z,t) - k_2\rho w_{tt}(z,t))\mathrm{d}z \\
&= \rho\int_0^L \tilde{d}_0(z,t)\big[k_2(Tw_{zz}(z,t) - \gamma w_t(z,t) + u_0(z,t)) \\
&\quad + k_2\hat{d}_0(z,t) - k_2\rho w_{tt}(z,t)\big]\mathrm{d}z \\
&= -\rho\int_0^L k_2\tilde{d}_0^2(z,t)\mathrm{d}z \\
&= -2k_2 V_{ob2}(t).
\end{aligned} \tag{5.13}
$$

By integrating Eq. (5.13) from 0 to t, the solution of Eq. (5.13) is

$$
V_{ob2}(t) = V_{ob2}(0)\exp(-2k_2 t). \tag{5.14}
$$

Therefore, from Eq. (5.14) we have $\tilde{d}_0(z,t) \to 0$ for $t \to \infty$.

Consequently, we can conclude that the boundary and distributed observers are exponentially stable and the uncertain disturbance of the flexible string can be observed effectively by the proposed observers.

5.4 Control Design and Stability Analysis

The control aim of this chapter is to design boundary and distributed controllers to restrain the elastic vibration of the string, that is, not only $w(z, t)$ but also $w_z(z, t)$ and $w_t(z, t)$ converge to zero for $t \to \infty$. In this section, based on LMI and the proposed observers in Sect. 5.3, we design boundary and distributed control laws and the closed-loop system asymptotic stability is verified through Lyapunov's direct method.

To realize control objective mentioned above, the boundary and distributed control laws are designed as

$$u_L(t) = -K_2 w_t(L, t) - \hat{d}_L(t) \tag{5.15}$$

$$u_0(z, t) = -K_1 w_z(z, t) - \hat{d}_0(z, t) \tag{5.16}$$

where K_1 and K_2 are two positive constants.

Theorem 5.2 *Define matrix $\boldsymbol{P}(z)$ as $\boldsymbol{P}(z) = \begin{bmatrix} \frac{T}{\rho} p & \beta z \\ \beta z & p \end{bmatrix}$ and matrix $\boldsymbol{\Phi}(z)$ as*

$$\boldsymbol{\Phi}(z) = \begin{bmatrix} \beta + \frac{1}{T\rho}\beta K_2^2 - 2K_2 p \frac{1}{L\rho} & 0 & 0 & -\frac{1}{T\rho}\beta K_2 + p\frac{1}{L\rho} & 0 \\ 0 & -\frac{T}{\rho}\beta + z\left(s_0 - 2\beta\frac{K_1}{\rho} + \beta^2 \frac{1}{\rho^2 s_1}\right) + 2\delta\frac{T}{\rho}p & -\frac{K_1}{\rho}p + 2\delta\beta z & 0 & 0 \\ 0 & -\frac{K_1}{\rho}p + 2\delta\beta z & -\beta + \beta^2\frac{\gamma^2}{\rho^2 s_0}z - \frac{\gamma}{\rho}2p + 2\delta p & 0 & \frac{1}{\rho}p \\ -\frac{1}{T\rho}\beta K_2 + p\frac{1}{L\rho} & 0 & 0 & \frac{1}{T\rho}\beta - \frac{k_1}{L} + \frac{\delta}{L} & 0 \\ 0 & 0 & \frac{1}{\rho}p & 0 & s_1 z - \rho k_2 + \delta\rho \end{bmatrix}$$

If there exist β, $p \in \Re^+$ and the following inequalities are satisfied

$$\boldsymbol{P}(z) \geq 0 \tag{5.17}$$

$$\boldsymbol{\Phi}(z) \leq 0 \tag{5.18}$$

the boundary and distributed control laws proposed by Eqs. (5.15) *and* (5.16) *can ensure that the flexible string is asymptotically stable in the following sense: $w(z, t)$, $w_z(z, t)$ and $w_t(z, t)$ converge to zero for $t \to \infty$.*

Proof Substituting $u_0(z, t)$ and $u_L(t)$ into Eqs. (5.1)–(5.3), we have the closed-loop system as follows

$$\rho w_{tt}(z, t) = T w_{zz}(z, t) - \gamma w_t(z, t) - K_1 w_z(z, t) + \tilde{d}_0(z, t) \tag{5.19}$$

$$Tw_z(L,t) = -K_2 w_t(L,t) + \tilde{d}_L(t) \tag{5.20}$$

$$w(0,t) = 0. \tag{5.21}$$

Define a Lyapunov function as

$$V(t) = V_1(t) + V_{ob1}(t) + V_{ob2}(t). \tag{5.22}$$

Let $\boldsymbol{x}(z,t) = \left[w_z(z,t)\; w_t(z,t)\right]^T$ and $V_1(t)$ is designed as

$$\begin{aligned} V_1(t) &= \int_0^L \boldsymbol{x}^T(z,t)\boldsymbol{P}(z)\boldsymbol{x}(z,t)\mathrm{d}z \\ &= \int_0^L \frac{T}{\rho} p w_z^2(z,t)\mathrm{d}z + \int_0^L 2\beta z w_z(z,t) w_t(z,t)\mathrm{d}z + \int_0^L p w_t^2(z,t)\mathrm{d}z \end{aligned} \tag{5.23}$$

where β and p are two positive constants.

Considering Eqs. (5.9), (5.10), (5.22) and (5.23), we can achieve

$$\begin{aligned} V(t) &= V_1(t) + V_{ob1}(t) + V_{ob2}(t) \\ &= \int_0^L \boldsymbol{x}^T(z,t)\boldsymbol{P}(z)\boldsymbol{x}(z,t)\mathrm{d}z + \frac{1}{2}\tilde{d}_L^2(t) + \frac{1}{2}\rho\int_0^L \tilde{d}_0^2(z,t)\mathrm{d}z \\ &= \int_0^L \frac{T}{\rho} p w_z^2(z,t)\mathrm{d}z + \int_0^L 2\beta z w_z(z,t) w_t(z,t)\mathrm{d}z + \int_0^L p w_t^2(z,t)\mathrm{d}z \\ &\quad + \frac{1}{2}\tilde{d}_L^2(t) + \frac{1}{2}\rho\int_0^L \tilde{d}_0^2(z,t)\mathrm{d}z. \end{aligned} \tag{5.24}$$

Differentiating $V_1(t)$ with respect to time yields

$$\begin{aligned} \dot{V}_1(t) &= \int_0^L 2\frac{T}{\rho} p w_z(z,t) w_{zt}(z,t)\mathrm{d}z + \int_0^L 2\beta z w_{zt}(z,t) w_t(z,t)\mathrm{d}z \\ &\quad + \int_0^L 2\beta z w_z(z,t) w_{tt}(z,t)\mathrm{d}z + \int_0^L 2p w_t(z,t) w_{tt}(z,t)\mathrm{d}z. \end{aligned} \tag{5.25}$$

Substituting Eqs. (5.19)–(5.21) into the middle terms of $\dot{V}_1(t)$, using integration by parts, we can achieve

$$\begin{aligned}
&\int_0^L 2\beta z w_{zt}(z,t)w_t(z,t)\mathrm{d}z + \int_0^L 2\beta z w_z(z,t)w_{tt}(z,t)\mathrm{d}z \\
&= \int_0^L 2\beta z w_{zt}(z,t)w_t(z,t)\mathrm{d}z + \int_0^L \frac{T}{\rho}\beta z\left(w_z^2(z,t)\right)_z\mathrm{d}z \\
&\quad + \int_0^L 2\beta z w_z(z,t)\left(-\frac{\gamma}{\rho}w_t(z,t) - \frac{K_1}{\rho}w_z(z,t) + \frac{1}{\rho}\tilde{d}_0(z,t)\right)\mathrm{d}z \\
&= \int_0^L 2\beta z w_{zt}(z,t)w_t(z,t)\mathrm{d}z + \frac{T}{\rho}\beta z w_z^2(z,t)\big|_0^L - \int_0^L \frac{T}{\rho}\beta w_z^2(z,t)\mathrm{d}z \\
&\quad + \int_0^L 2\beta z w_z(z,t)\left(-\frac{\gamma}{\rho}w_t(z,t) - \frac{K_1}{\rho}w_z(z,t) + \frac{1}{\rho}\tilde{d}_0(z,t)\right)\mathrm{d}z \\
&= \int_0^L 2\beta z w_{zt}(z,t)w_t(z,t)dz + \frac{T}{\rho}\beta L w_z^2(L,t) - \int_0^L \frac{T}{\rho}\beta w_z^2(z,t)\mathrm{d}z \\
&\quad + \int_0^L 2\beta z w_z(z,t)\left(-\frac{\gamma}{\rho}w_t(z,t) - \frac{K_1}{\rho}w_z(z,t) + \frac{1}{\rho}\tilde{d}_0(z,t)\right)\mathrm{d}z.
\end{aligned} \tag{5.26}$$

By virtue of

$$\begin{aligned}
&\int_0^L 2\beta z w_{zt}(z,t)w_t(z,t)\mathrm{d}z \\
&= 2\beta z w_t^2(z,t)\big|_0^L - \int_0^L 2\beta z w_t(z,t)w_{zt}(z,t)\mathrm{d}z - \int_0^L 2\beta w_t^2(z,t)\mathrm{d}z,
\end{aligned}$$

we can get

$$\int_0^L 2\beta z w_{zt}(z,t)w_t(z,t)\mathrm{d}z$$

$$= \beta z w_t^2(z,t)\Big|_0^L - \int_0^L \beta w_t^2(z,t)\mathrm{d}z$$

$$= \beta L w_t^2(L,t) - \int_0^L \beta w_t^2(z,t)\mathrm{d}z. \tag{5.27}$$

Substituting Eq. (5.27) into Eq. (5.26), we have that

$$\begin{aligned}
&\int_0^L 2\beta z w_{zt}(z,t) w_t(z,t)\mathrm{d}z + \int_0^L 2\beta z w_z(z,t) w_{tt}(z,t)\mathrm{d}z \\
&= \beta L w_t^2(L,t) - \int_0^L \beta w_t^2(z,t)\mathrm{d}z + \frac{T}{\rho}\beta L w_z^2(L,t) - \int_0^L \frac{T}{\rho}\beta w_z^2(z,t)\mathrm{d}z \\
&+ \int_0^L 2\beta z w_z(z,t)\left(-\frac{\gamma}{\rho} w_t(z,t) - \frac{K_1}{\rho} w_z(z,t) + \frac{1}{\rho}\tilde{d}_0(z,t)\right)\mathrm{d}z.
\end{aligned} \tag{5.28}$$

Substituting Eq. (5.28) into Eq. (5.25), we get

$$\begin{aligned}
\dot{V}_1(t) = &\int_0^L 2\frac{T}{\rho} p w_z(z,t) w_{zt}(z,t)\mathrm{d}z + \beta L w_t^2(L,t) - \int_0^L \beta w_t^2(z,t)\mathrm{d}z \\
&+ \frac{T}{\rho}\beta L w_z^2(L,t) - \int_0^L \frac{T}{\rho}\beta w_z^2(z,t)\mathrm{d}z \\
&+ \int_0^L 2\beta z w_z(z,t)\left(-\frac{\gamma}{\rho} w_t(z,t) - \frac{K_1}{\rho} w_z(z,t) + \frac{1}{\rho}\tilde{d}_0(z,t)\right)\mathrm{d}z \\
&+ \int_0^L 2 p w_t(z,t) w_{tt}(z,t)\mathrm{d}z.
\end{aligned} \tag{5.29}$$

Substituting Eqs. (5.29), (5.11) and (5.13) into $\dot{V}(t)$, we have

$$\begin{aligned}
\dot{V}(t) &= \dot{V}_1(t) + \dot{V}_{ob1}(t) + \dot{V}_{ob2}(t) \\
&= \int_0^L 2\frac{T}{\rho} p w_z(z,t) w_{zt}(z,t)\mathrm{d}z + \beta L w_t^2(L,t) - \int_0^L \beta w_t^2(z,t)\mathrm{d}z
\end{aligned}$$

$$
\begin{aligned}
&+\frac{T}{\rho}\beta L w_z^2(L,t)-\int_0^L \frac{T}{\rho}\beta w_z^2(z,t)\mathrm{d}z \\
&+\int_0^L 2\beta z w_z(z,t)\left(-\frac{\gamma}{\rho}w_t(z,t)-\frac{K_1}{\rho}w_z(z,t)+\frac{1}{\rho}\tilde{d}_0(z,t)\right)\mathrm{d}z \\
&+\int_0^L 2p w_t(z,t)w_{tt}(z,t)\mathrm{d}z-k_1\tilde{d}_L^2(t)-\rho\int_0^L k_2\tilde{d}_0^2(z,t)\mathrm{d}z.
\end{aligned}
\tag{5.30}
$$

Combining Eqs. (5.24) and (5.30) leads to

$$
\begin{aligned}
&\frac{\mathrm{d}V(t)}{\mathrm{d}t}+2\delta V(t) \\
&=\int_0^L 2\frac{T}{\rho}p w_z(z,t)w_{zt}(z,t)\mathrm{d}z+\beta L w_t^2(L,t)-\int_0^L \beta w_t^2(z,t)\mathrm{d}z \\
&+\frac{T}{\rho}\beta L w_z^2(L,t)-\int_0^L \frac{T}{\rho}\beta w_z^2(z,t)\mathrm{d}z \\
&+\int_0^L 2\beta z w_z(z,t)\left(-\frac{\gamma}{\rho}w_t(z,t)-\frac{K_1}{\rho}w_z(z,t)+\frac{1}{\rho}\tilde{d}_0(z,t)\right)\mathrm{d}z \\
&+\int_0^L 2p w_t(z,t)w_{tt}(z,t)\mathrm{d}z-k_1\tilde{d}_L^2(t)-\rho\int_0^L k_2\tilde{d}_0^2(z,t)\mathrm{d}z \\
&+2\delta\int_0^L \frac{T}{\rho}p w_z^2(z,t)\mathrm{d}z+2\delta\int_0^L 2\beta z w_z(z,t)w_t(z,t)\mathrm{d}z \\
&+2\delta\int_0^L p w_t^2(z,t)\mathrm{d}z+\delta\tilde{d}_L^2(t)+\delta\rho\int_0^L \tilde{d}_0^2(z,t)\mathrm{d}z
\end{aligned}
\tag{5.31}
$$

where δ is a positive constant.

Substituting Eqs. (5.19)–(5.21) into Eq. (5.31), we have

$$
\begin{aligned}
&\frac{\mathrm{d}V(t)}{\mathrm{d}t}+2\delta V(t) \\
&=\int_0^L 2\frac{T}{\rho}p w_z(z,t)w_{zt}(z,t)\mathrm{d}z+\beta L w_t^2(L,t)-\int_0^L \beta w_t^2(z,t)\mathrm{d}z
\end{aligned}
$$

$$
\begin{aligned}
&+\frac{T}{\rho}\beta L\Big(-\frac{K_2}{T}w_t(L,t)+\frac{1}{T}\tilde{d}_L(t)\Big)^2-\int_0^L\frac{T}{\rho}\beta w_z^2(z,t)\mathrm{d}z\\
&+\int_0^L 2\beta z w_z(z,t)\Big(-\frac{\gamma}{\rho}w_t(z,t)-\frac{K_1}{\rho}w_z(z,t)+\frac{1}{\rho}\tilde{d}_0(z,t)\Big)\mathrm{d}z\\
&+\int_0^L 2pw_t(z,t)\Big(\frac{T}{\rho}w_{zz}(z,t)-\frac{\gamma}{\rho}w_t(z,t)-\frac{K_1}{\rho}w_z(z,t)+\frac{1}{\rho}\tilde{d}_0(z,t)\Big)\mathrm{d}z\\
&-k_1\tilde{d}_L^2(t)-\rho\int_0^L k_2\tilde{d}_0^2(z,t)\mathrm{d}z+2\delta\int_0^L\frac{T}{\rho}pw_z^2(z,t)\mathrm{d}z\\
&+2\delta\int_0^L 2\beta z w_z(z,t)w_t(z,t)\mathrm{d}z+2\delta\int_0^L pw_t^2(z,t)\mathrm{d}z\\
&+\delta\tilde{d}_L^2(t)+\delta\rho\int_0^L\tilde{d}_0^2(z,t)\mathrm{d}z,
\end{aligned}
\tag{5.32}
$$

which can be further simplified as

$$
\begin{aligned}
&\frac{\mathrm{d}V(t)}{\mathrm{d}t}+2\delta V(t)\\
&=\beta L w_t^2(L,t)-\int_0^L\beta w_t^2(z,t)\mathrm{d}z-\int_0^L\frac{T}{\rho}\beta w_z^2(z,t)\mathrm{d}z\\
&\quad+\frac{1}{T\rho}\beta L\Big(K_2^2w_t^2(L,t)+\tilde{d}_L^2(t)-2K_2w_t(L,t)\tilde{d}_L(t)\Big)\\
&\quad+\int_0^L 2\beta z w_z(z,t)\Big(-\frac{\gamma}{\rho}w_t(z,t)-\frac{K_1}{\rho}w_z(z,t)+\frac{1}{\rho}\tilde{d}_0(z,t)\Big)\mathrm{d}z\\
&\quad+2\frac{1}{\rho}pw_t(L,t)\Big(-K_2w_t(L,t)+\tilde{d}_L(t)\Big)+\delta\tilde{d}_L^2(t)+\delta\rho\int_0^L\tilde{d}_0^2(z,t)\mathrm{d}z\\
&\quad+\int_0^L 2pw_t(z,t)\Big(-\frac{\gamma}{\rho}w_t(z,t)-\frac{K_1}{\rho}w_z(z,t)+\frac{1}{\rho}\tilde{d}_0(z,t)\Big)\mathrm{d}z\\
&\quad-k_1\tilde{d}_L^2(t)-\rho\int_0^L k_2\tilde{d}_0^2(z,t)\mathrm{d}z+2\delta\int_0^L\frac{T}{\rho}pw_z^2(z,t)\mathrm{d}z
\end{aligned}
$$

$$+ 2\delta \int_0^L 2\beta z w_z(z,t) w_t(z,t) \mathrm{d}z + 2\delta \int_0^L p w_t^2(z,t) \mathrm{d}z. \tag{5.33}$$

According to

$$\begin{aligned}
&\int_0^L 2\beta z w_z(z,t)\Big(-\frac{\gamma}{\rho} w_t(z,t) - \frac{K_1}{\rho} w_z(z,t) + \frac{1}{\rho}\tilde{d}_0(z,t)\Big)\mathrm{d}z \\
&= \int_0^L z\Big(-2\beta\frac{\gamma}{\rho} w_t(z,t) w_z(z,t) - 2\beta\frac{K_1}{\rho} w_z^2(z,t) + 2\beta\frac{1}{\rho}\tilde{d}_0(z,t) w_z(z,t)\Big)\mathrm{d}z \\
&\le \int_0^L z\Big(\beta^2\frac{\gamma^2}{\rho^2 s_0} w_t^2(z,t) + s_0 w_z^2(z,t) - 2\beta\frac{K_1}{\rho} w_z^2(z,t) \\
&\quad + \beta^2\frac{1}{\rho^2 s_1} w_z^2(z,t) + s_1\tilde{d}_0^2(z,t)\Big)\mathrm{d}z \\
&\le \int_0^L z\Big(\Big(s_0 - 2\beta\frac{K_1}{\rho} + \beta^2\frac{1}{\rho^2 s_1}\Big) w_z^2(z,t) + s_1\tilde{d}_0^2(z,t) + \beta^2\frac{\gamma^2}{\rho^2 s_0} w_t^2(z,t)\Big)\mathrm{d}z
\end{aligned} \tag{5.34}$$

where s_0 and s_1 are two positive constants, we can obtain

$$\begin{aligned}
&\frac{\mathrm{d}V(t)}{\mathrm{d}t} + 2\delta V(t) \\
&\le \beta L w_t^2(L,t) - \int_0^L \beta w_t^2(z,t)\mathrm{d}z + \frac{1}{T\rho}\beta L\Big(K_2^2 w_t^2(L,t) + \tilde{d}_L^2(t) - 2K_2 w_t(L,t)\tilde{d}_L(t)\Big) \\
&\quad - \int_0^L \frac{T}{\rho}\beta w_z^2(z,t)\mathrm{d}z + \int_0^L z\Big[\Big(s_0 - 2\beta\frac{K_1}{\rho} + \beta^2\frac{1}{\rho^2 s_1}\Big) w_z^2(z,t) + s_1\tilde{d}_0^2(z,t) \\
&\quad + \beta^2\frac{\gamma^2}{\rho^2 s_0} w_t^2(z,t)\Big]\mathrm{d}z - 2K_2 p\frac{1}{\rho} w_t^2(L,t) + 2p\frac{1}{\rho} w_t(L,t)\tilde{d}_L(t) \\
&\quad + \int_0^L 2p w_t(z,t)\Big(-\frac{\gamma}{\rho} w_t(z,t) - \frac{K_1}{\rho} w_z(z,t) + \frac{1}{\rho}\tilde{d}_0(z,t)\Big)\mathrm{d}z \\
&\quad - k_1\tilde{d}_L^2(t) - \rho\int_0^L k_2\tilde{d}_0^2(z,t)\mathrm{d}z + 2\delta\int_0^L \frac{T}{\rho} p w_z^2(z,t)\mathrm{d}z \\
&\quad + 2\delta\int_0^L 2\beta z w_z(z,t) w_t(z,t)\mathrm{d}z + 2\delta\int_0^L p w_t^2(z,t)\mathrm{d}z
\end{aligned}$$

$$+\delta\tilde{d}_L^2(t)+\delta\rho\int_0^L \tilde{d}_0^2(z,t)\mathrm{d}z. \tag{5.35}$$

Defining $\boldsymbol{\xi}(z,t) = \left[w_t(L,t)\ w_z(z,t)\ w_t(z,t)\ \tilde{d}_L(t)\ \tilde{d}_0(z,t)\right]^T$, we can get the following inequality from inequality (5.35)

$$\frac{\mathrm{d}V(t)}{\mathrm{d}t}+2\delta V(t) \le \int_0^L \boldsymbol{\xi}^T(z,t)\boldsymbol{\Phi}(z)\boldsymbol{\xi}(z,t)\mathrm{d}z. \tag{5.36}$$

If inequality (5.18) is satisfied, from inequality (5.36), we can obtain

$$\frac{\mathrm{d}V(t)}{\mathrm{d}t}+2\delta V(t) \le 0. \tag{5.37}$$

Solving inequality (5.37), we can achieve

$$V(t) \le V(0)e^{-2\delta t}. \tag{5.38}$$

According inequality (5.38), we can further obtain that $V(t) \to 0$ for $t \to \infty$.
If inequality (5.17) is satisfied, considering Eq. (5.24), we can obtain

$$\begin{aligned}
V(t) &= V_1(t)+V_{ob1}(t)+V_{ob2}(t)\\
&= \int_0^L \boldsymbol{x}^T(z,t)\boldsymbol{P}(z)\boldsymbol{x}(z,t)\mathrm{d}z+\frac{1}{2}\tilde{d}_L^2(t)+\frac{1}{2}\rho\int_0^L \tilde{d}_0^2(z,t)\mathrm{d}z\\
&\ge p_0\int_0^L \boldsymbol{x}^T(z,t)\boldsymbol{x}(z,t)\mathrm{d}z+\frac{1}{2}\tilde{d}_L^2(t)+\frac{1}{2}\rho\int_0^L \tilde{d}_0^2(z,t)\mathrm{d}z\\
&= p_0\int_0^L w_z^2(z,t)\mathrm{d}z+p_0\int_0^L w_t^2(z,t)\mathrm{d}z+\frac{1}{2}\tilde{d}_L^2(t)+\frac{1}{2}\rho\int_0^L \tilde{d}_0^2(z,t)\mathrm{d}z
\end{aligned} \tag{5.39}$$

where $p_0 = \min_{z\in[0,L]}\{\sigma(\boldsymbol{P}(z)\} > 0$.

From inequality (5.39), we can get $w_z(z,t) \to 0$ as well as $w_t(z,t) \to 0$ for $t \to \infty$.

According Lemma 2.2, we have

$$\frac{1}{L}w^2(z,t) \le \int_0^L w_z^2(z,t)\mathrm{d}z. \tag{5.40}$$

Hence, from inequalities (5.38)–(5.40), we can obtain

$$\frac{1}{L}w^2(z,t) \leq \int_0^L w_z^2(z,t)\mathrm{d}z \leq \frac{1}{p_0}V(t) \leq \frac{1}{p_0}V(0)e^{-2\delta t}. \tag{5.41}$$

Thus, we have

$$|w(z,t)| \leq \sqrt{\frac{L}{p_0}V(0)e^{-2\delta t}}. \tag{5.42}$$

From inequality (5.42), we can obtain that $w(z,t)$ exponentially converges to zero for $t \to \infty$.

The results of this section are presented as follows: boundary and distributed controllers are designed based on the disturbance observers proposed in Sect. 5.3. When there exist β, $p \in R^+$ and LMIs (5.17) and (5.18) are satisfied, the asymptotic stability of the closed-loop system can be verified by the Lyapunov's direct method.

5.5 Simulations

To verify the performance of the proposed observers-based boundary and distributed control laws, numerical simulations are provided in this section. The parameters of the string and the proposed observers are listed as: $L = 1\,\mathrm{m}$, $T = 2\,\mathrm{N}$, $\rho = 5\,\mathrm{kgm}^{-1}$, $\gamma = 0.5$, $k_1 = 16$, $k_2 = 65$.

The initial elastic displacement of the flexible sting is designed to be $w(z,0) = z^2$. The distributed disturbance is given as $d_0(z,t) = -0.3\sin(0.01\pi(z+t)) - 0.3\sin(0.02\pi(z+t))$ and the boundary disturbance is given as $d_L(t) = 1 + 0.2\sin(0.1\pi t) + 0.02\sin(0.2\pi t)$. The boundary and distributed disturbances and their estimations are shown in Figs. 5.2 and 5.3 respectively. The estimation errors of $d_L(t)$ and $d_0(z,t)$ are shown in Figs. 5.4 and 5.5.

Figures 5.4 and 5.5 illustrate that the estimated values of the boundary and distributed disturbances can fast and accurately converge to their true values, that is, when $t \to \infty$, $\tilde{d}_L(t) \to 0$ and $\tilde{d}_0(z,t) \to 0$. Therefore, the effectiveness of the disturbance observers is validated.

By Schur complements, inequality (5.18) holds if $\boldsymbol{\Psi}(z) \leq 0$, where

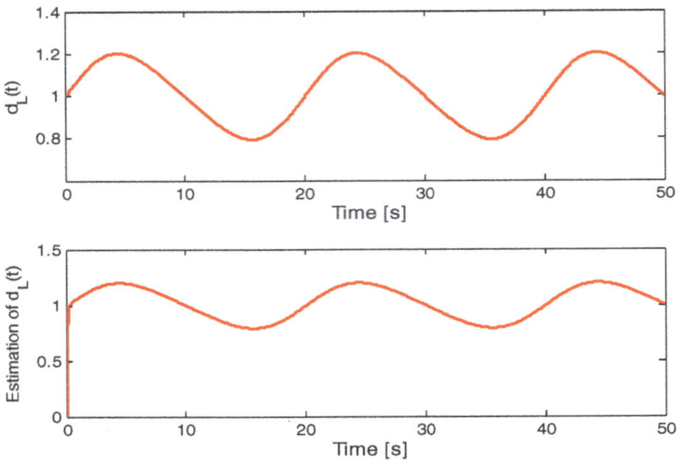

Fig. 5.2 $d_L(t)$ and $\hat{d}_L(t)$

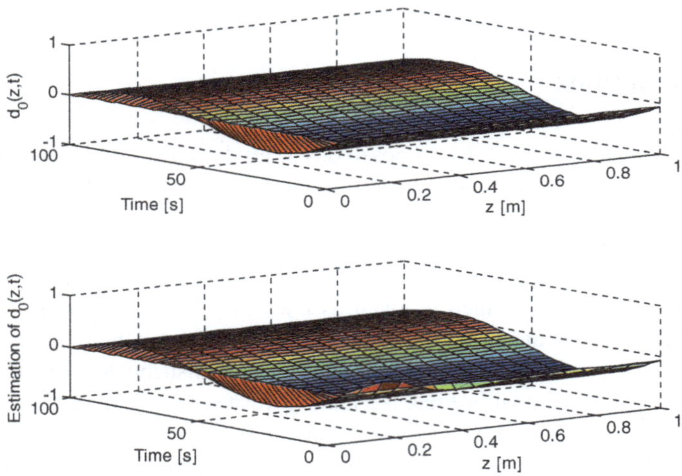

Fig. 5.3 $d_0(z, t)$ and $\hat{d}_0(z, t)$

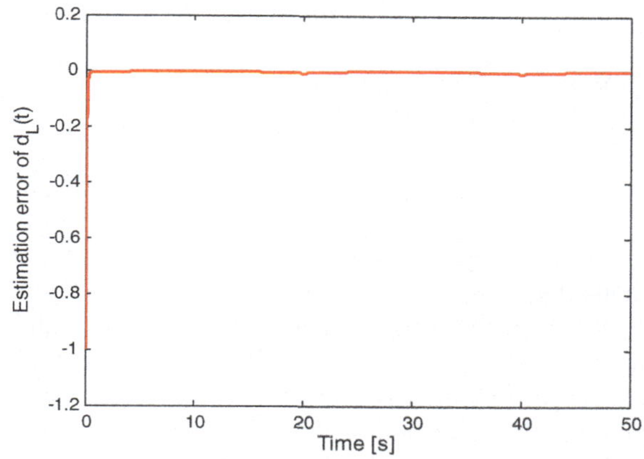

Fig. 5.4 Estimation error of $d_L(t)$

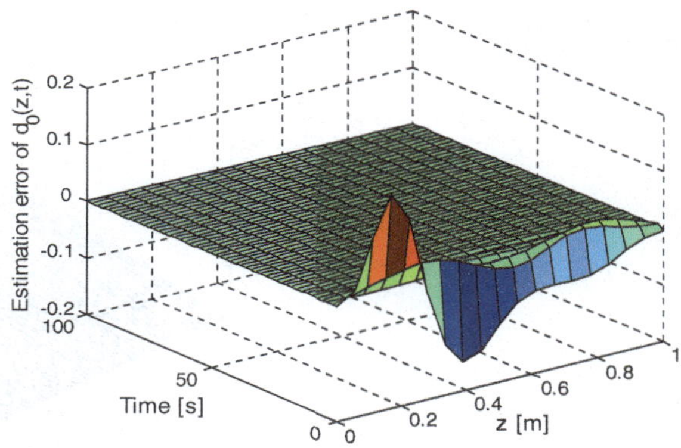

Fig. 5.5 Estimation error of $d_0(z, t)$

$$
\boldsymbol{\Psi}(z) = \begin{bmatrix}
\beta - 2K_2 p\frac{1}{L\rho} & 0 & 0 & -\frac{1}{T\rho}\beta K_2 + p\frac{1}{L\rho} & 0 & 0 & 0 & K_2 \\
0 & -\frac{T}{\rho}\beta + z\left(s_0 - 2\beta\frac{K_1}{\rho}\right) + 2\delta\frac{T}{\rho}p & -\frac{K_1}{\rho}p + 2\delta\beta z & 0 & 0 & z\frac{\beta}{\rho} & 0 & 0 \\
0 & -\frac{K_1}{\rho}p + 2\delta\beta z & -\beta - \frac{\gamma}{\rho}2p + 2\delta p & 0 & \frac{1}{\rho}p & 0 & \beta\frac{\gamma}{\rho}z & 0 \\
-\frac{1}{T\rho}\beta K_2 + p\frac{1}{L\rho} & 0 & 0 & \frac{1}{T\rho}\beta - \frac{k_1}{L} + \frac{\delta}{L} & 0 & 0 & 0 & 0 \\
0 & 0 & \frac{1}{\rho}p & 0 & s_1 z - \rho k_2 + \delta\rho & 0 & 0 & 0 \\
0 & z\frac{\beta}{\rho} & 0 & 0 & 0 & -zs_1 & 0 & 0 \\
0 & 0 & \beta\frac{\gamma}{\rho}z & 0 & 0 & 0 & -zs_0 & 0 \\
K_2 & 0 & 0 & 0 & 0 & 0 & 0 & -\frac{T\rho}{\beta}
\end{bmatrix}
$$

By LMI techniques, the feasible solutions to inequality (5.17) can be obtained as:

$$p = 52.4250, \beta = 15.7275.$$

With the result p and β, we can obtain the control gain as follows by solving the LMI $\boldsymbol{\Psi}(z) \le 0$

$$K_1 = 0.5824, K_2 = 8.3681.$$

In this chapter, the spatial step size and temporal step size of the finite difference method are chosen as $\Delta t = 0.01$ s and $\Delta h = 0.05$ m.

When no control input is applied to the system, the dynamic responses of $w(z, t)$, $w_z(z, t)$ and $w_t(z, t)$ are shown in Figs. 5.6, 5.7 and 5.8 respectively where the system is disturbed by the environmental disturbance. We can notice that in Figs. 5.6, 5.7 and 5.8, the deflection of the string is large without control. From Figs. 5.9, 5.10 and 5.11, we can see the dynamic responses of the flexible string system with the

Fig. 5.6 $w(z, t)$ without control

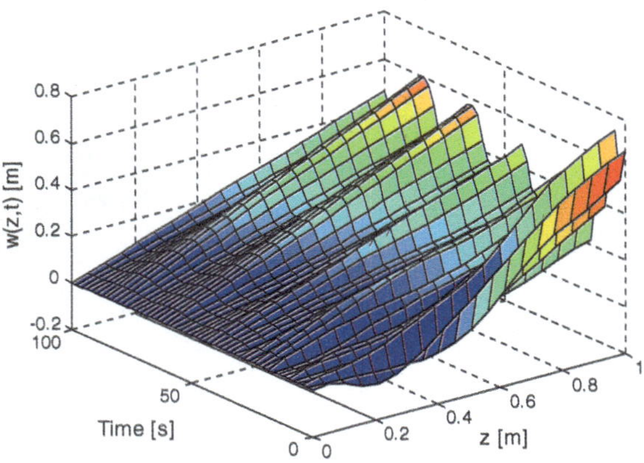

Fig. 5.7 $w_z(z, t)$ without control

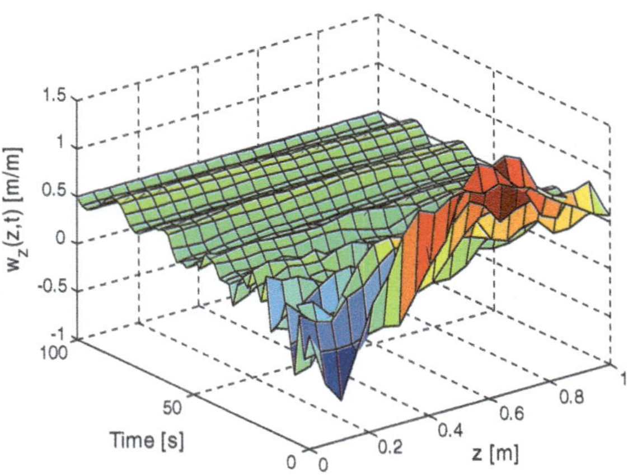

Fig. 5.8 $w_t(z, t)$ without control

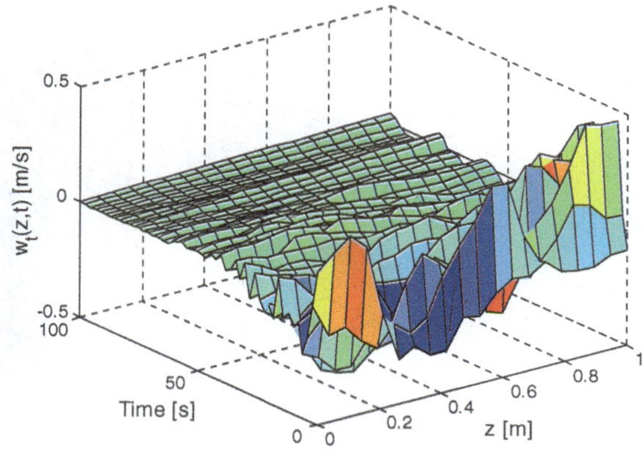

Fig. 5.9 $w(z, t)$ with the proposed control laws

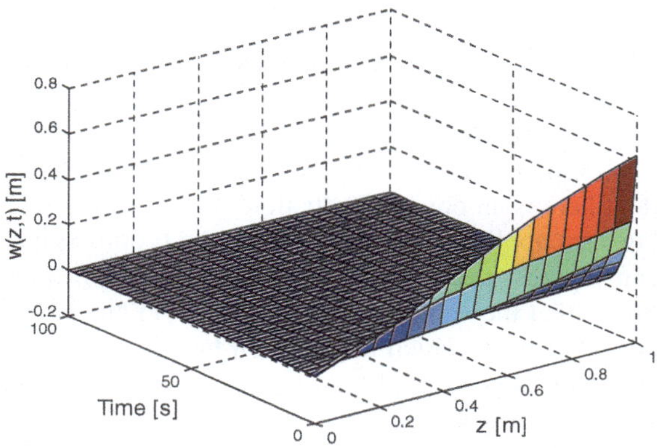

Fig. 5.10 $w_z(z, t)$ with the proposed control laws

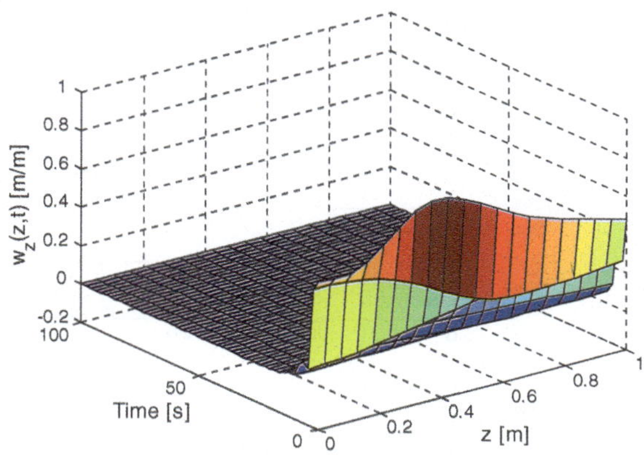

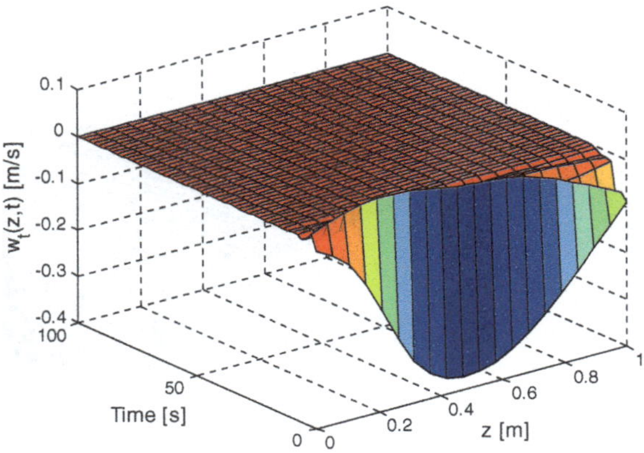

Fig. 5.11 $w_t(z, t)$ with the proposed control laws

boundary and distributed controllers based on the proposed disturbance observers. $w(L, t)$, $w_z(L, t)$ and $w_t(L, t)$ of the closed-loop system are shown in Fig. 5.12. The boundary and distributed control laws are shown in Figs. 5.13 and 5.14 respectively.

Comparing Figs. 5.9, 5.10 and 5.11 with Figs. 5.6, 5.7 and 5.8, it is observed that there is an obvious reduction of $w(z, t)$, $w_z(z, t)$ and $w_t(z, t)$ when the control laws are applied. It is apparent that the proposed boundary and distributed controls can suppress the vibration of the flexible string effectively. From Fig. 5.12, we can notice that the values of $w(L, t)$, $w_z(L, t)$ and $w_t(L, t)$ converge to zero in a short period of time, which implies that based on the observers, the proposed boundary and distributed control laws can reduce the vibration quickly and make the system asymptotically stable.

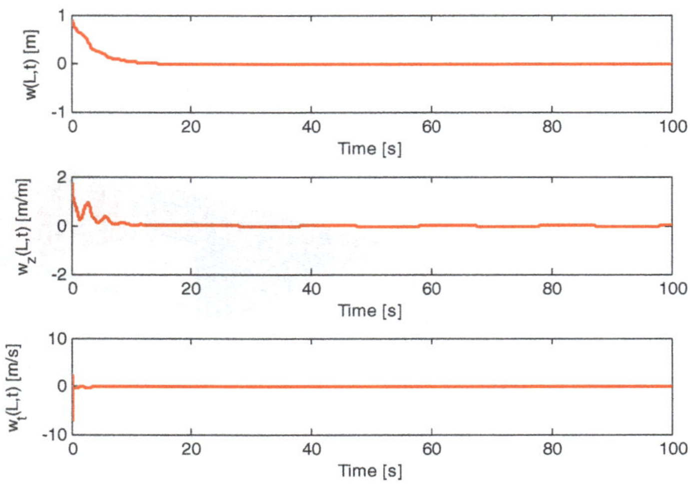

Fig. 5.12 $w(L, t)$, $w_z(L, t)$ and $w_t(L, t)$ with the proposed control laws

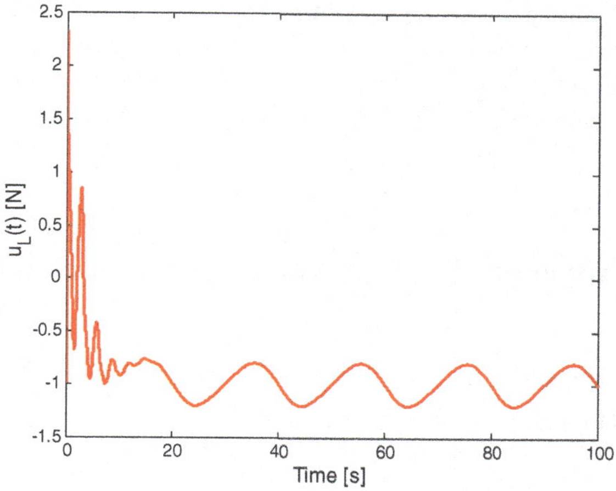

Fig. 5.13 Boundary control input

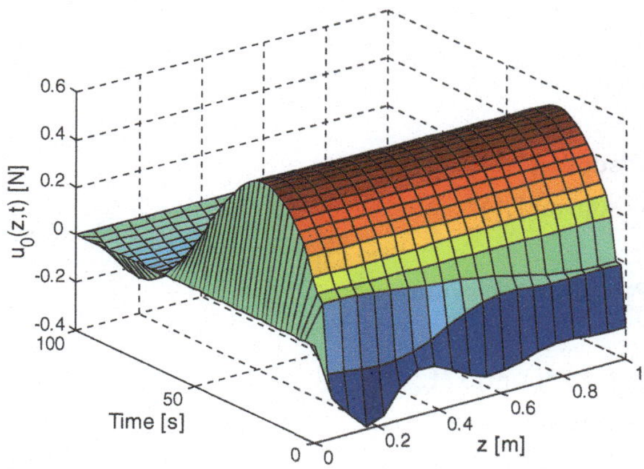

Fig. 5.14 Distributed control input

Therefore, we can conclude that the simulation results in Figs. 5.2, 5.3, 5.4, 5.5, 5.6, 5.7, 5.8, 5.9, 5.10, 5.11, 5.12, 5.13 and 5.14 agree well with the stability analysis and prove the effectiveness of the proposed control scheme.

The simulation programs are given in the Appendixes 5.1–5.3, where chapter5_LMI1.m in Appendix 5.1 is used to solve LMI (5.17) to obtain p and β, chapter5_LMI2.m in Appendix 5.2 is used to solve LMI (5.18) to obtain K_1 and K_2, and chapter5.m in Appendix 5.3 is the main program containing the proposed control laws (5.15), (5.16) and observers in Eqs. (5.6)–(5.8).

5.6 Conclusions

In this chapter, exponentially stable disturbance observers have been proposed to compensate the boundary and distributed disturbances of the flexible string. Based on the observers, LMIs have been introduced to design boundary and distributed control laws. The closed-loop stability has been analyzed by Lyapunov's direct method. With the proposed controls, the vibration suppression is achieved and the deformation of the string converges to zero in a short time. In addition, the effectiveness of the proposed disturbance observers and control laws has been proved by numerical simulations.

Appendix 5.1: Simulation Program: chapter5_LMI1.m

```
clear all;
close all;

L=1;
m=5;
T=2;

p=sdpvar(1,1);
x=sdpvar(1,1);

P=sdpvar(2,2,'symmetric');

P=[p*T*(1/m) x*L;x*L p];
L1=(P>0);
L2=(x>0);
L3=(p>0);

L=L1+L2+L3;
solvesdp(L);

M=[p x];

p=double(M(1,1));
x=double(M(1,2));
```

Appendix 5.2: Simulation Program: chapter5_LMI2.m

```
clear all;
close all;

z1=0;
z=1;
L=1;
m=5;
T=2;
b=0.5;
p=52.4250;
x=15.7275;
K3=16;
K4=65;

K1=sdpvar(1,1);
K2=sdpvar(1,1);

s0=sdpvar(1,1);
s1=sdpvar(1,1);
a=sdpvar(1,1);

N=sdpvar(8,8,'symmetric');
```

```
N=[-2*K2*p*(1/m)*(1/L)+x 0 0 p*(1/m)*(1/L)-(1/(T*m))*x*K2 0 0 0 K2;0 -
    x*(T/m)+2*a*(T/m)*p+z*(s0-2*x*K1*(1/m)) 2*a*x*z-p*K1*(1/m) 0 0 x*(1/m)*z 0 0;0
    2*a*x*z-p*K1*(1/m) -x-2*p*b*(1/m)+2*a*p 0 p*(1/m) 0 z*x*b*(1/m) 0;p*(1/m)*(1/L)-
    (1/(T*m))*x*K2 0 0 -K3*(1/L)+a*(1/L)+(1/(T*m))*x 0 0 0 0;0 0 p*(1/m) 0 a*p-K4*m+s1*z
    0 0 0;0 x*(1/m)*z 0 0 0 -z*s1 0 0;0 0 x*(1/m)*z*b 0 0 0 -z*s0 0;K2 0 0 0 0 0 0 -T*m*(1/(x))];
N1=-[-2*K2*p*(1/m)*(1/L)+x 0 0 p*(1/m)*(1/L)-(1/(T*m))*x*K2 0 ;0 -
    x*(T/m)+2*a*(T/m)*p+z1*(s0-2*x*K1*(1/m)) 2*a*x*z1-p*K1*(1/m) 0 0 ;0 2*a*x*z1-
    p*K1*(1/m) -x-2*p*b*(1/m)+2*a*p 0 p*(1/m);p*(1/m)*(1/L)-(1/(T*m))*x*K2 0 0 -
    K3*(1/L)+a*(1/L)+(1/(T*m))*x 0 ;0 0 p*(1/m) 0 a*p-K4*m+s1*z1];

L0=(N1>0);
L1=(N<0);
L2=(K1>0);
L3=(K2>0);

L6=(a>0);
L7=(s0>0);
L8=(s1>0);

L=L0+L1+L2+L3+L6+L7+L8;

solvesdp(L);

K1=double(K1);
K2=double(K2);

a=double(a);
s0=double(s0);
s1=double(s1);
```

Appendix 5.3: Simulation Program: chapter5.m

```
close all;
clear all;
N=20;
n=10000;

TIME=100;
L=1;

h=L/N;
t=TIME/n;

T=2;
b=0.5;
rho=5;
K1=0.582416350481034;
K2=8.368133782636251;
K3=16;
K4=65;

A=-2*T*t^2/(rho*h^2)-b*t/rho+2;
B=T*t^2/(rho*h^2);
C=T*t^2/(rho*h^2);
D=(b*t/rho)-1;

for j=1:n+1
    for i=2:N-1
    d0(j,i)=-0.3*sin(0.01*pi*(j+i)*t)-0.3*sin(0.02*pi*(j+i)*t);
    dp(j,i)=0;
    ds(j,i)=0;
    u0(j,i)=0;
    end

    d1(j)=1+0.2*sin(0.1*pi*j*t)+0.02*sin(0.2*pi*j*t);

    dpL(j)=0;
    dLs(j)=0;
    u1(j)=0;
end
```

```
z=zeros(n+1,N+1);

for j=1:n+1
  z(j,1)=0;
end

for i=2:N+1
   z(1,i)=1*((i-1)*h)^2;
   dz0=0.1;
   z(2,i)=z(1,i)+t*dz0;
end


z(2,1)=0;


dtL(2)=(z(2,N)-z(1,N))/t;


for i=2:N-1
dz(2,i)=(z(2,i)-z(2,i-1))/h;
end

z(2,N+1)=(h/T)*(d1(1)+u1(1))+z(2,N);

dLs(2)=dLs(1)+t*(-K3*(-T*(z(2,N+1)-z(2,N))/h+u1(1))-K3*dpL(1));
dpL(2)=dLs(2);

for i=2:N-1
ds(2,i)=ds(1,i)+t*(-K4*(T*(z(2,i-1)-2*z(2,i)+z(2,i+1))/(h^2)-b*(z(2,i)-z(1,i))/t+u0(1,i))-
K4*dp(1,i));
dp(2,i)=ds(2,i)+K4*rho*(z(2,i)-z(1,i))/t;
end

       for i=2:N+1
              dzt(1,i-1)=(z(2,i)-z(1,i))/t;
              dz1(1,i-1)=(z(2,i)-z(2,i-1))/h;
       end

u1(2)=-K2*dtL(2)-dpL(2);
```

```
    for i=2:N-1
    u0(2,i)=-K1*dz(2,i)-dp(2,i);
    end

    %u1(2)=0;

    %for i=2:N-1
    %u0(2,i)=0;
    %end

    for j=3:n+1
        for i=2:N-1
        z(j,i)=A*z(j-1,i)+B*z(j-1,i-1)+C*z(j-1,i+1)+D*z(j-2,i)+(t^2/rho)*d0(j-
1,i)+(t^2/rho)*u0(j-1,i);
    end

    z(j,N)=A*z(j-1,N)+B*z(j-1,N-1)+C*z(j-1,N+1)+D*z(j-2,N);

    z(j,N+1)=(h/T)*(u1(j-1)+d1(j-1))+z(j,N);

    for i=2:N+1
        dzt(j-1,i-1)=(z(j,i)-z(j-1,i))/t;
        dz1(j-1,i-1)=(z(j,i)-z(j,i-1))/h;
    end
        z(j,1)=0;

       dtL(j)=(z(j,N)-z(j-1,N))/t;

        for i=2:N-1
          dz(j,i)=(z(j,i)-z(j,i-1))/h;
        end

        dLs(j)=dLs(j-1)+t*(-K3*(-T*(z(j,N+1)-z(j,N))/h+u1(j-1))-K3*dpL(j-1));
        dpL(j)=dLs(j);

         for i=2:N-1
              ds(j,i)=ds(j-1,i)+t*(-K4*(T*(z(j,i-1)-2*z(j,i)+z(j,i+1))/(h^2)-b*(z(j,i)-z(j-
```

```
1,i))/t+u0(j-1,i))-K4*dp(j-1,i));
            dp(j,i)=ds(j,i)+K4*rho*(z(j,i)-z(j-1,i))/t;
        end

        u1(j)=-K2*dtL(j)-dpL(j);% with boundary control
        %u1(j)=0;% without boundary control
        for i=2:N-1
         u0(j,i)=-K1*dz(j,i)-dp(j,i);% with distributed control
         %u0(j,i)=0;% without distributed control
        end

end

zshort=zeros(n/200,N);
dzshort=zeros(n/200,N);
dztshort=zeros(n/200,N);

for i=1:N
    for j=1:n/200
        zshort(j,i)=z(j*200,i);
        dztshort(j,i)=dzt(j*200,i);
        dzshort(j,i)=dz1(j*200,i);

    end
end

for i=1:N-1
    for j=1:n/200
        d0short(j,i)=d0(j*200,i);
        dpshort(j,i)=dp(j*200,i);
    end
end
for i=1:N-1
    for j=1:n/200
       u0short(j,i)=u0(j*200,i);

    end
```

```
end
for j=1:n

            z1(j)=z(j,20);
            z2(j)=dzt(j,20);
            z3(j)=dz1(j,18);
        end
tshort=linspace(0,TIME,n/200);
y=linspace(0,L,N);
[y1,tshort1]=meshgrid(y,tshort);

y0=linspace(0,L,N-1);
[y2,tshort2]=meshgrid(y0,tshort);

figure(1);

surf(y1,tshort1,zshort);
xlabel('z [m]'); ylabel('Time [s]');zlabel('w(z,t) [m]');

figure(2);
surf(y1,tshort1,dztshort);
xlabel('z [m]'); ylabel('Time [s]');zlabel('w_t(z,t)[m/s]');

figure(3);
surf(y1,tshort1,dzshort);
xlabel('z [m]'); ylabel('Time [s]');zlabel('w_z(z,t) [m/m]');

figure(4);
subplot(311);
plot(z1,'r','linewidth',2);
xlabel('Time [s]');ylabel('w(L,t) [m]');
set(gca,'xticklabel',[0 10 20 30 40 50 60 70 80 90 100]);
subplot(312);
plot(z3,'r','linewidth',2);
xlabel('Time [s]');ylabel('w_z(L,t) [m/m]');
set(gca,'xticklabel',[0 10 20 30 40 50 60 70 80 90 100]);
subplot(313);
```

```
plot(z2,'r','linewidth',2);
xlabel('Time [s]');ylabel('w_t(L,t) [m/s]');
set(gca,'xticklabel',[0 10 20 30 40 50 60 70 80 90 100]);

figure(5);
subplot(211);
plot(d1(1:5000),'r','linewidth',2);
xlabel('Time [s]');ylabel('d_L(t)');
set(gca,'xticklabel',[0 5 10 15 20 25 30 35 40 45 50]);
subplot(212);
plot(dpL(1:5000),'r','linewidth',2);
xlabel('Time [s]');ylabel('Estimation of d_L(t)');
set(gca,'xticklabel',[0 5 10 15 20 25 30 35 40 45 50]);
hold off

figure(6);
plot(dpL(1:5000)-d1(1:5000),'r','linewidth',2);
xlabel('Time [s]');ylabel('Estimation error of d_L(t)');
set(gca,'xticklabel',[0 5 10 15 20 25 30 35 40 45 50]);

figure(7);
subplot(211);
surf(y2,tshort2,d0short);
xlabel('z [m]'); ylabel('Time [s]');zlabel('d_0(z,t)');
subplot(212);
surf(y2,tshort2,dpshort);
xlabel('z [m]'); ylabel('Time [s]');zlabel('Estimation of d_0(z,t)');

figure(8);
surf(y2,tshort2,dpshort-d0short);
xlabel('z [m]'); ylabel('Time [s]');zlabel('Estimation error of d_0(z,t)');
figure(9);
plot(u1,'r','linewidth',2);
xlabel('Time [s]');ylabel('u_L(t) [N]');
axis([0 10000 -1.5 2.5])
set(gca,'xticklabel',[0 10 20 30 40 50 60 70 80 90 100]);

figure(10);
surf(y2,tshort2,u0short);
xlabel('z [m]'); ylabel('Time [s]');zlabel('u_0(z,t) [N]');
```

References

1. W. He, S.S. Ge, Vibration Control of a Flexible String With Both Boundary Input and Output Constraints. IEEE Trans. Control Syst. Technol. **23**(4), 1245–1254 (2015)
2. W.H. Chen, D.J. Balance, P.J. Gawthrop, J.J. Gribble, J. OReilly, A nonlinear disturbance observer for two link robotic manipulators. 38th IEEE Conference on Decision and Control, Phoenix, Arizona (1999)

Chapter 6
PDE Modeling and Basic Control of Vehicle-Mounted Flexible Manipulator

6.1 Introduction

With the development of the modern mechanized production, robots play an important role of improving productivity in industry manufacture. These robots generally have complex dynamical behaviors and long rigid or flexible arms which have been widely applied in different kinds of advanced fields in the modern society, such as marine engineering, aerospace and aviation. Compared with rigid manipulators, the flexible ones have strengths of wider operating ranges, faster responses, and longer working lives. As a result, quite a few studies on modeling and controlling for flexible manipulators have been carried out all over the world. Although flexible manipulators have advantages mentioned above, there also exist some problems in their application. In the actual process, the elastic deformation of flexible manipulator is inevitable because of its elastic characteristic. The vibration will affect the control effect.

In this chapter, the control law for a vehicle-mounted flexible manipulator is designed. Based on the PDE model established in this chapter, the boundary control is proposed aiming at position regulation and vibration suppression. The asymptotic stability of the closed-loop system is proved using the extending LaSalle's invariance principle.

6.2 Problem Formulation

In this chapter, a vehicle-mounted flexible manipulator which moves in the horizontal direction is considered and for simplicity, the simplified system is shown in Fig. 6.1. The whole system is composed of a vehicle, a flexible manipulator and a payload at the free end without damping. We suppose that flexible manipulator rotates in the

X. Xing and J. Liu, *Vibration Control Methods of Mechanical Distributed Parameter Systems*, Springer Tracts in Mechanical Engineering,
https://doi.org/10.1007/978-981-16-1532-0_6

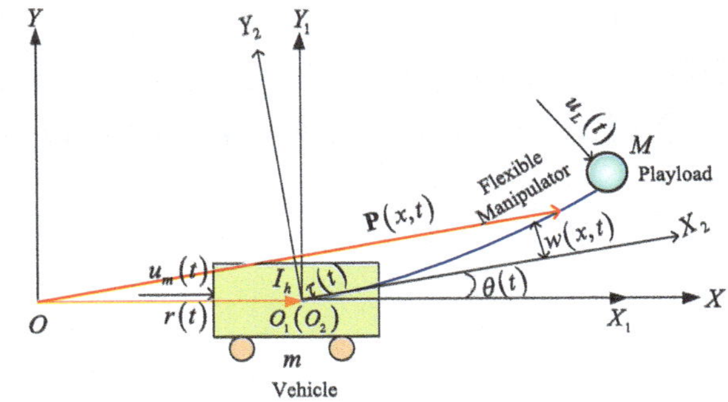

Fig. 6.1 Configuration of the vehicle-mounted flexible manipulator

horizontal direction and three actuators respectively focus on the joint, vehicle, and payload at the end of the flexible arm.

According to Fig. 6.1, XOY represents the global inertial coordinate and $X_i O_i Y_i$ represents the moving coordinates fixed on the robot arm where O_i is the center of mass of the vehicle, $i = 1, 2$. The system variables are listed as follows. The distance traveled by the vehicle is denoted as $r(t)$ in XOY. $\theta(t)$ is the rotation angle of the link in $X_1 O_1 Y_1$. The elastic deformation of beam measured from $X_2 O_2 Y_2$ is defined as $w(x, t)$. $\boldsymbol{P}(x, t)$ is a vector which denotes the position of a particular point on the flexible manipulator relative to the inertial frame. As a result, according to Fig. 6.1, vector $\boldsymbol{P}(x, t)$ can be expressed as

$$\boldsymbol{P}(x,t) = \begin{bmatrix} P_x(x,t) \\ P_y(x,t) \end{bmatrix} = \begin{bmatrix} x\cos\theta(t) - w(x,t)\sin\theta(t) + r(t) \\ x\sin\theta(t) + w(x,t)\cos\theta(t) \end{bmatrix} \tag{6.1}$$

where $P_x(x, t)$ and $P_y(x, t)$ are components of the position vector related to X axis and Y axis respectively, x denotes the position of the point on the flexible arm along X_2 axis.

Let $x = L$ and $\boldsymbol{P}(L, t)$ can be written as

$$\boldsymbol{P}(L,t) = \begin{bmatrix} P_x(L,t) \\ P_y(L,t) \end{bmatrix} = \begin{bmatrix} L\cos\theta(t) - w(L,t)\sin\theta(t) + r(t) \\ L\sin\theta(t) + w(L,t)\cos\theta(t) \end{bmatrix}. \tag{6.2}$$

Remark 6.1 (Applicable for this chapter, Chaps. 7, 8 and 9). For clarity, the following notations are introduced throughout this chapter: $(*)_x = \frac{\alpha(*)}{\partial x}$, $(*)_{xx} = \frac{\alpha(*)}{\partial^2 x}$, $(*)_{xxx} = \frac{\alpha(*)}{\partial^3 x}$, $(*)_{xxxx} = \frac{\alpha(*)}{\partial^4 x}$, $(\dot{*}) = \frac{\alpha(*)}{\partial t}$, $(\ddot{*}) = \frac{\alpha(*)}{\partial^2 t}$.

To obtain the PDE dynamic model of the system, the energy of the system and the work done by the non-conservative force should be achieved.

The expression of kinetic energy $E_k(t)$ can be represented as

$$E_k(t) = \frac{1}{2} I_h \dot{\theta}^2(t) + \frac{1}{2}\rho \int_0^L \dot{\boldsymbol{P}}^{\mathrm{T}}(x,t)\dot{\boldsymbol{P}}(x,t)\mathrm{d}x + \frac{1}{2} M \dot{\boldsymbol{P}}^{\mathrm{T}}(L,t)\dot{\boldsymbol{P}}(L,t) + \frac{1}{2} m\dot{r}^2(t) \tag{6.3}$$

where L is the length of the flexible link, I_h is the moment of inertia of the flexible link, ρ is the mass of the flexible link per unit span, M is the mass of the payload and m is the mass of the vehicle.

The potential energy $E_p(t)$ can be written as

$$E_p(t) = \frac{EI}{2}\int_0^L w_{xx}^2(x,t)\mathrm{d}x + g\rho\int_0^L P_y(x,t)\mathrm{d}x + MgP_y(L,t) \tag{6.4}$$

where EI is the uniform flexural rigidity of the flexible link.

The non-conservative force $W(t)$ is given as

$$W(t) = \tau(t)\theta(t) + u_L(t)w(L,t) + u_m(t)r(t) + \int_0^L F(x,t)w(x,t)\mathrm{d}x \tag{6.5}$$

where $\tau(t)$ is the joint torque, $u_L(t)$ represents the boundary control input force at the tip of flexible link, and $u_m(t)$ is the control input force on the vehicle.

Taking the derivative of Eqs. (6.1) and (6.2), after making a series of calculation, we have

$$\begin{aligned}
\dot{\boldsymbol{P}}^{\mathrm{T}}(x,t)\dot{\boldsymbol{P}}(x,t) ={}& x^2\sin^2\theta(t)\dot{\theta}^2(t) + \cos^2\theta(t)\dot{\theta}^2(t)w^2(x,t) + \sin^2\theta(t)\dot{w}^2(x,t) + \dot{r}^2(t)\\
&+ 2x\sin\theta(t)\dot{\theta}^2(t)\cos\theta(t)w(x,t) + 2x\sin^2\theta(t)\dot{\theta}(t)\dot{w}(x,t)\\
&- 2x\sin\theta(t)\dot{\theta}(t)\dot{r}(t) + 2\cos\theta(t)\dot{\theta}(t)w(x,t)\sin\theta(t)\dot{w}(x,t)\\
&- 2\cos\theta(t)\dot{\theta}(t)w(x,t)\dot{r}(t) - 2\sin\theta(t)\dot{w}(x,t)\dot{r}(t)\\
&+ x^2\cos^2\theta(t)\dot{\theta}^2(t) + \dot{w}^2(x,t)\cos^2\theta(t) + w^2(x,t)\sin^2\theta(t)\dot{\theta}^2(t)\\
&+ 2x\cos^2\theta(t)\dot{\theta}(t)\dot{w}(x,t) - 2x\cos\theta(t)\dot{\theta}^2(t)w(x,t)\sin\theta(t)\\
&- 2\dot{w}(x,t)\cos\theta(t)w(x,t)\sin\theta(t)\dot{\theta}(t)\\
={}& x^2\dot{\theta}^2(t) + \dot{\theta}^2(t)w^2(x,t) + \dot{w}^2(x,t) + \dot{r}^2(t) + 2x\dot{\theta}(t)\dot{w}(x,t)\\
&- 2x\sin\theta(t)\dot{\theta}(t)\dot{r}(t) - 2\cos\theta(t)\dot{\theta}(t)w(x,t)\dot{r}(t) - 2\sin\theta(t)\dot{w}(x,t)\dot{r}(t)
\end{aligned} \tag{6.6}$$

$$\begin{aligned}
\dot{\boldsymbol{P}}^{\mathrm{T}}(L,t)\dot{\boldsymbol{P}}(L,t) ={}& L^2\dot{\theta}^2(t) + \dot{\theta}^2(t)w^2(L,t) + \dot{w}^2(L,t) + \dot{r}^2(t) + 2L\dot{\theta}(t)\dot{w}(L,t)\\
&- 2L\sin\theta(t)\dot{\theta}(t)\dot{r}(t) - 2\cos\theta(t)\dot{\theta}(t)w(L,t)\dot{r}(t) - 2\sin\theta(t)\dot{w}(L,t)\dot{r}(t).
\end{aligned} \tag{6.7}$$

Then the Hamilton principle in Eq. (2.5) is applied as

$$\int_{t_1}^{t_2}\big(\delta E_k(t) - \delta E_p(t) + \delta W(t)\big)\mathrm{d}t = 0. \tag{6.8}$$

The time integration of variation of $E_k(t)$, $E_p(t)$, and $W(t)$ are given as

$$
\begin{aligned}
\int_{t_1}^{t_2} \delta E_k(t)\mathrm{d}t = {} & I_h \int_{t_1}^{t_2} \dot{\theta}(t)\delta\dot{\theta}(t)\mathrm{d}t + \rho \int_{t_1}^{t_2}\int_0^L \Big(x^2\dot{\theta}(t)\delta\dot{\theta}(t) + \dot{\theta}(t)^2 w(x,t)\delta w(x,t) \\
& + w^2(x,t)\dot{\theta}(t)\delta\dot{\theta}(t) + \dot{w}(x,t)\delta\dot{w}(x,t) + \dot{r}(t)\delta r(t) + x\dot{\theta}(t)\delta\dot{w}(x,t) \\
& + x\dot{w}(x,t)\delta\dot{\theta}(t) - x\cos\theta(t)\dot{\theta}(t)\dot{r}(t)\delta\theta(t) - x\sin\theta(t)\dot{r}(t)\delta\dot{\theta}(t) \\
& - x\sin\theta(t)\dot{\theta}(t)\delta\dot{r}(t) + \sin\theta(t)\dot{\theta}(t)w(x,t)\dot{r}(t)\delta\theta(t) \\
& - \cos\theta(t)w(x,t)\dot{r}(t)\delta\theta(t) - \cos\theta(t)\dot{\theta}(t)\dot{r}(t)\delta w(x,t) \\
& - \cos\theta(t)\dot{\theta}(t)w(x,t)\delta(t)\dot{r}(t) - \cos\theta(t)\dot{w}(x,t)\dot{r}(t)\delta\theta(t) \\
& - \sin\theta(t)\dot{r}(t)\delta\dot{w}(x,t) - \sin\theta(t)\dot{w}(x,t)\delta\dot{r}(t))\mathrm{d}x\mathrm{d}t \\
& + M\int_{t_1}^{t_2} \Big(L^2\dot{\theta}(t)\delta\dot{\theta}(t) + \dot{\theta}(t)w^2(L,t)\delta\dot{\theta}(t) + \dot{\theta}^2(t)w(L,t)\delta w(L,t) \\
& + \dot{w}(L,t)\delta\dot{w}(L,t) + \dot{r}(t)\delta r(t) + L\dot{\theta}(t)\delta\dot{w}(L,t) + L\dot{w}(L,t)\delta\dot{\theta}(t) \\
& - L\cos\theta(t)\dot{\theta}(t)\dot{r}(t)\delta\theta(t) - L\sin\theta(t)\dot{r}(t)\delta\dot{\theta}(t) \\
& - L\sin\theta(t)\dot{\theta}(t)\delta\dot{r}(t) + \sin\theta(t)\dot{\theta}(t)w(L,t)\dot{r}(t)\delta\theta(t) \\
& - \cos\theta(t)w(L,t)\dot{r}(t)\delta\theta(t) - \cos\theta(t)\dot{\theta}(t)w(L,t)\delta\dot{r}(t) \\
& - \cos\theta(t)\dot{\theta}(t)\dot{r}(t)\delta w(L,t) - \cos\theta(t)\dot{w}(L,t)\dot{r}(t)\delta\theta(t) \\
& - \sin\theta(t)\dot{r}(t)\delta\dot{w}(L,t) - \sin\theta(t)\dot{w}(L,t)\delta\dot{r}(t))\mathrm{d}t + m\int_{t_1}^{t_2}\dot{r}(t)\delta\dot{r}(t)\mathrm{d}t
\end{aligned} \tag{6.9}
$$

$$
\begin{aligned}
\int_{t_1}^{t_2} \delta E_p(t)\mathrm{d}t = {} & EI\int_{t_1}^{t_2}\int_0^L w_{xx}(x,t)\delta w_{xx}(x,t)\mathrm{d}x\mathrm{d}t + g\rho\int_{t_1}^{t_2}\int_0^L (x\cos\theta(t)\delta\theta(t) \\
& - w(x,t)\sin\theta(t)\delta\theta(t) + \cos\theta(t)\delta w(x,t))\mathrm{d}x\mathrm{d}t \\
& + Mg\int_{t_1}^{t_2} (L\cos\theta(t)\delta\theta(t) - w(L,t)\sin\theta(t)\delta\theta(t) + \cos\theta(t)\delta w(L,t))\mathrm{d}t
\end{aligned} \tag{6.10}
$$

$$
\int_{t_1}^{t_2} \delta W(t)\mathrm{d}t = \int_{t_1}^{t_2} (\tau(t)\delta\theta(t) + u_L(t)\delta w(L,t) + u_m(t)\delta r(t))\mathrm{d}t + \int_{t_1}^{t_2}\int_0^L F(x,t)\delta w(x,t)\mathrm{d}x\mathrm{d}t. \tag{6.11}
$$

Using integration by parts, Eqs. (6.9) and (6.10) can be respectively simplified as

$$
\begin{aligned}
\int_{t_1}^{t_2} \delta E_p(t)\mathrm{d}t = {} & EI\int_{t_1}^{t_2} (w_{xx}(L,t)\delta w_x(L,t) - w_{xx}(0,t)\delta w_x(0,t) - w_{xxx}(L,t)\delta w(L,t) \\
& + w_{xxx}(0,t)\delta w(0,t))\mathrm{d}t + EI\int_{t_1}^{t_2}\int_0^L w_{xxxx}(x,t)\delta w(x,t)\mathrm{d}x\mathrm{d}t \\
& + g\rho\int_{t_1}^{t_2}\int_0^L (x\cos\theta(t)\delta\theta(t) - w(x,t)\sin\theta(t)\delta\theta(t) + \cos\theta(t)\delta w(x,t))\mathrm{d}x\mathrm{d}t
\end{aligned}
$$

$$+ Mg\int_{t_1}^{t_2}(L\cos\theta(t)\delta\theta(t) - w(L,t)\sin\theta(t)\delta\theta(t) + \cos\theta(t)\delta w(L,t))\mathrm{d}t \tag{6.12}$$

$$\int_{t_1}^{t_2}\delta E_k(t)\mathrm{d}t = -I_h\int_{t_1}^{t_2}\ddot{\theta}(t)\delta\dot{\theta}(t)\mathrm{d}t + \rho\int_{t_1}^{t_2}\int_0^L\Big(-x^2\ddot{\theta}(t)\delta\theta(t) + \dot{\theta}^2(t)w(x,t)\delta w(x,t)$$
$$- 2w(x,t)\dot{w}(x,t)\dot{\theta}(t)\delta\theta(t) - w^2(x,t)\ddot{\theta}(t)\delta\theta(t) - \ddot{w}(x,t)\delta w(x,t)$$
$$- \ddot{r}(t)\delta r(t) - x\ddot{\theta}(t)\delta w(x,t) - x\ddot{w}(x,t)\delta\theta(t) + x\sin\theta(t)\ddot{r}(t)\delta\theta(t)$$
$$+ x\cos\theta(t)\dot{\theta}^2(t)\delta r(t) + x\sin\theta(t)\ddot{\theta}(t)\delta r(t) + \cos\theta(t)w(x,t)\ddot{r}(t)\delta\theta(t)$$
$$- \sin\theta(t)\dot{\theta}^2(t)w(x,t)\delta r(t) + \cos\theta(t)\ddot{\theta}(t)w(x,t)\delta r(t)$$
$$+ 2\cos\theta(t)\dot{\theta}(t)\dot{w}(x,t)\delta r(t) + \sin\theta(t)\ddot{r}(t)\delta w(x,t)$$
$$+ \sin\theta(t)\ddot{w}(x,t)\delta r(t))\mathrm{d}x\mathrm{d}t + M\int_{t_1}^{t_2}\Big(-L^2\ddot{\theta}(t)\delta\theta(t) - \ddot{\theta}(t)w^2(L,t)\delta\theta(t)$$
$$- 2w(L,t)\dot{w}(L,t)\dot{\theta}(t)\delta\theta(t) + \dot{\theta}^2(t)w(L,t)\delta w(L,t) - \ddot{w}(L,t)\delta w(L,t)$$
$$- \ddot{r}(t)\delta r(t) - L\ddot{\theta}(t)\delta w(L,t) - L\ddot{w}(L,t)\delta\theta(t) + L\sin\theta(t)\ddot{r}(t)\delta\theta(t)$$
$$+ L\cos\theta(t)\dot{\theta}^2(t)\delta r(t) + L\sin\theta(t)\ddot{\theta}(t)\delta r(t) + \cos\theta(t)w(L,t)\ddot{r}(t)\delta\theta(t)$$
$$- \sin\theta(t)\dot{\theta}^2(t)w(L,t)\delta r(t) + \cos\theta(t)\ddot{\theta}(t)w(L,t)\delta r(t)$$
$$+ 2\cos\theta(t)\dot{\theta}(t)\dot{w}(L,t)\delta r(t) + \sin\theta(t)\ddot{r}(t)\delta w(L,t)$$
$$+ \sin\theta(t)\ddot{w}(L,t)\delta r(t))\mathrm{d}t - m\int_{t_1}^{t_2}\ddot{r}(t)\delta r(t)\mathrm{d}t. \tag{6.13}$$

Natural boundary conditions of flexible manipulator are given as

$$w(0,t) = w_x(0,t) = 0. \tag{6.14}$$

By substituting Eqs. (6.11)–(6.14) into Eq. (6.8), we can further obtain the following PDE model of the system

$$\left(I_h + \frac{1}{3}\rho L^3 + ML^2\right)\ddot{\theta}(t) + 2M\dot{\theta}(t)w(L,t)\dot{w}(L,t) - \left(ML + \frac{1}{2}\rho L^2\right)\sin\theta(t)\ddot{r}(t)$$
$$- M\cos\theta(t)w(L,t)\ddot{r}(t) + M\ddot{\theta}(t)w^2(L,t) + ML\ddot{w}(L,t)$$
$$+ \frac{1}{2}g\rho L^2\cos\theta(t) + MgL\cos\theta(t) - Mgw(L,t)\sin\theta(t) + \rho\int_0^L(2w(x,t)\dot{w}(x,t)\dot{\theta}(t)$$
$$+ w^2(x,t)\ddot{\theta}(t) + x\ddot{w}(x,t) - \cos\theta(t)w(x,t)\ddot{r}(t) - gw(x,t)\sin\theta(t)\Big)\mathrm{d}x - \tau(t) = 0 \tag{6.15}$$

$$-M\dot{\theta}^2(t)w(L,t) + M\ddot{w}(L,t) + ML\ddot{\theta}(t) - M\sin\theta(t)\ddot{r}(t) + Mg\cos\theta(t) - EIw_{xxx}(L,t) - u_L(t) = 0 \tag{6.16}$$

$$(M + m + \rho L)\ddot{r}(t) - \left(\rho\frac{1}{2}L^2 + ML\right)\cos\theta(t)\dot{\theta}^2(t)$$

$$
\begin{aligned}
&-\left(\rho\frac{1}{2}L^2+ML\right)\sin\theta(t)\ddot{\theta}(t)+M\sin\theta(t)\dot{\theta}^2(t)w(L,t)-M\cos\theta(t)\ddot{\theta}(t)w(L,t)\\
&-2M\cos\theta(t)\dot{\theta}(t)\dot{w}(L,t)-M\sin\theta(t)\ddot{w}(L,t)+\rho\int_0^L\left(\sin\theta(t)\dot{\theta}^2(t)w(x,t)\right.\\
&\left.-\cos\theta(t)\ddot{\theta}(t)w(x,t)-2\cos\theta(t)\dot{\theta}(t)\dot{w}(x,t)-\sin\theta(t)\ddot{w}(x,t)\right)\mathrm{d}x-u_m(t)=0
\end{aligned}
\tag{6.17}
$$

$$
\rho\ddot{w}(x,t)+\rho x\ddot{\theta}(t)-\rho\sin\theta(t)\ddot{r}(t)+EIw_{xxxx}(x,t)-\rho\dot{\theta}^2(t)w(x,t)+g\rho\cos\theta(t)-F(x,t)=0.
\tag{6.18}
$$

The corresponding boundary conditions are given by the following set of equations

$$
w(0,t)=0 \tag{6.19}
$$

$$
w_x(0,t)=0 \tag{6.20}
$$

$$
w_{xx}(L,t)=0. \tag{6.21}
$$

Assumption 6.1 The desired angle, expected position of the vehicle and all parameters are bounded.

6.3 Control Design and Stability Analysis

In this section, a boundary PD control of the vehicle-mounted flexible manipulator system is developed. Based on Assumption 6.1, the designed controllers have the following objectives: (1) Regulating the positions of the joint and vehicle; (2) Suppressing the elastic vibration of the flexible manipulator. To summarize, the final objective of the system can be described as $\theta(t)\to\theta_d$, $\dot{\theta}(t)\to 0$, $r(t)\to r_d$, $\dot{r}(t)\to 0$, $w(x,t)\to 0$ and $\dot{w}(x,t)\to 0$ when $t\to\infty$, where θ_d is the desired angle of the flexible link and r_d is the desired position of the vehicle. Both θ_d and r_d are invariant constants.

To realize the control target, we define tracking errors as $e_1(t)=\theta(t)-\theta_d$, $e_2(t)=r(t)-r_d$ and control inputs are proposed as following.

$$
\tau(t)=-k_{p1}e_1(t)-k_{d1}\dot{e}_1(t)+\frac{1}{2}g\rho L^2\cos\theta(t)+MgL\cos\theta(t)
$$

$$- Mgw(L,t)\sin\theta(t) - \rho\int_0^L gw(x,t)\sin\theta(t)\mathrm{d}x \tag{6.22}$$

$$u_L(t) = -k_{p2}w(L,t) - k_{d2}\dot{w}(L,t) + Mg\cos\theta(t) \tag{6.23}$$

$$u_m(t) = -k_{p3}e_3(t) - k_{d3}\dot{e}_3(t) \tag{6.24}$$

$$F(x,t) = -k_{d4}\dot{w}(x,t) + g\rho\cos\theta(t) \tag{6.25}$$

where k_{p1}, k_{d1}, k_{p2}, k_{d2}, k_{p3} and k_{d3} are positive control coefficients.

Theorem 6.1 *Considering the system described by Eqs.* (6.15)–(6.21) *under Assumption* 6.1, *the asymptotical stability of the closed-loop system can be guaranteed under the proposed control laws in Eqs.* (6.22)–(6.24) *in the sense of* $\theta(t) \to \theta_d(t)$, $\dot{\theta}(t) \to 0$, $r(t) \to r_d(t)$, $\dot{r}(t) \to 0$, $w(x,t) \to 0$ *and* $\dot{w}(x,t) \to 0$ *when* $t \to \infty$.

Proof In the first step, a Lyapunov function is taken to be

$$V(t) = V_1(t) + V_2(t) \tag{6.26}$$

where

$$V_1(t) = \frac{1}{2}I_h\dot{\theta}^2(t) + \frac{1}{2}\rho\int_0^L \dot{\boldsymbol{P}}^{\mathrm{T}}(x,t)\dot{\boldsymbol{P}}(x,t)\mathrm{d}x + \frac{1}{2}M\dot{\boldsymbol{P}}^{\mathrm{T}}(L,t)\dot{\boldsymbol{P}}(L,t)$$
$$+ \frac{1}{2}m\dot{r}^2(t) + \frac{EI}{2}\int_0^L w_{xx}^2(x,t)\mathrm{d}x \tag{6.27}$$

$$V_2(t) = \frac{1}{2}k_{p1}e_1^2(t) + \frac{1}{2}k_{p2}w^2(L,t) + \frac{1}{2}k_{p3}e_2^2(t) \tag{6.28}$$

in which $V_1(t)$ is a term of the system energy and $V_2(t)$ is a term of tracking errors.

Differentiating Eq. (6.26) with respect to time, we can obtain

$$\dot{V}(t) = I_h\dot{\theta}(t)\ddot{\theta}(t) + \frac{1}{2}\rho\int_0^L \big(2x^2\dot{\theta}(t)\ddot{\theta}(t) + 2w(x,t)\dot{w}(x,t)\dot{\theta}^2(t)$$
$$+ 2w^2(x,t)\dot{\theta}(t)\ddot{\theta}(t) + 2\dot{w}(x,t)\ddot{w}(x,t) + 2\dot{r}(t)\ddot{r}(t) + 2x\ddot{\theta}(t)\dot{w}(x,t)$$
$$+ 2x\dot{\theta}(t)\ddot{w}(x,t) - 2x\cos\theta(t)\dot{\theta}^2(t)\dot{r}(t) - 2x\sin\theta(t)\ddot{\theta}(t)\dot{r}(t)$$

$$
\begin{aligned}
&-2x\sin\theta(t)\dot{\theta}(t)\ddot{r}(t)+2\sin\theta(t)\dot{\theta}^2(t)w(x,t)\dot{r}(t)\\
&-2\cos\theta(t)\ddot{\theta}(t)w(x,t)\dot{r}(t)-4\cos\theta(t)\dot{\theta}(t)\dot{w}(x,t)\dot{r}(t)\\
&-2\cos\theta(t)\dot{\theta}(t)w(x,t)\ddot{r}(t)-2\sin\theta(t)\ddot{w}(x,t)\dot{r}(t)\\
&-2\sin\theta(t)\dot{w}(x,t)\ddot{r}(t))\mathrm{d}x+k_{p3}(r(t)-r_d)\dot{r}(t)+\frac{1}{2}M\big(2L^2\dot{\theta}(t)\ddot{\theta}(t)\\
&+2w(L,t)\dot{w}(L,t)\dot{\theta}^2(t)+2w^2(L,t)\dot{\theta}(t)\ddot{\theta}(t)+2\dot{w}(L,t)\ddot{w}(L,t)\\
&+2\dot{r}(t)\ddot{r}(t)+2L\ddot{\theta}(t)\dot{w}(L,t)+2L\dot{\theta}(t)\ddot{w}(L,t)-2L\cos\theta(t)\dot{\theta}^2(t)\dot{r}(t)\\
&-2L\sin\theta(t)\ddot{\theta}(t)\dot{r}(t)-2L\sin\theta(t)\dot{\theta}(t)\ddot{r}(t)\\
&+2\sin\theta(t)\dot{\theta}^2(t)w(L,t)\dot{r}(t)-2\cos\theta(t)\ddot{\theta}(t)w(L,t)\dot{r}(t)\\
&-4\cos\theta(t)\dot{\theta}(t)\dot{w}(L,t)\dot{r}(t)-2\cos\theta(t)\dot{\theta}(t)w(L,t)\ddot{r}(t)\\
&-2\sin\theta(t)\ddot{w}(L,t)\dot{r}(t)-2\sin\theta(t)\dot{w}(L,t)\ddot{r}(t))+m\dot{r}(t)\ddot{r}(t)\\
&+EI\int_0^L w_{xx}(x,t)\dot{w}_{xx}(x,t)\mathrm{d}x+k_{p1}(\theta(t)-\theta_d)\dot{\theta}(t)+k_{p2}w(L,t)\dot{w}(L,t).
\end{aligned}
\tag{6.29}
$$

Multiplying Eq. (6.15) by $\dot{\theta}(t)$, Eq. (6.16) by $\dot{w}(L,t)$ and Eq. (6.17) by $\dot{r}(t)$, we can get

$$
\begin{aligned}
\dot{\theta}(t)\tau(t)=&\left(I_h+\frac{1}{3}\rho L^3+ML^2\right)\ddot{\theta}(t)\dot{\theta}(t)+2M\dot{\theta}^2(t)w(L,t)\dot{w}(L,t)\\
&-\left(ML+\frac{1}{2}\rho L^2\right)\sin\theta(t)\ddot{r}(t)\dot{\theta}(t)-M\cos\theta(t)w(L,t)\ddot{r}(t)\dot{\theta}(t)\\
&+M\ddot{\theta}(t)w^2(L,t)\dot{\theta}(t)+ML\ddot{w}(L,t)\dot{\theta}(t)+\frac{1}{2}g\rho L^2\cos\theta(t)\dot{\theta}(t)\\
&+MgL\cos\theta(t)\dot{\theta}(t)-Mgw(L,t)\sin\theta(t)\dot{\theta}(t)\\
&+\rho\int_0^L\Big(2w(x,t)\dot{w}(x,t)\dot{\theta}^2(t)+w^2(x,t)\ddot{\theta}(t)\dot{\theta}(t)+x\ddot{w}(x,t)\dot{\theta}(t)\\
&-\cos\theta(t)w(x,t)\ddot{r}(t)\dot{\theta}(t)-gw(x,t)\sin\theta(t)\dot{\theta}(t)\Big)\mathrm{d}x
\end{aligned}
\tag{6.30}
$$

$$
\begin{aligned}
\dot{w}(L,t)u_L(t)=&-M\dot{\theta}^2(t)w(L,t)\dot{w}(L,t)+M\ddot{w}(L,t)\dot{w}(L,t)+ML\ddot{\theta}(t)\dot{w}(L,t)\\
&-M\sin\theta(t)\ddot{r}(t)\dot{w}(L,t)-EIw_{xxx}(x,t)\dot{w}(L,t)+Mg\cos\theta(t)\dot{w}(L,t)
\end{aligned}
\tag{6.31}
$$

$$
\begin{aligned}
\dot{r}(t)u_m(t)=&(M+m+\rho L)\ddot{r}(t)\dot{r}(t)-\left(\rho\frac{1}{2}L^2+ML\right)\cos\theta(t)\dot{\theta}^2(t)\dot{r}(t)\\
&-M\sin\theta(t)\ddot{w}(L,t)\dot{r}(t)+M\sin\theta(t)\dot{\theta}^2(t)w(L,t)\dot{r}(t)\\
&-M\cos\theta(t)\ddot{\theta}(t)w(L,t)\dot{r}(t)-2M\cos\theta(t)\dot{\theta}(t)\dot{w}(L,t)\dot{r}(t)\\
&+\rho\int_0^L\Big(\sin\theta(t)\dot{\theta}^2(t)w(x,t)\dot{r}(t)-\cos\theta(t)\ddot{\theta}(t)w(x,t)\dot{r}(t)
\end{aligned}
$$

$$-2\cos\theta(t)\dot{\theta}(t)\dot{w}(x,t)\dot{r}(t)-\sin\theta(t)\ddot{w}(x,t)\dot{r}(t)\big)\mathrm{d}x$$
$$-\left(\rho\frac{1}{2}L^2+ML\right)\sin\theta(t)\ddot{\theta}(t)\dot{r}(t). \tag{6.32}$$

Substituting Eqs. (6.30)–(6.32) into Eq. (6.29) gives

$$\begin{aligned}
\dot{V}(t) &= \dot{\theta}(t)\tau(t)-\rho\int_0^L w(x,t)\dot{w}(x,t)\dot{\theta}^2(t)\mathrm{d}x+\dot{w}(L,t)u_L(t)+EIw_{xxx}(L,t)\dot{w}(L,t)+\dot{r}(t)u_m(t)\\
&+\rho\int_0^L \dot{w}(x,t)\ddot{w}(x,t)\mathrm{d}x+\rho\int_0^L x\ddot{\theta}(t)\dot{w}(x,t)\mathrm{d}x-\rho\int_0^L \sin\theta(t)\dot{w}(x,t)\ddot{r}(t)\mathrm{d}x\\
&+EI\int_0^L w_{xx}(x,t)\dot{w}_{xx}(x,t)\mathrm{d}x+k_{p1}(\theta(t)-\theta_d)\dot{\theta}(t)+k_{p2}w(L,t)\dot{w}(L,t)+k_{p3}(r(t)-r_d)\dot{r}(t)\\
&+\int_0^L EIw_{xxxx}(x,t)\dot{w}(x,t)\mathrm{d}x-\int_0^L g\rho\cos\theta(t)\dot{w}(x,t)\mathrm{d}x\\
&-\frac{1}{2}g\rho L^2\cos\theta(t)\dot{\theta}(t)-MgL\cos\theta(t)\dot{\theta}(t)+Mgw(L,t)\sin\theta(t)\dot{\theta}(t)\\
&+\rho\int_0^L gw(x,t)\sin\theta(t)\dot{\theta}(t)\mathrm{d}x-Mg\cos\theta(t)\dot{w}(L,t)+\int_0^L g\rho\cos\theta(t)\dot{w}(x,t)\mathrm{d}x.
\end{aligned} \tag{6.33}$$

Multiplying Eq. (6.18) by $\dot{w}(x,t)$ and integrating it respect to x from 0 to L, we have

$$\begin{aligned}
\int_0^L \dot{w}(x,t)F(x,t)\mathrm{d}x &= \int_0^L \rho\ddot{w}(x,t)\dot{w}(x,t)\mathrm{d}x+\int_0^L \rho x\ddot{\theta}(t)\dot{w}(x,t)\mathrm{d}x-\int_0^L \rho\sin\theta(t)\ddot{r}(t)\dot{w}(x,t)\mathrm{d}x\\
&+\int_0^L EIw_{xxxx}(x,t)\dot{w}(x,t)\mathrm{d}x-\int_0^L \rho\dot{\theta}^2(t)w(x,t)\dot{w}(x,t)\mathrm{d}x+\int_0^L g\rho\cos\theta(t)\dot{w}(x,t)\mathrm{d}x.
\end{aligned} \tag{6.34}$$

Substituting Eq. (6.34) into Eq. (6.33), we can get

$$\begin{aligned}
\dot{V}(t) &= \dot{\theta}(t)\tau(t)+\dot{w}(L,t)u_L(t)+\dot{r}(t)u_m(t)+\int_0^L F(x,t)\dot{w}(x,t)\mathrm{d}x-\frac{1}{2}g\rho L^2\cos\theta(t)\dot{\theta}(t)\\
&-MgL\cos\theta(t)\dot{\theta}(t)+Mgw(L,t)\sin\theta(t)\dot{\theta}(t)+\rho\int_0^L gw(x,t)\sin\theta(t)\dot{\theta}(t)\mathrm{d}x\\
&-Mg\cos\theta(t)\dot{w}(L,t)-\int_0^L g\rho\cos\theta(t)\dot{w}(x,t)\mathrm{d}x+EI\int_0^L w_{xx}(x,t)\dot{w}_{xx}(x,t)\mathrm{d}x
\end{aligned}$$

$$
\begin{aligned}
&-\int_0^L EIw_{xxxx}(x,t)\dot{w}(x,t)\mathrm{d}x + k_{p1}(\theta(t)-\theta_d)\dot{\theta}(t) + k_{p2}w(L,t)\dot{w}(L,t)\\
&+k_{p3}(r(t)-r_d)\dot{r}(t) + EIw_{xxx}(L,t)\dot{w}(L,t).
\end{aligned} \tag{6.35}
$$

Since

$$
\begin{aligned}
&-\int_0^L EIw_{xxxx}(x,t)\dot{w}(x,t)\mathrm{d}x + EIw_{xxx}(L,t)\dot{w}(L,t) + EI\int_0^L w_{xx}(x,t)\dot{w}_{xx}(x,t)\mathrm{d}x\\
&=-\int_0^L EIw_{xxxx}(x,t)\dot{w}(x,t)\mathrm{d}x + EIw_{xxx}(L,t)\dot{w}(L,t) + EIw_{xx}(x,t)\dot{w}_x(x,t)\Big|_0^L\\
&-EI\int_0^L w_{xxx}(x,t)\dot{w}_x(x,t)\mathrm{d}x\\
&=EIw_{xxx}(L,t)\dot{w}(L,t) + EIw_{xx}(L,t)\dot{w}_x(L,t) - EIw_{xx}(0,t)\dot{w}_x(0,t) - EIw_{xxx}(x,t)\dot{w}(x,t)\Big|_0^L\\
&=EIw_{xxx}(L,t)\dot{w}(L,t) + EIw_{xx}(L,t)\dot{w}_x(L,t) - EIw_{xx}(0,t)\dot{w}_x(0,t)\\
&\quad-\big(EIw_{xxx}(L,t)\dot{w}(L,t) - EIw_{xxx}(0,t)\dot{w}(0,t)\big)\\
&=EIw_{xx}(L,t)\dot{w}_x(L,t) - EIw_{xx}(0,t)\dot{w}_x(0,t) + EIw_{xxx}(0,t)\dot{w}(0,t)\\
&=0,
\end{aligned} \tag{6.36}
$$

we can achieve the following equation by Eq. (6.35)

$$
\begin{aligned}
\dot{V}(t) &= \dot{\theta}(t)\tau(t) + \dot{w}(L,t)u_L(t) + \dot{r}(t)u_m(t) + \int_0^L F(x,t)\dot{w}(x,t)\mathrm{d}x - \frac{1}{2}g\rho L^2\cos\theta(t)\dot{\theta}(t)\\
&-MgL\cos\theta(t)\dot{\theta}(t) + Mgw(L,t)\sin\theta(t)\dot{\theta}(t) + \rho\int_0^L gw(x,t)\sin\theta(t)\dot{\theta}(t)\mathrm{d}x\\
&-Mg\cos\theta(t)\dot{w}(L,t) - \int_0^L g\rho\cos\theta(t)\dot{w}(x,t)\mathrm{d}x + k_{p1}(\theta(t)-\theta_d)\dot{\theta}(t)\\
&+k_{p2}w(L,t)\dot{w}(L,t) + k_{p3}(r(t)-r_d)\dot{r}(t).
\end{aligned} \tag{6.37}
$$

Substituting the proposed control laws in Eqs. (6.22)–(6.25), we obtain

$$
\dot{V}(t) = -k_{d1}\dot{\theta}^2(t) - k_{d2}\dot{w}^2(L,t) - k_{d3}\dot{r}^2(t) - k_{d4}\int_0^L \dot{w}^2(x,t)\mathrm{d}x \le 0. \tag{6.38}
$$

Consequently, the derivation of Lyapunov function in Eq. (6.26) is negative semi-definite. Then we will use the extended LaSalle's invariance principle [1] to prove the asymptotic stability of the system.

If $\dot{V}(t) \equiv 0$, we have

$$\dot{\theta}(t) = \dot{w}(L, t) = \dot{r}(t) \equiv \dot{w}(x, t) = 0 \tag{6.39}$$

which implies

$$\ddot{\theta}(t) = \ddot{w}(L, t) = \ddot{r}(t) = \ddot{w}(x, t) \equiv 0. \tag{6.40}$$

Applying Eqs. (6.39) and (6.40) into Eqs. (6.15)–(6.21) yields

$$\tau(t) = \frac{1}{2}g\rho L^2 \cos\theta(t) + MgL\cos\theta(t) - Mgw(L, t)\sin\theta(t) - \rho\int_0^L gw(x, t)\sin\theta(t)\mathrm{d}x \tag{6.41}$$

$$u_L(t) = -EIw_{xxx}(L, t) + Mg\cos\theta(t) \tag{6.42}$$

$$u_m(t) = 0 \tag{6.43}$$

$$EIw_{xxxx}(x, t) + \rho g\cos\theta(t) = F(x, t). \tag{6.44}$$

Referring back to Eqs. (6.39) and (6.44), it is clear that

$$EIw_{xxxx}(x, t) = 0. \tag{6.45}$$

It is noteworthy that Eq. (6.45) is separable and the technique of separation of variables [2] can be applied on it. It is assumed that $w(x, t)$ can be separated as

$$w(x, t) = \Phi(x) \cdot q(t) \tag{6.46}$$

where $\Phi(x)$ and $q(t)$ are two uncertain functions related to space and time to be determined respectively.

Solving Eq. (6.46), we can obtain

$$\Phi(x) = b_1 + b_2x + b_3x^2 + b_4x^3 \tag{6.47}$$

where $b_i \in R, i = 1, 2, 3, 4$ are four undetermined constants.

Using boundary conditions in Eqs. (6.19)–(6.21) and Eq. (6.42), we have

$$b_1 = 0 \tag{6.48}$$

$$b_2 = 0 \tag{6.49}$$

$$b_3 + 3b_4 L = 0 \tag{6.50}$$

$$6EIb_4 = k_{p2}\big(b_1 + b_2 L + b_3 L^2 + b_4 L^3\big). \tag{6.51}$$

By solving Eqs. (6.48)–(6.51), we can derive that $b_i = 0, i = 1, 2, 3, 4$, is the unique solution. Therefore, $\Phi(x) = 0$, $w(x, t) = 0$, $\ddot{w}(x, t) = 0$, $w_{xxx}(L, t) = 0$ can be obtained. Substituting Eqs. (6.22)–(6.25) into Eqs. (6.41)–(6.44), we can get

$$\theta(t) - \theta_d = 0 \tag{6.52}$$

$$r(t) - r_d = 0 \tag{6.53}$$

$$w(L, t) = 0. \tag{6.54}$$

Therefore, applying the extended LaSalle's invariance principle, $\theta(t) \to \theta_d$, $\dot{\theta}(t) \to 0$, $r(t) \to r_d$, $\dot{r}(t) \to 0$, $w(x, t) \to 0$, $\dot{w}(x, t) \to 0$ when $t \to \infty$ can be guaranteed by the PD boundary control proposed in Eqs. (6.22)–(6.25) and the asymptotical stability of the closed-loop system is verified.

6.4 Simulations

The numerical simulation results of the proposed control scheme for the vehicle-mounted flexible manipulator are presented by MATLAB and the effectiveness of the proposed boundary control laws in Eqs. (6.22)–(6.25) is verified in this section. The desired positions are set as $\theta_d = 0.2$ rad, $r_d = 1$ m, and the final signals of the system are expected to be $\theta(t) \to \theta_d$, $\dot{\theta}(t) \to 0$, $r(t) \to r_d$, $\dot{r}(t) \to 0$, $w(x, t) \to 0$, $\dot{w}(x, t) \to 0$ with time. The system parameters are given in Table 6.1.

Table 6.1 Parameters of the vehicle-mounted flexible manipulator

Parameter	Description	Value
L	Length of the flexible link	0.6 m
ρ	Uniform mass per unit length of the flexible link	0.8 kg/m
M	Mass of the tip payload	0.3 kg
EI	Bending rigidity of the flexible link	20 Nm^2
I_h	Moment of inertia of the flexible link	2 kgm^2
m	Mass of the vehicle	1.5 kg

The control gains of the system are set as $k_{p1} = 30, k_{d1} = 10, k_{p2} = 10, k_{d2} = 60, k_{p3} = 40$, and $k_{d3} = 15$.

System states without control are reflected by Figs. 6.2, 6.3, 6.4, 6.5 and 6.6, from which it can be observed that the flexible manipulator vibrates and the open-loop system is unstable.

The simulation results for the closed-loop system are shown in Figs. 6.7, 6.8, 6.9, 6.10, 6.11 and 6.12. Figure 6.7 gives the target and real position of the joint. Figure 6.8 shows target and real position of the vehicle. The speeds of the joint and vehicle are represented in Figs. 6.9 and 6.10 respectively. The elastic deflection of the flexible manipulator $w(x, t)$ under control laws is depicted in Fig. 6.11 and the speed of elastic deflection $\dot{w}(x, t)$ in the closed-loop system is shown in Fig. 6.12.

Control inputs $\tau(t)$, $u_m(t)$, $u_L(t)$ and $F(x, t)$ are shown in Figs. 6.13, 6.14, 6.15 and 6.16 as follows.

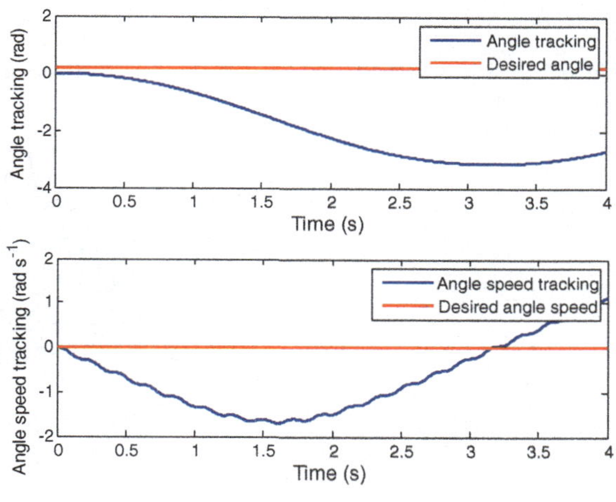

Fig. 6.2 Angular position and speed of the joint without control

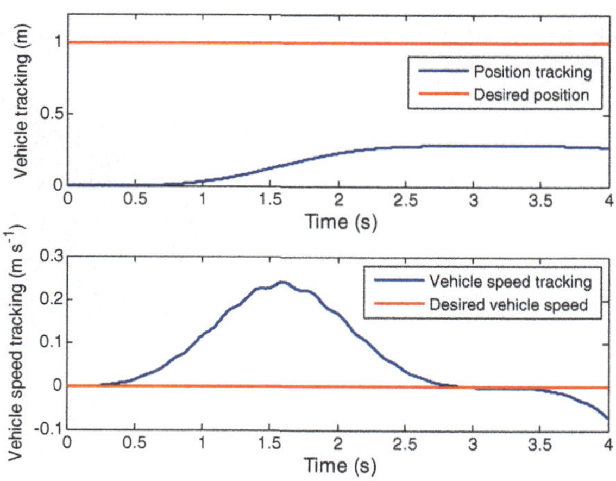

Fig. 6.3 Position and speed of the vehicle without control

Fig. 6.4 Elastic deflection and deformation speed of the flexible manipulator without control

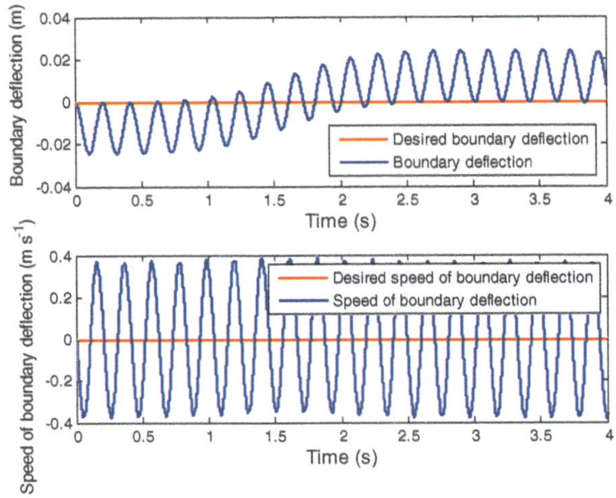

Fig. 6.5 Elastic deflection of the flexible manipulator without control

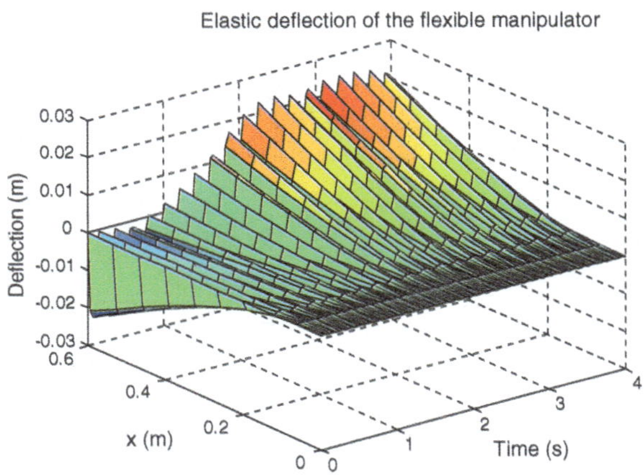

Fig. 6.6 Speed of elastic deflection without control

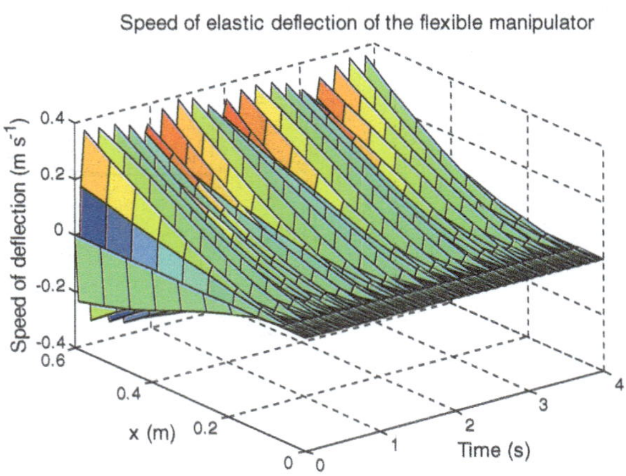

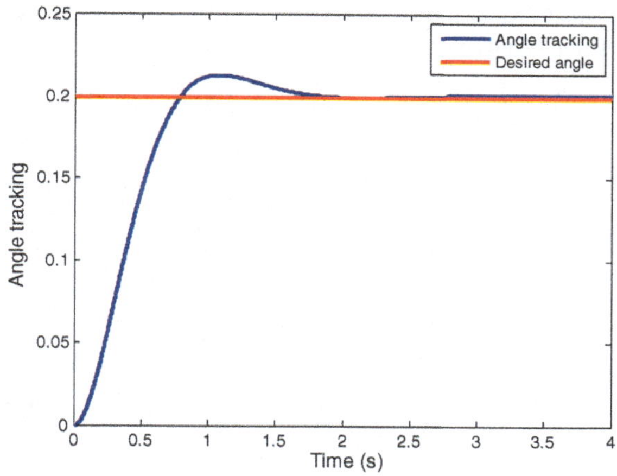

Fig. 6.7 Angular position of the joint with control

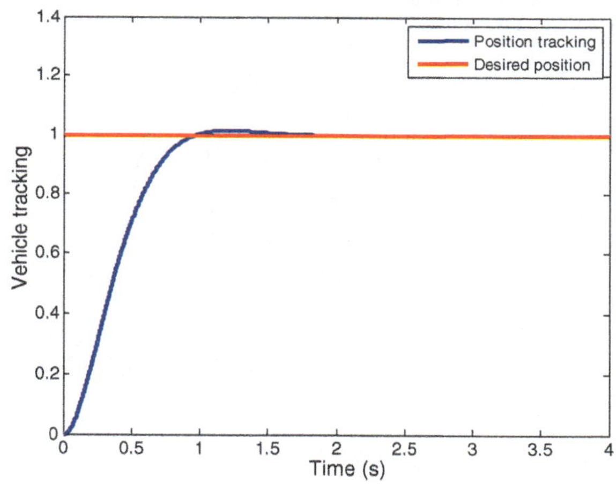

Fig. 6.8 Position of the vehicle with control

From the simulation results, we can find that the system can track the desired position and eliminate the vibration simultaneously within 2 s under the proposed control scheme. Besides that, $\theta(t) \to \theta_d, \dot{\theta}(t) \to 0, r(t) \to r_d, \dot{r}(t) \to 0, w(x,t) \to 0$, $\dot{w}(x,t) \to 0$ can be realized when $t \to \infty$. Therefore, the proposed control laws are effective and the closed-loop system is asymptotic stable.

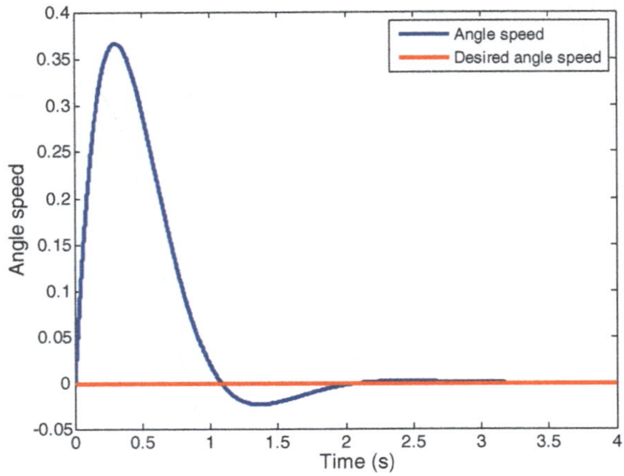

Fig. 6.9 Angular speed of the joint with control

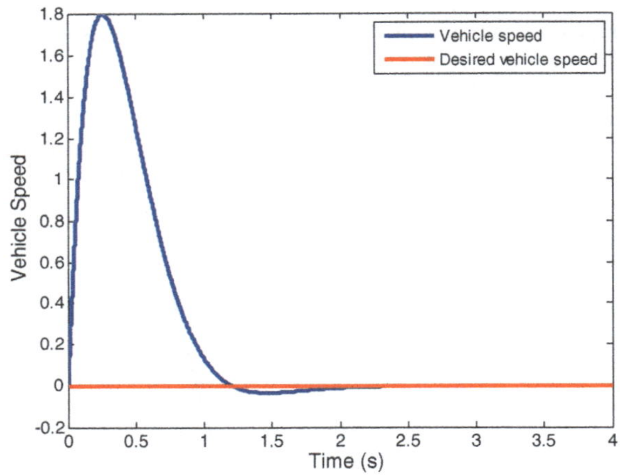

Fig. 6.10 Speed of the vehicle with control

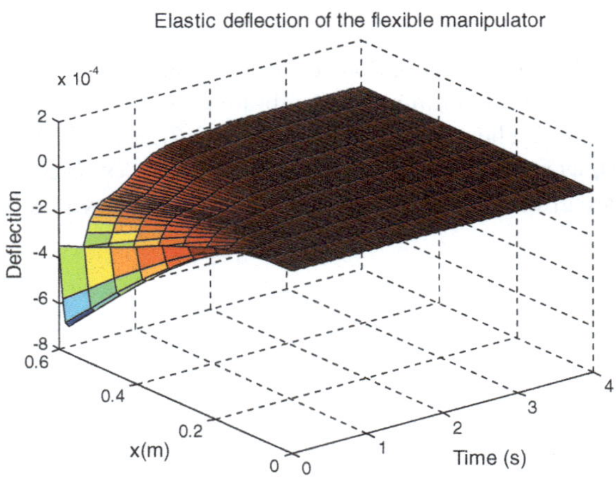

Fig. 6.11 Elastic deflection of the flexible manipulator with control

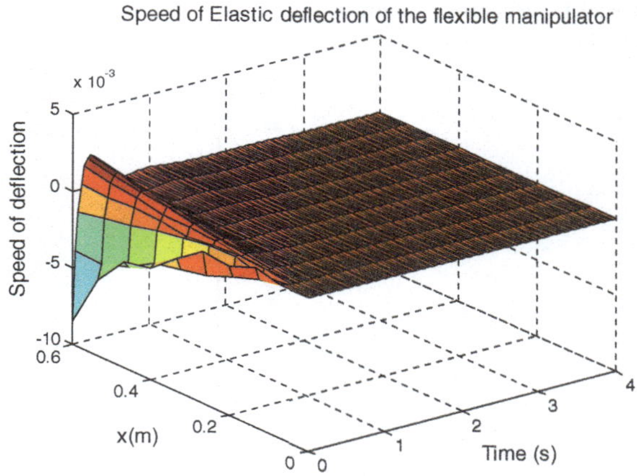

Fig. 6.12 Speed of the elastic deflection with control

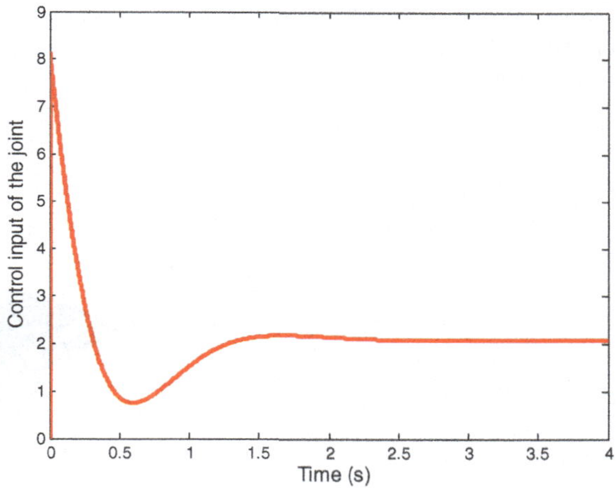

Fig. 6.13 Control input $\tau(t)$

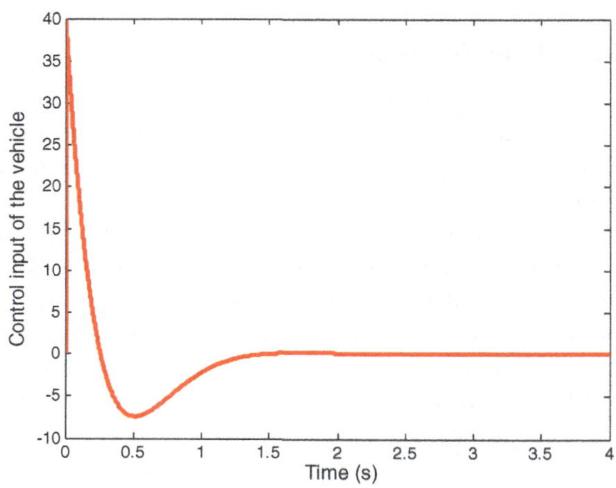

Fig. 6.14 Control input $u_m(t)$

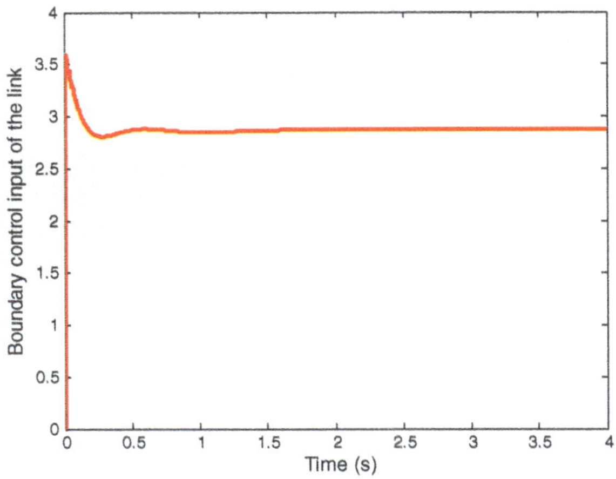

Fig. 6.15 Boundary control input $u_L(t)$

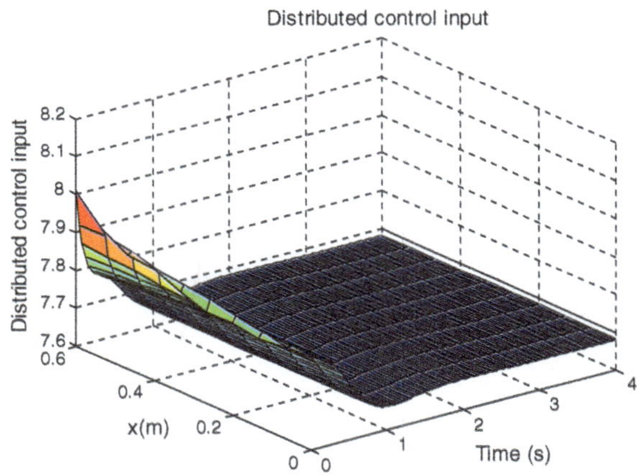

Fig. 6.16 Distributed control input $F(x, t)$

6.5 Conclusions

In this chapter, a boundary PD control has been designed for the vehicle-mounted flexible manipulator. A PDE model of the system has been derived by Hamilton principle without neglecting any high frequency mode. Based on the original PDE model, the vehicle motion of the vehicle-mounted flexible manipulator is considered. The proposed controller can not only regulate the positions of joint and vehicle, but also can suppress the vibration of the system. By extending LaSalle's invariance principle to infinite dimensions, the asymptotic stability of the closed-loop system has been rigorously demonstrated. Numerical simulations have been also provided to illustrate the effectiveness of the proposed control.

References

1. F. Cao, J. Liu, An adaptive iterative learning algorithm for boundary control of a coupled ODE-PDE two-link rigid-flexible manipulator. J. Franklin Inst. **354**(1), 277–297 (2016)
2. W.H. Ray, *Advanced process control* (McGraw-Hill Companies, New York, NY, 1981)

Chapter 7
Adaptive Fault-Tolerant Control of Vehicle-Mounted Flexible Manipulator

7.1 Introduction

In the actual control process, control systems inevitably suffer from actuator faults due to the widespread uncertainty. Consequently, the fault-tolerant control is developed to deal with actuator failures and ensure the control performance.

A neural network based adaptive fault-tolerant control scheme is developed in this chapter for the vehicle-mounted flexible manipulator system in the presence of actuator failures and uncertainty. The control is proposed based on a hybrid PDE model. Under the proposed control laws, it guarantees that the vibration of the flexible manipulator can be suppressed with actuation redundancy. The boundary uncertainty of the vehicle-mounted flexible manipulator can be approached using RBF neural networks. Moreover, the position regulation of the vehicle and joint can also be achieved with the proposed control. The stability of the closed-loop system is demonstrated by extending LaSalle's invariance principle. Illustrative simulation results are provided to confirm the correctness of the proposed control strategy.

7.2 Problem Formulation

In this chapter, we consider a vehicle-mounted flexible manipulator system with uncertainties. The flexible manipulator is depicted in Fig. 7.1. Compared the model given in Chap. 6, the gravity of the flexible manipulator is neglected. Parameters and variables of the system are defined as same as the ones given in Chap. 6.

Assumption 7.1 The flexible manipulator moves in the horizontal direction and thus the potential energy only depends on the flexural deflection of the link.

X. Xing and J. Liu, *Vibration Control Methods of Mechanical Distributed Parameter Systems*, Springer Tracts in Mechanical Engineering,
https://doi.org/10.1007/978-981-16-1532-0_7

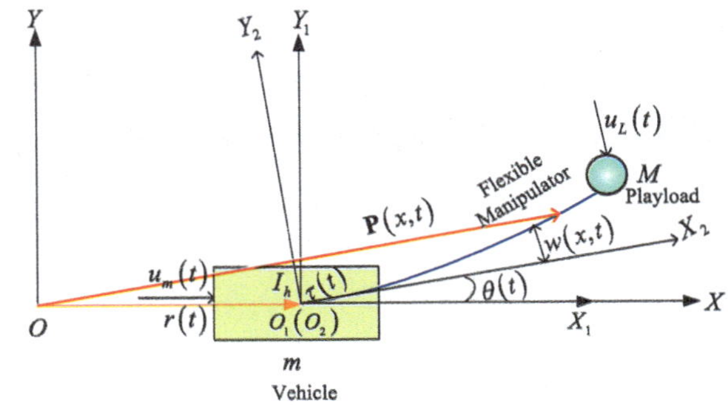

Fig. 7.1 Configuration of the vehicle-mounted flexible manipulator

Remark 7.1 Since the flexible manipulator is usually made of lightweight materials, if the flexible manipulator only moves in the horizontal plane, the influence of the gravity can be ignored [1–3].

The potential energy $E_p(t)$ of the system is described as

$$E_p(t)=\frac{EI}{2}\int_0^L w_{xx}^2(x,t)dx. \tag{7.1}$$

The $W(t)$ done by the non-conservative force is given as

$$W(t)=\tau(t)\theta(t)+(u_L(t)+d(\boldsymbol{\omega}))w(L,t)+u_m(t)r(t) \tag{7.2}$$

where $d(\boldsymbol{\omega})$ is a function of the modeling uncertainty occurring at the boundary conditions of the system.

The formulation of the kinetic energy $E_k(t)$ is given in Eq. (6.3).

Remark 7.2 In this chapter, $d(\boldsymbol{\omega})$ is the boundary modeling uncertainty, where $\boldsymbol{\omega}$ is an independent variable defined as $\boldsymbol{\omega}=\left[\, w(L,t)\ \dot{w}(L,t)\,\right]^{\mathrm{T}}$. It is noteworthy that $\boldsymbol{\omega}$ is also the input signal of the RBF neural network which will be designed later.

Considering the modeling uncertainty at the boundary of the flexible manipulator, we get the following dynamic PDE model of the system by using the Hamilton principle

$$\begin{aligned}&\left(I_h+\frac{1}{3}\rho L^3+ML^2\right)\ddot{\theta}(t)+2M\dot{\theta}(t)w(L,t)\dot{w}(L,t)-\left(ML+\frac{1}{2}\rho L^2\right)\sin\theta(t)\ddot{r}(t)\\&-M\cos\theta(t)w(L,t)\ddot{r}(t)+M\ddot{\theta}(t)w^2(L,t)+ML\ddot{w}(L,t)\\&+\rho\int_0^L\left(2w(x,t)\dot{w}(x,t)\dot{\theta}(t)+w^2(x,t)\ddot{\theta}(t)+x\ddot{w}(x,t)-\cos\theta(t)w(x,t)\ddot{r}(t)\right)\\&\mathrm{d}x-\tau(t)=0\end{aligned} \tag{7.3}$$

$$
\begin{aligned}
&(M + m + \rho L)\ddot{r}(t) - \left(\rho\frac{1}{2}L^2 + ML\right)\cos\theta(t)\dot{\theta}^2(t) \\
&-\left(\rho\frac{1}{2}L^2 + ML\right)\sin\theta(t)\ddot{\theta}(t) + M\sin\theta(t)\dot{\theta}^2(t)w(L,t) - M\cos\theta(t)\ddot{\theta}(t)w(L,t) \\
&-2M\cos\theta(t)\dot{\theta}(t)\dot{w}(L,t) - M\sin\theta(t)\ddot{w}(L,t) + \rho\int_0^L \big(\sin\theta(t)\dot{\theta}^2(t)w(x,t) \\
&-\cos\theta(t)\ddot{\theta}(t)w(x,t) - 2\cos\theta(t)\dot{\theta}(t)\dot{w}(x,t) - \sin\theta(t)\ddot{w}(x,t)\big)\mathrm{d}x - u_m(t) = 0
\end{aligned}
\tag{7.4}
$$

$$
\rho\ddot{w}(x,t) + \rho x\ddot{\theta}(t) - \rho\sin\theta(t)\ddot{r}(t) + EIw_{xxxx}(x,t) - \rho\dot{\theta}^2(t)w(x,t) = 0 \tag{7.5}
$$

where $\forall(x,t) \in (0,L) \times [0,\infty)$, and the boundary conditions of the flexible manipulator at $x = 0$ and $x = L$ are given as

$$
\begin{aligned}
&-M\dot{\theta}^2(t)w(L,t) + M\ddot{w}(L,t) + ML\ddot{\theta}(t) - M\sin\theta(t)\ddot{r}(t) \\
&\quad - EIw_{xxx}(L,t) - u_L(t) - d(\boldsymbol{\omega}) = 0
\end{aligned}
\tag{7.6}
$$

$$
w(0,t) = 0 \tag{7.7}
$$

$$
w_x(0,t) = 0 \tag{7.8}
$$

$$
w_{xx}(L,t) = 0. \tag{7.9}
$$

Assumption 7.2 It is reasonable to assume that the uncertainty $d(\boldsymbol{\omega})$ is bounded by $|d(\boldsymbol{\omega})| \leq (b_5\sigma_5 + b_6\sigma_6)\overline{d}$, where $\overline{d}$ is an unknown positive constant and b_5, σ_5, b_6, σ_6 are constants which are defined in the following control design process.

7.3 Control Design and Stability Analysis

7.3.1 RBF Neural Network Design

In order to facilitate the stability analysis, tracking errors of the joint and vehicle are denoted as $e_1(t) = \theta(t) - \theta_d$, $e_2(t) = r(t) - r_d$, respectively, where θ_d and r_d are constants. We also define $e_3(t) = w(L,t)$ as the deflection error of the end of the flexible manipulator. Then $\dot{e}_1(t) = \dot{\theta}(t)$, $\dot{e}_2(t) = \dot{r}(t)$, and $\dot{e}_3(t) = \dot{w}(L,t)$ can be achieved.

First all of, the neural network is designed to approximate the uncertain function $d(\boldsymbol{\omega})$.

Let

$$d(\boldsymbol{\omega}) = \mathbf{Q}^{*\mathrm{T}}\mathbf{h}(\boldsymbol{\omega}) + \varepsilon \tag{7.10}$$

where $\mathbf{Q}^* \in R^{n\times 1}$ is the ideal weight vector; $\mathbf{h}(\boldsymbol{\omega}) = \left[h_j\right]^{\mathrm{T}} \in R^{n\times 1}$, $j = 1, 2, \cdots n$, is the vector of the basis function; n is the network node number of the hidden layer; ε is the bounded approximation error in which $|\varepsilon| \leq \eta$ and η is a positive constant.

Then the adaptive RBF neural network algorithm can be designed as

$$h_j = \exp\left(\frac{\left\|\boldsymbol{\omega} - \mathbf{f}_j\right\|^2}{2g_j^2}\right) \tag{7.11}$$

$$\hat{d}(\boldsymbol{\omega}) = \hat{\mathbf{Q}}^{\mathrm{T}}\mathbf{h}(\boldsymbol{\omega}) \tag{7.12}$$

where $\mathbf{f}_j \in R^{2\times 1}$, $g_j \in R$, $j = 1, 2, \cdots n$, are the parameters of the Gaussian function, and h_j is the output signal of the Gaussian function, and adaptive law is proposed as

$$\dot{\hat{\mathbf{Q}}} = \gamma \dot{e}_3(t)\mathbf{h}(\boldsymbol{\omega}) \tag{7.13}$$

in which $\hat{\mathbf{Q}} \in R^{n\times 1}$ is a time related vector and γ is a positive constant.

Define $\tilde{\mathbf{Q}} = \mathbf{Q}^* - \hat{\mathbf{Q}}$ and $\tilde{d}(\boldsymbol{\omega}) = d(\boldsymbol{\omega}) - \hat{d}(\boldsymbol{\omega})$. Thus we have

$$\tilde{d}(\boldsymbol{\omega}) = \mathbf{Q}^{*\mathrm{T}}\mathbf{h}(\boldsymbol{\omega}) + \varepsilon - \hat{\mathbf{Q}}^{\mathrm{T}}\mathbf{h}(\boldsymbol{\omega}) = \tilde{\mathbf{Q}}^{\mathrm{T}}\mathbf{h}(\boldsymbol{\omega}) + \varepsilon. \tag{7.14}$$

7.3.2 Adaptive Fault-Tolerant Control Design and Stability Analysis

In the practical engineering, actuator faults are inevitable which may cause the system to be out of control and even lead to unrecoverable damage to the system. In this part, a redundancy fault-tolerant control of the vehicle-mounted flexible manipulator system with boundary uncertainty is proposed to guarantee the stability of the closed-loop system even when one of the actuators completely fails. With uncertainty and actuator failures, the control objective of this chapter is to try to control the position of the vehicle and joint, and suppress the vibration of the flexible link with the proposed control laws, which means that $\theta(t) \to \theta_d$, $r(t) \to r_d$, $w(L,t) \to 0$, and $w(x,t) \to 0$ are expected to be realized when $t \to \infty$.

To make sure the remaining actuator can continue operating during the complete fault occurring to one of the existing actuators, a dual-actuator redundancy control strategy is applied in this chapter, and two actuators are needed for each control object since failures may occur to the actuators of the joint, vehicle, and tip of the flexible link. Hence the output signals of the dual-actuators can be presented as follows

$$\tau(t) = b_1\tau_1(t) + b_2\tau_2(t) \tag{7.15}$$

$$u_m(t) = b_3u_{m1}(t) + b_4u_{m2}(t) \tag{7.16}$$

$$u_L(t) = b_5u_{L1}(t) + b_6u_{L2}(t) \tag{7.17}$$

where b_k and $b_{k+1}, i = 1, 3, 5$, are unknown constants with the same signs; $\tau_1(t)$ and $\tau_2(t)$, $u_{m1}(t)$ and $u_{m2}(t)$, $u_{L1}(t)$ and $u_{L2}(t)$ are the output signals of two actuators of the joint, vehicle and tip of the flexible link respectively, with the same direction; $\tau_2(t)$, $u_{m2}(t)$, $u_{L2}(t)$ are the actuation redundancy to provide the desired control performance even when a complete failure occurs to one of these two actuators.

Then the controller failures can be described by the following equations

$$\tau_1(t) = \sigma_1\tau_c(t),\ \ \tau_2(t) = \sigma_2\tau_c(t) \tag{7.18}$$

$$u_{m1}(t) = \sigma_3u_{mc}(t),\ \ u_{m2}(t) = \sigma_4u_{mc}(t) \tag{7.19}$$

$$u_{L1}(t) = \sigma_5u_{Lc}(t),\ \ u_{L2}(t) = \sigma_6u_{Lc}(t) \tag{7.20}$$

where $0 \leq \sigma_i \leq 1$, σ_1 and σ_2, σ_3 and σ_4, σ_5 and σ_6 cannot be zero simultaneously which means $\max\{\sigma_1, \sigma_2\} > 0$, $\max\{\sigma_3, \sigma_4\} > 0$, $\max\{\sigma_5, \sigma_6\} > 0$.

Remark 7.3 $\tau_c(t), u_{mc}(t), u_{Lc}(t)$ are actual input signals of actuators and $\tau(t), u_m(t)$, $u_L(t)$ are signals feeding into the control system. Here we use the actuator of the joint as an example: when $\sigma_i = 1$, $i = 1, 2$, the actuator of $\tau_i(t)$ is regarded as fault-free; $0 < \sigma_i < 1$ means that the actuator of $\tau_i(t)$ is with partial degradation; $\sigma_i = 0$ implies that the actuator of $\tau_i(t)$ completely fails and cannot work at all. The description can also be applied to the actuator of $u_{mi}(t)$ with σ_3, σ_4 and the actuator of $u_{Li}(t)$ with σ_5, σ_6.

The flow chart of the proposed adaptive fault-tolerant control is shown in Fig. 7.2. Control inputs of actuators in Eqs. (7.18)–(7.20) are proposed as

$$\tau_c(t) = -k_{p1}e_1(t) - k_{d1}\dot{e}_1(t) - k_1\hat{p}_1\dot{e}_1(t) \tag{7.21}$$

$$u_{mc}(t) = -k_{p2}e_2(t) - k_{d2}\dot{e}_2(t) - k_2\hat{p}_2\dot{e}_2(t) \tag{7.22}$$

$$u_{Lc}(t) = -k_{p3}e_3(t) - k_{d3}\dot{e}_3(t) - k_3\hat{p}_3\dot{e}_3(t) - \hat{d}(\boldsymbol{\omega}) - \eta\,\mathrm{sgn}(\dot{e}_3(t)) \tag{7.23}$$

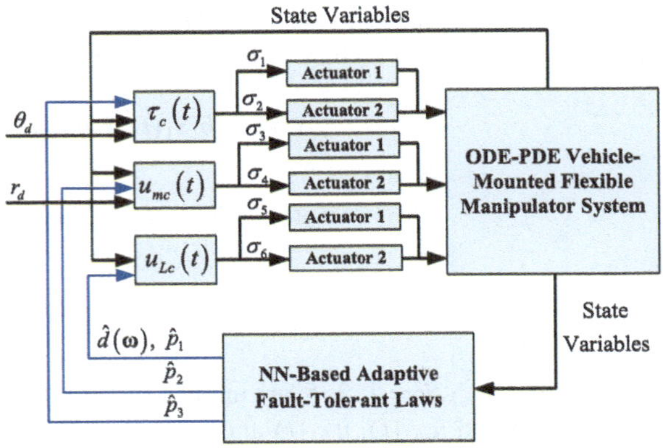

Fig. 7.2 Structure of the closed-loop system where NN denotes the neural network

where $k_{p1}, k_{d1}, k_{p2}, k_{d2}, k_{p3}, k_{d3}, k_1, k_2, k_3, c_1, c_2, c_3$ are positive control coefficients, and $\hat{p}_i$ is the estimation of p_i, $i = 1, 2, 3$, defined as follows

$$p_1 = b_1\sigma_1 + b_2\sigma_2 \tag{7.24}$$

$$p_2 = b_3\sigma_3 + b_4\sigma_4 \tag{7.25}$$

$$p_3 = b_5\sigma_5 + b_6\sigma_6. \tag{7.26}$$

Define the estimation error of p_i as $\tilde{p}_i$ where $\tilde{p}_i = \hat{p}_i - p_i$ and the adaptive control law for p_i is proposed as follows

$$\dot{\hat{p}}_i = k_i\lambda_i\dot{e}_i^2(t) \tag{7.27}$$

where λ_i is a positive constant and $i = 1, 2, 3$.

Theorem 7.1 Under Assumptions 7.1 and 7.2, the following properties of the closed-loop system should be satisfied with the proposed neural network-based fault-tolerant control law in Eqs. (7.11)–(7.13) and Eqs. (7.21)–(7.23): the asymptotical stability of the system can be guaranteed in the presence of the uncertainty in the sense of $\theta(t) \to \theta_d(t)$, $r(t) \to r_d(t)$, $w(L, t) \to 0$, $w(x, t) \to 0$ when $t \to \infty$, even as one of the actuators of $\tau_i(t)$, $u_{mi}(t)$, $u_{Li}(t)$, $i = 1, 2$, completely fails, respectively.

Proof A Lyapunov function is taken to be

$$V(t) = V_1(t) + V_2(t) + V_3(t) + V_4(t) \tag{7.28}$$

where

$$V_1(t) = \frac{1}{2}I_h\dot{\theta}^2(t) + \frac{1}{2}\rho\int_0^L \dot{\mathbf{P}}^{\mathrm{T}}(x,t)\dot{\mathbf{P}}(x,t)\mathrm{d}x + \frac{1}{2}M\dot{\mathbf{P}}^{\mathrm{T}}(L,t)\dot{\mathbf{P}}(L,t) + \frac{1}{2}m\dot{r}^2(t) + \frac{EI}{2}\int_0^L w_{xx}^2(x,t)\mathrm{d}x \tag{7.29}$$

$$V_2(t) = \frac{1}{2}k_{p1}(b_1\sigma_1 + b_2\sigma_2)e_1^2(t) + \frac{1}{2}k_{p2}(b_3\sigma_3 + b_4\sigma_4)e_2^2(t) + \frac{1}{2}k_{p3}(b_5\sigma_5 + b_6\sigma_6)e_3^2(t) \tag{7.30}$$

$$V_3(t) = \frac{b_1\sigma_1 + b_2\sigma_2}{2\lambda_1}\tilde{p}_1^2 + \frac{b_3\sigma_3 + b_4\sigma_4}{2\lambda_2}\tilde{p}_2^2 + \frac{b_5\sigma_5 + b_6\sigma_6}{2\lambda_3}\tilde{p}_3^2 \tag{7.31}$$

$$V_4(t) = \frac{b_5\sigma_5 + b_6\sigma_6}{2\gamma}\tilde{\mathbf{Q}}^{\mathrm{T}}\tilde{\mathbf{Q}} \tag{7.32}$$

where $V_1(t)$ is a term of the system energy; $V_2(t)$ is a term of tracking errors; $V_3(t)$ is a term of the estimation error of actuator failures; $V_4(t)$ is a term of the network approximation error; $\mathbf{P}(x,t)$ and $\mathbf{P}(L,t)$ are respectively defined in Eqs. (6.1) and (6.2).

Differentiating Eq. (7.29) with respect to time and substituting Eqs. (7.3)–(7.5), we can obtain

$$\dot{V}_1(t) = \dot{e}_1(t)\tau(t) + \dot{e}_2(t)u_m(t) + \dot{e}_3(t)(u_L(t) + d(\omega)) + EIw_{xxx}(L,t)\dot{w}(L,t) + EI\int_0^L w_{xx}(x,t)\dot{w}_{xx}(x,t)\mathrm{d}x - \int_0^L EIw_{xxxx}(x,t)\dot{w}(x,t)\mathrm{d}x. \tag{7.33}$$

With Eq. (7.33) and substituting Eqs. (7.21)–(7.23), we can obtain

$$\begin{aligned}\dot{V}_1(t) = {} & (b_1\sigma_1 + b_2\sigma_2)\dot{e}_1(t)\big(-k_{p1}e_1(t) - k_{d1}\dot{e}_1(t) - k_1\hat{p}_1\dot{e}_1(t)\big)\\ & + (b_3\sigma_3 + b_4\sigma_4)\dot{e}_2(t)\big(-k_{p2}e_2(t) - k_{d2}\dot{e}_2(t) - k_2\hat{p}_2\dot{e}_2(t)\big)\\ & + (b_5\sigma_5 + b_6\sigma_6)\dot{e}_3(t)\Big(-k_{p3}e_3(t) - k_{d3}\dot{e}_3(t) - k_3\hat{p}_3\dot{e}_3(t) - \eta\,\mathrm{sgn}(\dot{e}_3(t)) + \tilde{d}(\omega)\Big).\end{aligned} \tag{7.34}$$

Differentiating Eq. (7.30) with respect to time, we have

$$\begin{aligned}\dot{V}_2(t) = {} & k_{p1}(b_1\sigma_1 + b_2\sigma_2)e_1(t)\dot{e}_1(t) + k_{p2}(b_3\sigma_3 + b_4\sigma_4)e_2(t)\dot{e}_2(t)\\ & + k_{p3}(b_5\sigma_5 + b_6\sigma_6)e_3(t)\dot{e}_3(t).\end{aligned} \tag{7.35}$$

Taking the derivative of Eq. (7.31) and using Eq. (7.27), we have

$$\dot{V}_3(t) = (b_1\sigma_1 + b_2\sigma_2)\tilde{p}_1k_1\dot{e}_1^2(t) + (b_3\sigma_3 + b_4\sigma_4)\tilde{p}_2k_2\dot{e}_2^2(t) + (b_5\sigma_5 + b_6\sigma_6)\tilde{p}_3k_3\dot{e}_3^2(t). \tag{7.36}$$

Substituting Eq. (7.13) into the derivative of $V_4(t)$

$$\dot{V}_4(t) = -\frac{b_5\sigma_5 + b_6\sigma_6}{\gamma}\tilde{\mathbf{Q}}^{\mathrm{T}}\dot{\hat{\mathbf{Q}}} = -(b_5\sigma_5 + b_6\sigma_6)\dot{e}_3(t)\tilde{\mathbf{Q}}^{\mathrm{T}}\mathbf{h}(\boldsymbol{\omega}). \tag{7.37}$$

Combining Eqs. (7.34)–(7.37), we obtain

$$\begin{aligned}\dot{V}(t) &= -k_{d1}(b_1\sigma_1 + b_2\sigma_2)\dot{e}_1^2(t) - k_{d2}(b_3\sigma_3 + b_4\sigma_4)\dot{e}_2^2(t)\\ &-k_{d3}(b_5\sigma_5 + b_6\sigma_6)\dot{e}_3^2(t) + (b_5\sigma_5 + b_6\sigma_6)\dot{e}_3(t)\Big(-\eta\,\mathrm{sgn}(\dot{e}_3(t)) + \tilde{\mathbf{Q}}^{\mathrm{T}}\mathbf{h}(\boldsymbol{\omega}) + \varepsilon\Big)\\ &-k_1\dot{e}_1^2(t) - k_2\dot{e}_2^2(t) - k_3\dot{e}_3^2(t)\\ &-(b_5\sigma_5 + b_6\sigma_6)\dot{e}_3(t)\tilde{\mathbf{Q}}^{\mathrm{T}}\mathbf{h}(\boldsymbol{\omega})\\ &\le -[k_{d1}(b_1\sigma_1 + b_2\sigma_2) + k_1]\dot{e}_1^2(t) - [k_{d2}(b_3\sigma_3 + b_4\sigma_4) + k_2]\dot{e}_2^2(t)\\ &-[k_{d3}(b_5\sigma_5 + b_6\sigma_6) + k_3]\dot{e}_3^2(t).\end{aligned} \tag{7.38}$$

From inequality (7.38), we can know that $\dot{V}(t) \le 0$ and thus according to the extended LaSalle's invariance principle in Chap. 6, it follows that the closed-loop PDE system is asymptotically stable with actuator faults and $\theta(t) \to \theta_d$, $r(t) \to r_d$, $w(L,t) \to 0$, $w(x,t) \to 0$ are assured when $t \to \infty$.

As a result, Theorem 7.1 has been proved.

7.4 Simulations

In this section, the effectiveness of the proposed adaptive fault-tolerant control scheme is demonstrated by numerical simulations. Under actuator failures, $\theta(t) \to \theta_d, r(t) \to r_d, w(L,t) \to 0, w(x,t) \to 0$ are expected to be obtained when $t \to \infty$.

The initial conditions of the vehicle-mounted flexible manipulator are chosen as $\theta(0) = 0.1$, $\dot{\theta}(0) = 1$, $r(0) = 0.5$, $\dot{r}(0) = 1$, $w(x,0) = 0.04\sin(0.01x)$, $\dot{w}(x,0) = 1$, and the desired positions of the joint and vehicle are $\theta_d = 0.2$ rad and $r_d = 1$ m. The boundary uncertainty is given as $d(\boldsymbol{\omega}) = w^2(L,t) + \dot{w}^2(L,t)$.

Parameters of the system are listed in Table 7.1.

Control gains of $\tau_c(t)$, $u_{mc}(t)$, $u_{Lc}(t)$ in Eqs. (7.21)–(7.23) are set as $k_{p1} = 100$, $k_{d1} = 15$, $k_{p2} = 100$, $k_{d2} = 15$, $k_{p3} = 10$, $k_{d3} = 60$, $k_1 = 50$, $k_2 = 10$, $k_3 = 2$, and $\eta = 0.3$.

Parameters of the RBF neural network are set as $n = 5$, $\mathbf{f}_1 = [-2,-2]^{\mathrm{T}}$, $\mathbf{f}_2 = [-1,-1]^{\mathrm{T}}$, $\mathbf{f}_3 = [0,0]^{\mathrm{T}}$, $\mathbf{f}_4 = [1,1]^{\mathrm{T}}$, $\mathbf{f}_5 = [2,2]^{\mathrm{T}}$, $g_1 = 0.3$, $g_2 = 0.3$, $g_3 = 0.3$, $g_4 = 0.3$, $g_5 = 0.3$ and $\gamma = 1$. The parameters of the adaptive tolerant laws are given as $\lambda_1 = 2$, $\lambda_2 = 2$, $\lambda_3 = 2$.

For simulating total and partial actuator failures, let $b_1 = b_2 = 1$, and σ_i, $i = 1,2,3,4,5,6$, in Eqs. (7.18)–(7.20) are given as: $\sigma_1 = \sigma_2 = \sigma_3 = \sigma_4 = \sigma_5 = \sigma_6 = 1$

Table 7.1 Parameters of the system

Parameter	Description	Value
L	Length of the flexible link	0.6 m
ρ	Uniform mass per unit length of the flexible link	0.8 kg m^{-1}
M	Mass of the tip payload	0.3 kg
EI	Bending rigidity of the flexible link	2 N m^2
I_h	Moment of inertia of the flexible link	2 kg m^2
m	Mass of the vehicle	1.5 kg

for $t \leq 0.45$s, $\sigma_1 = 0.01$, $\sigma_2 = 0.02$, $\sigma_3 = 0.02$, $\sigma_4 = 0.05$, $\sigma_5 = 0.1$, $\sigma_6 = 0.2$ for $0.45\text{s} < t \leq 0.9$s, and $\sigma_1 = 0$, $\sigma_2 = 0.02$, $\sigma_3 = 0$, $\sigma_4 = 0.05$, $\sigma_5 = 0$, $\sigma_6 = 0.2$ for $t > 0.9$s, which means when $t \leq 0.45$s, actuators $\tau(t)$, $u_m(t)$, $u_L(t)$ are fault-free; when $0.45\text{s} < t \leq 0.9$s, all the actuators lose partial effectiveness, for instance, a 98% reduction of control effectiveness in $\tau_1(t)$ and a 99% loss of control effectiveness in $\tau_2(t)$ occur at $t = 0.45$s; when $t > 0.9$s, the actuators of $\tau_1(t)$, $u_{m1}(t)$, $u_{L1}(t)$ completely fault, respectively, and the redundant actuators of $\tau_2(t)$, $u_{m2}(t)$, $u_{L2}(t)$ continue working with partial actuator failures.

The system outputs with the proposed adaptive fault-tolerant control are given in Figs. 7.3, 7.4, 7.5 and 7.6, where the trajectories of $\theta(t)$, $r(t)$, $w(L, t)$, and $w(x, t)$ with the proposed adaptive fault-tolerant control algorithm are shown respectively. From Figs. 7.3, 7.4, 7.5 and 7.6 we can see that the proposed control laws can realize the precise position regulation even under complete actuator failures and uncertainty. The vibration is also greatly reduced with time. That is to say, $\theta(t) \rightarrow \theta_d$, $r(t) \rightarrow r_d$, $w(L, t) \rightarrow 0$, and $w(x, t) \rightarrow 0$ can be ensured even when one of the two actuators completely fails (Figs. 7.7, 7.8 and 7.9).

Through the simulation, it can prove that the proposed neural network-based adaptive fault-tolerant control offers exact position regulation in the event of complete and

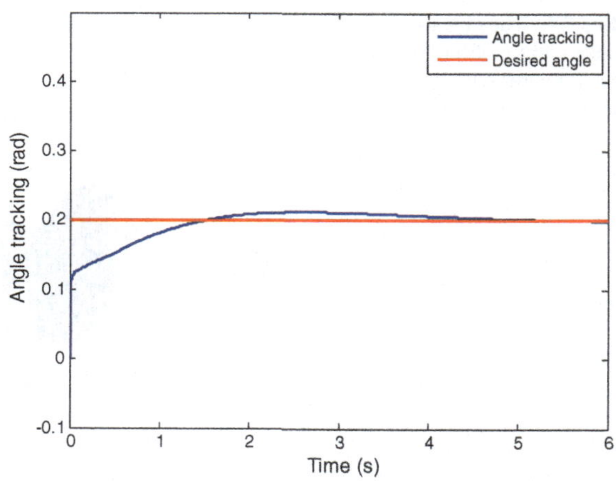

Fig. 7.3 Angle of joint with the proposed control laws

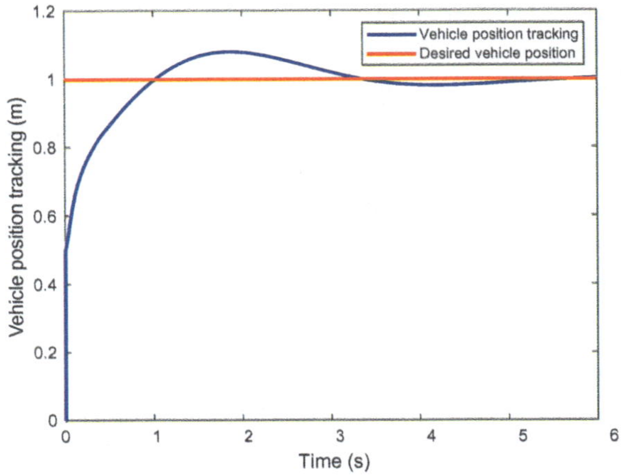

Fig. 7.4 Position tracking of the vehicle with the proposed control laws

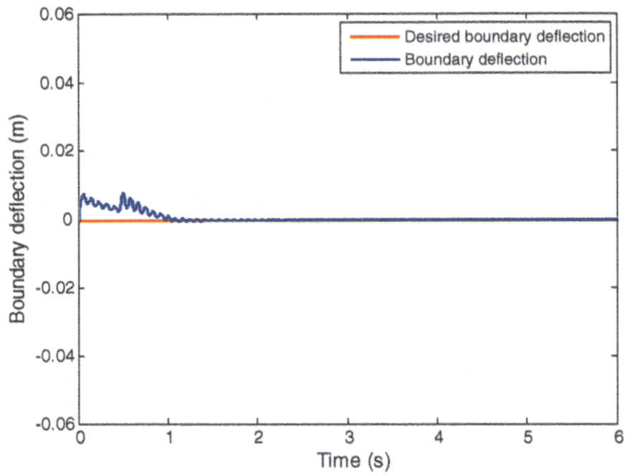

Fig. 7.5 Vibration of the end of the flexible manipulator with the proposed control laws

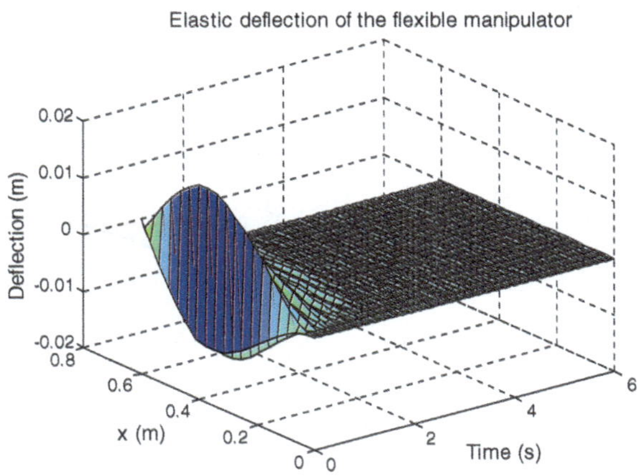

Fig. 7.6 The elastic deformation of the flexible manipulator with the proposed control laws

Fig. 7.7 Control input $\tau(t)$

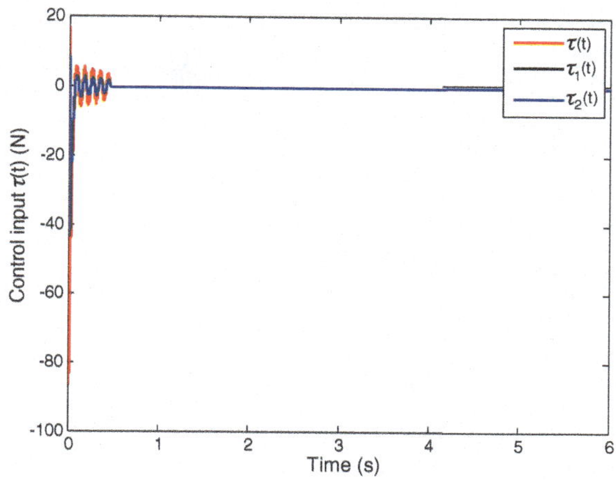

Fig. 7.8 Control input $u_m(t)$

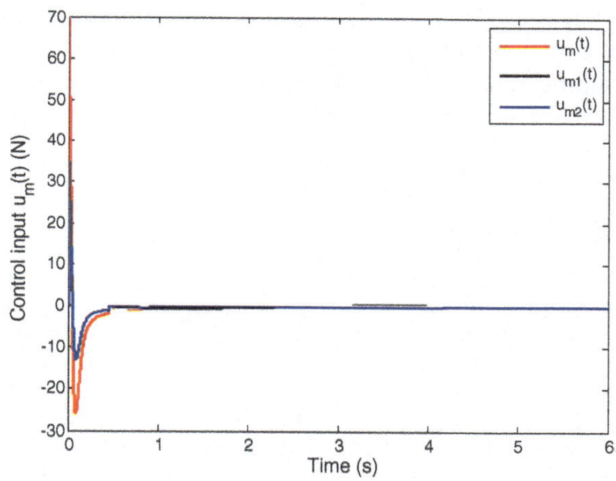

partial actuator failures by compensating failure coefficients. Moreover, the proposed control scheme also has the ability to compensate the uncertainty appearing at the free end of the flexible link.

7.5 Conclusions

In this chapter, a problem of neural network-based adaptive fault-tolerant control design has been investigated for a vehicle-mounted flexible manipulator. Based on the PDE model without neglecting any mode of the system, the neural network-based adaptive fault-tolerant control scheme has been developed to preserve system stability

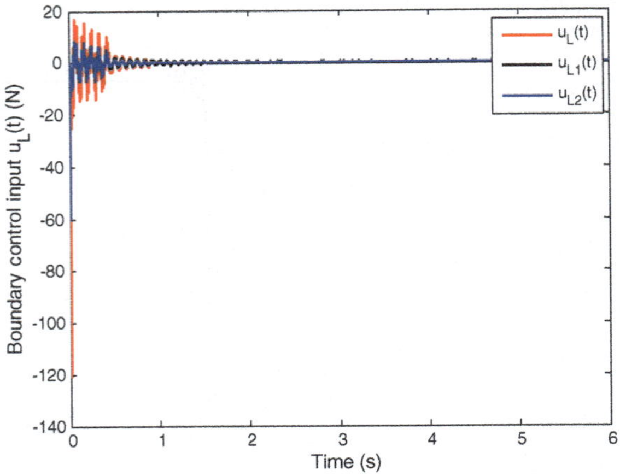

Fig. 7.9 Boundary control input $u_L(t)$

with boundary uncertainty and actuator failures. With the actuator redundancy and the proposed control laws, position regulation of the joint and vehicle and vibration suppression of the flexible manipulator can be achieved simultaneously under the occurrence of both total and partial controller failures.

References

1. L. Zhang, J. Liu, Observer-based partial differential equation boundary control for a flexible two-link manipulator in task space. IET Control Theory Appl **6**(13), 2120–2133 (2012)
2. S.S. Ge, T.H. Lee, G. Zhu, A nonlinear feedback controller for a single-link flexible manipulator based on a finite element model. J. Robot. Syst.v **14**(3), 165–178 (1997)
3. Z. Liu, J. Liu, W. He, An adaptive iterative learning algorithm for boundary control of a flexible manipulator. Int. J. Adapt. Control Signal Process. **31**(6), 903–916 (2017)

Chapter 8
Switching Fault-Tolerant Control of Vehicle-Mounted Flexible Manipulator System with Output Constraints

8.1 Introduction

Considering that output constraints must be strictly met during the control process, rigorous handling constraints in the control design has aroused widespread concern in recent years. Once the constraints of physical systems are violated during operation, serious consequences, such as system instability, control performance degradation, or even systematic destruction may be resulted. Besides, systems inevitably suffer actuator faults due to external stochastic factors or internal structural abrasion in the actual control process. To avoid adverse effects caused by the actuator failure, fault-tolerant control and switching strategies are developed as a result of safety concerns so that the actuators can furthest continue functioning when a portion of the system fails [1–4].

Different from Chap. 7, this chapter considers the influence of gravity of the flexible manipulator and proposes a switching fault-tolerant control for the vehicle-mounted flexible manipulator with output constraints. The joint angle, vehicle position, and payload vibration satisfy all the necessary constraints by using BLF. The fault-free closed-loop system is asymptotically stable with vibration suppression and position regulation. Further, output constraints can be preserved effectively even under the occurrence of actuator failures by introducing the switching strategy. All the closed-loop signals are kept bounded indefinitely when the initial condition requirement is satisfied.

X. Xing and J. Liu, *Vibration Control Methods of Mechanical Distributed Parameter Systems*, Springer Tracts in Mechanical Engineering,
https://doi.org/10.1007/978-981-16-1532-0_8

8.2 Problem Formulation

In this chapter, a vehicle-mounted flexible manipulator system which consists of a vehicle, a flexible manipulator and a payload at the free end is discussed in the vertical plane so that the influence of gravity is considered. The simplified system configuration is introduced in Fig. 8.1. As shown in Fig. 8.1, XOY is the inertia frame; $X_1O_1Y_1$ is a moving coordinate fixed on the vehicle; $X_2O_2Y_2$ is a rotating reference frame with the hub O_2.

The modeling process of the system is as same as the one given in Chap. 6. Without considering the external interference, the following PDE model of system dynamics can be derived as

$$
\begin{aligned}
&\left(I_h + \frac{1}{3}\rho L^3 + ML^2\right)\ddot{\theta}(t) + 2M\dot{\theta}(t)w(L,t)\dot{w}(L,t) - \left(ML + \frac{1}{2}\rho L^2\right)\sin\theta(t)\ddot{r}(t) \\
&- M\cos\theta(t)w(L,t)\ddot{r}(t) + M\ddot{\theta}(t)w^2(L,t) + ML\ddot{w}(L,t) \\
&+ \frac{1}{2}g\rho L^2\cos\theta(t) + MgL\cos\theta(t) - Mgw(L,t)\sin\theta(t) \\
&+ \rho\int_0^L \Big(2w(x,t)\dot{w}(x,t)\dot{\theta}(t) + w^2(x,t)\ddot{\theta}(t) + x\ddot{w}(x,t) \\
&- \cos\theta(t)w(x,t)\ddot{r}(t) - gw(x,t)\sin\theta(t)\Big)dx - \tau_C(t) = 0
\end{aligned} \tag{8.1}
$$

$$
\begin{aligned}
&(M + m + \rho L)\ddot{r}(t) - \left(\rho\frac{1}{2}L^2 + ML\right)\cos\theta(t)\dot{\theta}^2(t) \\
&- \left(\rho\frac{1}{2}L^2 + ML\right)\sin\theta(t)\ddot{\theta}(t) + M\sin\theta(t)\dot{\theta}^2(t)w(L,t) \\
&- M\cos\theta(t)\ddot{\theta}(t)w(L,t) \\
&- 2M\cos\theta(t)\dot{\theta}(t)\dot{w}(L,t) - M\sin\theta(t)\ddot{w}(L,t)
\end{aligned}
$$

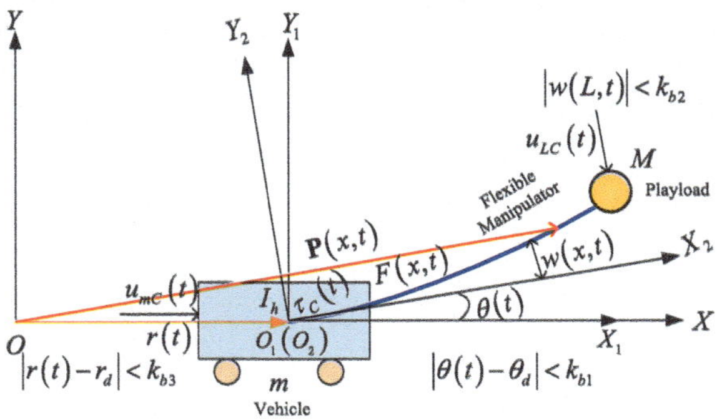

Fig. 8.1 Configuration of the vehicle-mounted flexible manipulator

$$+\rho\int_0^L\big(\sin\theta(t)\dot{\theta}^2(t)w(x,t)-\cos\theta(t)\ddot{\theta}(t)w(x,t)$$
$$-2\cos\theta(t)\dot{\theta}(t)\dot{w}(x,t)-\sin\theta(t)\ddot{w}(x,t)\big)dx-u_{mC}(t)=0 \quad (8.2)$$

$$\rho\ddot{w}(x,t)+\rho x\ddot{\theta}(t)-\rho\sin\theta(t)\ddot{r}(t)+EIw_{xxxx}(x,t)$$
$$-\rho\dot{\theta}^2(t)w(x,t)+g\rho\cos\theta(t)-F(x,t)=0 \quad (8.3)$$

with corresponding boundary conditions

$$-M\dot{\theta}^2(t)w(L,t)+M\ddot{w}(L,t)+ML\ddot{\theta}(t)-M\sin\theta(t)\ddot{r}(t)$$
$$+Mg\cos\theta(t)-EIw_{xxx}(L,t)-u_{LC}(t)=0 \quad (8.4)$$

$$w(0,t)=0 \quad (8.5)$$

$$w_x(0,t)=0 \quad (8.6)$$

$$w_{xx}(L,t)=0 \quad (8.7)$$

where $\tau_C(t)$ denotes the output signal of the actuator on the joint generated by the joint motor; $u_{LC}(t)$ denotes the output of the actuator on the end of the flexible manipulator; $u_{mC}(t)$ is the output of the actuator on the vehicle; $F(x,t)$ is a distributed control applied on the flexible arm.

The detailed modeling process can be found in Chap. 6.

Assumption 8.1 The desired angle, the expected position of the vehicle and all parameters in this chapter are bounded.

8.3 Switching Fault-Tolerant Control Design with Output Constraints

8.3.1 Control Design with Output Constraints Under Non-fault Conditions

In practical engineering process, vehicle and joint positions are generally required to be strictly limited. Furthermore, the elastic deflection of the payload always needs to be confined to a small limit. In this section, a switching fault-tolerant control of a vehicle-mounted flexible manipulator system with output constraints is proposed by introducing a BLF. Considering the system output constraints, control objectives of this chapter are summarized as follows:

(1) Control the vehicle and joint so that they can achieve their desired positions without actuator failures. The vibration of the flexible arm can also be suppressed with healthy actuators.
(2) Under the supervision of the proposed monitoring functions, actuator faults can be detected in time and the output constraints of $\theta(t)$, $r(t)$, $w(L,t)$ can always be guaranteed within the given bounds by actuator switchings even under unknown actuator failures.

In this part, we discuss the case in which no faults occur to the actuators, which paves the way for the following switching fault-tolerant control design. We define tracking errors as $e_1(t) = \theta(t) - \theta_d$, $e_2(t) = w(L,t)$, $e_3(t) = r(t) - r_d$, where the desired angle position is given as θ_d and the desired vehicle position is denoted as r_d. Both θ_d and r_d are invariant constants, then $\dot{e}_1(t) = \dot{\theta}(t)$, $\dot{e}_2(t) = \dot{w}(L,t)$, $\dot{e}_3(t) = \dot{r}(t)$.

Provided that

$$|e_1(0)| = |\theta(0) - \theta_d| < k_{b1} \tag{8.8}$$

$$|e_2(0)| = |w(L,0)| < k_{b2} \tag{8.9}$$

$$|e_3(0)| = |r(0) - r_d| < k_{b3} \tag{8.10}$$

where k_{b1}, k_{b2}, k_{b3} are three positive constants, we define $\tau(t), u_L(t), u_m(t)$ as input signals which are applied to the actuators on the joint, end of the flexible manipulator, and vehicle, respectively. The model-based inputs of actuators are given as follows to make $\theta(t) \to \theta_d$, $\dot{\theta}(t) \to 0$, $r(t) \to r_d$, $\dot{r}(t) \to 0$, $w(L,t) \to 0$, $\dot{w}(L,t) \to 0$, $w(x,t) \to 0$, $\dot{w}(x,t) \to 0$ with output constraints $|e_1(t)| < k_{b1}$, $|e_2(t)| < k_{b2}$, $|e_3(t)| < k_{b3}$ when $t \to \infty$.

The control laws are proposed as

$$\begin{aligned}\tau(t) =& - k_{p1}e_1(t) - k_{d1}\dot{e}_1(t) + \frac{1}{2}g\rho L^2\cos\theta(t)\\ &+ MgL\cos\theta(t) - Mgw(L,t)\sin\theta(t)\\ &- \rho\int_0^L gw(x,t)\sin\theta(t)dx - \frac{e_1(t)}{k_{b1}^2 - e_1^2(t)}\end{aligned} \tag{8.11}$$

$$u_L(t) = -k_{p2}w(L,t) - k_{d2}\dot{w}(L,t) + Mg\cos\theta(t) - \frac{e_2(t)}{k_{b2}^2 - e_2^2(t)} \tag{8.12}$$

$$u_m(t) = -k_{p3}e_3(t) - k_{d3}\dot{e}_3(t) - \frac{e_3(t)}{k_{b3}^2 - e_3^2(t)} \tag{8.13}$$

$$F(x,t) = -k_{p4}w(x,t) - k_{d4}\dot{w}(x,t) + g\rho\cos\theta(t) \tag{8.14}$$

where $k_{p1}, k_{d1}, k_{p2}, k_{d2}, k_{p3}, k_{d3}, k_{p4}$ and k_{d4} are positive control coefficients.

Remark 8.1 In the control system, $\tau(t)$, $u_m(t)$, $u_L(t)$ are input signals of the actuators; $\tau_C(t)$, $u_{mC}(t)$, $u_{LC}(t)$ are actual output signals from the actuators to the system. If actuators are all fault-free, then we have $\tau_C(t) = \tau(t)$, $u_{mC}(t) = u_m(t)$, $u_{LC}(t) = u_L(t)$.

The stability of the closed-loop system without actuator failure is proved by Lyapunov theory and extended LaSalle's invariance principle.

Theorem 8.1 *Suppose that Assumption* 8.1 *holds and all actuators are fault-free on* $t \in [0, \infty)$. *Then the resulting closed-loop system with fault-free actuators has following properties:*

(1) *The output errors of system indefinitely remain in* $|e_1(t)| < k_{b1}$, $|e_2(t)| < k_{b2}$, $|e_3(t)| < k_{b3}$.
(2) *The output tracking errors and elastic deflection of the flexible manipulator converge to zero asymptotically, that is,* $\theta(t) \to \theta_d$, $\dot{\theta}(t) \to 0$, $r(t) \to r_d$, $\dot{r}(t) \to 0$, $w(L,t) \to 0$, $\dot{w}(L,t) \to 0$, $w(x,t) \to 0$, $\dot{w}(x,t) \to 0$ *with time.*

Proof Construct the Lyapunov function candidate as

$$V(t) = V_1(t) + V_2(t) + V_3(t) \tag{8.15}$$

where $V_1(t)$ is a term of the system energy; $V_2(t)$ is a term of tracking errors; $V_3(t)$ is a term of output constraints.

Define $V_1(t)$, $V_2(t)$ and $V_3(t)$ as follows

$$\begin{aligned} V_1(t) = {} & \frac{1}{2}I_h\dot{\theta}^2(t) + \frac{1}{2}\rho\int_0^L \dot{\boldsymbol{P}}^T(x,t)\dot{\boldsymbol{P}}(x,t)dx + \frac{1}{2}M\dot{\boldsymbol{P}}^T(L,t)\dot{\boldsymbol{P}}(L,t) \\ & + \frac{1}{2}m\dot{r}^2(t) + \frac{EI}{2}\int_0^L w_{xx}^2(x,t)dx \end{aligned} \tag{8.16}$$

$$\begin{aligned} V_2(t) = {} & \frac{1}{2}k_{p1}e_1^2(t) + \frac{1}{2}k_{p2}e_2^2(t) \\ & + \frac{1}{2}k_{p3}e_3^2(t) + \frac{1}{2}k_{p4}\int_0^L w^2(x,t)\mathrm{d}x \end{aligned} \tag{8.17}$$

$$V_3(t) = \frac{1}{2}\ln\frac{k_{b1}^2}{k_{b1}^2 - e_1^2(t)} + \frac{1}{2}\ln\frac{k_{b2}^2}{k_{b2}^2 - e_2^2(t)} + \frac{1}{2}\ln\frac{k_{b3}^2}{k_{b3}^2 - e_3^2(t)} \tag{8.18}$$

where $\mathbf{P}(x,t)$ and $\mathbf{P}(L,t)$ are defined in Eqs. (6.1) and (6.2).

The derivative of Eq. (8.15) is given as

$$\begin{aligned}
\dot{V}(t) =& I_h\dot{\theta}(t)\ddot{\theta}(t) + \frac{1}{2}\rho\int_0^L \Big(2x^2\dot{\theta}(t)\ddot{\theta}(t) + 2w(x,t)\dot{w}(x,t)\dot{\theta}^2(t) \\
&+ 2w^2(x,t)\dot{\theta}(t)\ddot{\theta}(t) + 2\dot{w}(x,t)\ddot{w}(x,t) + 2\dot{r}(t)\ddot{r}(t) \\
&+ 2x\ddot{\theta}(t)\dot{w}(x,t) + 2x\dot{\theta}(t)\ddot{w}(x,t) - 2x\cos\theta(t)\dot{\theta}^2(t)\dot{r}(t) \\
&- 2x\sin\theta(t)\ddot{\theta}(t)\dot{r}(t) - 2x\sin\theta(t)\dot{\theta}(t)\ddot{r}(t) \\
&+ 2\sin\theta(t)\dot{\theta}^2(t)w(x,t)\dot{r}(t) - 2\cos\theta(t)\ddot{\theta}(t)w(x,t)\dot{r}(t) \\
&- 4\cos\theta(t)\dot{\theta}(t)\dot{w}(x,t)\dot{r}(t) - 2\cos\theta(t)\dot{\theta}(t)w(x,t)\ddot{r}(t) \\
&-2\sin\theta(t)\ddot{w}(x,t)\dot{r}(t) - 2\sin\theta(t)\dot{w}(x,t)\ddot{r}(t))\mathrm{d}x \\
&+ \frac{1}{2}M\Big(2L^2\dot{\theta}(t)\ddot{\theta}(t) + 2w(L,t)\dot{w}(L,t)\dot{\theta}^2(t) \\
&+ 2w^2(L,t)\dot{\theta}(t)\ddot{\theta}(t) + 2\dot{w}(L,t)\ddot{w}(L,t) + 2\dot{r}(t)\ddot{r}(t) + 2L\ddot{\theta}(t)\dot{w}(L,t) \\
&+ 2L\dot{\theta}(t)\ddot{w}(L,t) - 2L\cos\theta(t)\dot{\theta}^2(t)\dot{r}(t) - 2L\sin\theta(t)\ddot{\theta}(t)\dot{r}(t) \\
&- 2L\sin\theta(t)\dot{\theta}(t)\ddot{r}(t) + 2\sin\theta(t)\dot{\theta}^2(t)w(L,t)\dot{r}(t) \\
&- 2\cos\theta(t)\ddot{\theta}(t)w(L,t)\dot{r}(t) - 4\cos\theta(t)\dot{\theta}(t)\dot{w}(L,t)\dot{r}(t) \\
&-2\cos\theta(t)\dot{\theta}(t)w(L,t)\ddot{r}(t) - 2\sin\theta(t)\ddot{w}(L,t)\dot{r}(t) - 2\sin\theta(t)\dot{w}(L,t)\ddot{r}(t)\Big) \\
&+ m\dot{r}(t)\ddot{r}(t) + EI\int_0^L w_{xx}(x,t)\dot{w}_{xx}(x,t)\mathrm{d}x + k_{p1}e_1(t)\dot{e}_1(t) + k_{p2}e_2(t)\dot{e}_2(t) \\
&+ k_{p3}e_3(t)\dot{e}_3(t) + k_{p4}\int_0^L w(x,t)\dot{w}(x,t)\mathrm{d}x + \frac{e_1(t)\dot{e}_1(t)}{k_{b1}^2 - e_1^2(t)} + \frac{e_2(t)\dot{e}_2(t)}{k_{b2}^2 - e_2^2(t)} + \frac{e_3(t)\dot{e}_3(t)}{k_{b3}^2 - e_3^2(t)}.
\end{aligned} \tag{8.19}$$

Making some arrangements and substituting Eqs. (8.11)–(8.14) into Eq. (8.19) yield

$$\begin{aligned}
\dot{V}(t) = \dot{\theta}(t)\Big[&\Big(I_h + \frac{1}{3}\rho L^3 + ML^2\Big)\ddot{\theta}(t) + 2M\dot{\theta}(t)w(L,t)\dot{w}(L,t) \\
&- \Big(ML + \frac{1}{2}\rho L^2\Big)\sin\theta(t)\ddot{r}(t) - M\cos\theta(t)w(L,t)\ddot{r}(t) + M\ddot{\theta}(t)w^2(L,t) \\
&+ ML\ddot{w}(L,t) + \frac{1}{2}g\rho L^2\cos\theta(t) + MgL\cos\theta(t) - Mgw(L,t)\sin\theta(t) \\
&+ \rho\int_0^L \Big(2w(x,t)\dot{w}(x,t)\dot{\theta}(t) + w^2(x,t)\ddot{\theta}(t) + x\ddot{w}(x,t) - \cos\theta(t)w(x,t)\ddot{r}(t)
\end{aligned}$$

$$
\begin{aligned}
&-gw(x,t)\sin\theta(t))\mathrm{d}x] + \dot{w}(L,t)\Big(-M\dot{\theta}^2(t)w(L,t) + M\ddot{w}(L,t) + ML\ddot{\theta}(t)\\
&-M\sin\theta(t)\ddot{r}(t) + Mg\cos\theta(t) - EIw_{xxx}(x,t)) + \dot{r}(t)[(M+m+\rho L)\ddot{r}(t)\\
&-\left(\rho\frac{1}{2}L^2 + ML\right)\cos\theta(t)\dot{\theta}^2(t) - \left(\rho\frac{1}{2}L^2 + ML\right)\sin\theta(t)\ddot{\theta}(t)\\
&+M\sin\theta(t)\dot{\theta}^2(t)w(L,t) - M\cos\theta(t)\ddot{\theta}(t)w(L,t) - 2M\cos\theta(t)\dot{\theta}(t)\dot{w}(L,t)\\
&-M\sin\theta(t)\ddot{w}(L,t) + \rho\int_0^L\Big(\sin\theta(t)\dot{\theta}^2(t)w(x,t) - \cos\theta(t)\ddot{\theta}(t)w(x,t)\\
&-2\cos\theta(t)\dot{\theta}(t)\dot{w}(x,t) - \sin\theta(t)\ddot{w}(x,t)\Big)\mathrm{d}x\Big] + \int_0^L \dot{w}(x,t)\big(\rho\ddot{w}(x,t) + \rho x\ddot{\theta}(t)\\
&-\rho\sin\theta(t)\ddot{r}(t) + EIw_{xxxx}(x,t) - \rho\dot{\theta}^2(t)w(x,t) + g\rho\cos\theta(t)\big)\mathrm{d}x\\
&-\dot{\theta}(t)\left(\frac{1}{2}g\rho L^2\cos\theta(t) + MgL\cos\theta(t) - Mgw(L,t)\sin\theta(t) - \rho\int_0^L gw(x,t)\sin\theta(t)\mathrm{d}x\right)\\
&-Mg\cos\theta(t)\dot{w}(L,t) - \int_0^L g\rho\cos\theta(t)\dot{w}(x,t)\mathrm{d}x + EIw_{xxx}(L,t)\dot{w}(L,t)\\
&+EI\int_0^L w_{xx}(x,t)\dot{w}_{xx}(x,t)\mathrm{d}x - \int_0^L EIw_{xxxx}(x,t)\dot{w}(x,t)\mathrm{d}x + k_{p1}e_1(t)\dot{e}_1(t)\\
&+k_{p2}e_2(t)\dot{e}_2(t) + k_{p3}e_3(t)\dot{e}_3(t) + k_{p4}\int_0^L w(x,t)\dot{w}(x,t)\mathrm{d}x\\
&+\frac{e_1(t)\dot{e}_1(t)}{k_{b1}^2 - e_1^2(t)} + \frac{e_2(t)\dot{e}_2(t)}{k_{b2}^2 - e_2^2(t)} + \frac{e_3(t)\dot{e}_3(t)}{k_{b3}^2 - e_3^2(t)}
\end{aligned}
\tag{8.20}
$$

which means

$$
\begin{aligned}
\dot{V}(t) = {} & \dot{\theta}(t)\tau_C(t) + \dot{w}(L,t)u_{LC}(t) + \dot{r}(t)u_{mC}(t) \\
& + \int_0^L \dot{w}(x,t)F(x,t)\mathrm{d}x + EIw(L,t)\dot{w}(L,t) \\
& - \frac{1}{2}g\rho L^2 \cos\theta(t)\dot{\theta}(t) - MgL\cos\theta(t)\dot{\theta}(t) \\
& + Mgw(L,t)\sin\theta(t)\dot{\theta}(t) + \rho\int_0^L gw(x,t)\sin\theta(t)\dot{\theta}(t)\mathrm{d}x \\
& - Mg\cos\theta(t)\dot{w}(L,t) - \int_0^L g\rho\cos\theta(t)\dot{w}(x,t)\mathrm{d}x \\
& + EI\int_0^L w_{xx}(x,t)\dot{w}_{xx}(x,t)\mathrm{d}x - \int_0^L EIw_{xxxx}(x,t)\dot{w}(x,t)\mathrm{d}x \\
& + k_{p1}e_1(t)\dot{e}_1(t) + k_{p2}e_2(t)\dot{e}_2(t) + k_{p3}e_3(t)\dot{e}_3(t) \\
& + k_{p4}\int_0^L w(x,t)\dot{w}(x,t)\mathrm{d}x \\
& + \frac{e_1(t)\dot{e}_1(t)}{k_{b1}^2 - e_1^2(t)} + \frac{e_2(t)\dot{e}_2(t)}{k_{b2}^2 - e_2^2(t)} + \frac{e_3(t)\dot{e}_3(t)}{k_{b3}^2 - e_3^2(t)}.
\end{aligned} \tag{8.21}
$$

By integration by part, we have

$$
\begin{aligned}
& -\int_0^L EIw_{xxxx}(x,t)\dot{w}(x,t)\mathrm{d}x + EIw_{xxx}(L,t)\dot{w}(L,t) \\
& + EI\int_0^L w_{xx}(x,t)\dot{w}_{xx}(x,t)\mathrm{d}x = 0.
\end{aligned} \tag{8.22}
$$

Substituting Eq. (8.22) into Eq. (8.21), we can obtain

$$
\begin{aligned}
\dot{V}(t) = {} & \dot{\theta}(t)\tau_C(t) + \dot{w}(L,t)u_{LC}(t) + \dot{r}(t)u_{mC}(t) \\
& + \int_0^L \dot{w}(x,t)F(x,t)\mathrm{d}x - \frac{1}{2}g\rho L^2\cos\theta(t)\dot{\theta}(t) \\
& - MgL\cos\theta(t)\dot{\theta}(t) + Mgw(L,t)\sin\theta(t)\dot{\theta}(t)
\end{aligned}
$$

$$
\begin{aligned}
&+\rho\int_0^L gw(x,t)\sin\theta(t)\dot{\theta}(t)\mathrm{d}x - Mg\cos\theta(t)\dot{w}(L,t)\\
&-\int_0^L g\rho\cos\theta(t)\dot{w}(x,t)\mathrm{d}x + k_{p1}e_1(t)\dot{e}_1(t) + k_{p2}e_2(t)\dot{e}_2(t)\\
&+k_{p3}e_3(t)\dot{e}_3(t) + k_{p4}\int_0^L w(x,t)\dot{w}(x,t)\mathrm{d}x\\
&+\frac{e_1(t)\dot{e}_1(t)}{k_{b1}^2-e_1^2(t)}+\frac{e_2(t)\dot{e}_2(t)}{k_{b2}^2-e_2^2(t)}+\frac{e_3(t)\dot{e}_3(t)}{k_{b3}^2-e_3^2(t)}.
\end{aligned}
\tag{8.23}
$$

Since actuators are fault-free and $\tau_C(t)=\tau(t), u_{mC}(t)=u_m(t), u_{LC}(t)=u_L(t)$, substituting control laws in Eqs. (8.11)–(8.14) to Eq. (8.23), we obtain

$$
\begin{aligned}
\dot{V}(t) = {}& \dot{e}_1(t)\bigg(-k_{p1}e_1(t) - k_{d1}\dot{e}_1(t) + \frac{1}{2}g\rho L^2\cos\theta(t) + MgL\cos\theta(t)\\
&-Mgw(L,t)\sin\theta(t) - \rho\int_0^L gw(x,t)\sin\theta(t) - \frac{e_1(t)}{k_{b1}^2-e_1^2(t)}\bigg)\\
&+\dot{e}_2(t)\bigg(-k_{p2}w(L,t) - k_{d2}\dot{w}(L,t) + Mg\cos\theta(t) - \frac{e_2(t)}{k_{b2}^2-e_2^2(t)}\bigg)\\
&+\dot{e}_3(t)\bigg(-k_{p3}e_3(t) - k_{d3}\dot{e}_3(t) - \frac{e_3(t)}{k_{b3}^2-e_3^2(t)}\bigg)\\
&+\int_0^L \dot{w}(x,t)\big(-k_{p4}w(x,t) - k_{d4}\dot{w}(x,t) + g\rho\cos\theta(t)\big)\mathrm{d}x\\
&-\frac{1}{2}g\rho L^2\cos\theta(t)\dot{\theta}(t) - MgL\cos\theta(t)\dot{\theta}(t) + Mgw(L,t)\sin\theta(t)\dot{\theta}(t)\\
&+\rho\int_0^L gw(x,t)\sin\theta(t)\dot{\theta}(t)dx - Mg\cos\theta(t)\dot{w}(L,t)\\
&-\int_0^L g\rho\cos\theta(t)\dot{w}(x,t)\mathrm{d}x + k_{p1}e_1(t)\dot{e}_1(t) + k_{p2}e_2(t)\dot{e}_2(t)\\
&+k_{p3}e_3(t)\dot{e}_3(t) + k_{p4}\int_0^L w(x,t)\dot{w}(x,t)\mathrm{d}x\\
&+\frac{e_1(t)\dot{e}_1(t)}{k_{b1}^2-e_1^2(t)}+\frac{e_2(t)\dot{e}_2(t)}{k_{b2}^2-e_2^2(t)}+\frac{e_3(t)\dot{e}_3(t)}{k_{b3}^2-e_3^2(t)}
\end{aligned}
$$

$$
\begin{aligned}
&= -k_{d1}\dot{\theta}^2(t) - k_{d2}\dot{w}^2(L,t) - k_{d3}\dot{r}^2(t) - k_{d4}\int_0^L \dot{w}^2(x,t)\mathrm{d}x \\
&\leq 0.
\end{aligned}
\tag{8.24}
$$

Then from inequality (8.24) we can know that $V(t)$ is negative semi-definite. By extending LaSalle's invariance principle to the infinite dimensional space, the proof of asymptotic stability of the closed-loop system is given as follows.

If $\dot{V}(t) \equiv 0$, then from inequality (8.24), we get

$$
\dot{\theta}(t) = \dot{w}(L,t) = \dot{r}(t) = \dot{w}(x,t) \equiv 0 \tag{8.25}
$$

which implies

$$
\ddot{\theta}(t) = \ddot{w}(L,t) = \ddot{r}(t) = \ddot{w}(x,t) \equiv 0. \tag{8.26}
$$

Substituting Eqs. (8.25) and (8.26) into Eqs. (8.1)–(8.7) yields

$$
\begin{aligned}
\tau_C(t) =& \frac{1}{2}g\rho L^2\cos\theta(t) + MgL\cos\theta(t) \\
&- Mgw(L,t)\sin\theta(t) - \rho\int_0^L gw(x,t)\sin\theta(t)\mathrm{d}x
\end{aligned}
\tag{8.27}
$$

$$
u_{LC}(t) = -EIw_{xxx}(L,t) + Mg\cos\theta(t) \tag{8.28}
$$

$$
u_{mC}(t) = 0 \tag{8.29}
$$

$$
EIw_{xxxx}(x,t) + \rho g\cos\theta(t) = F(x,t). \tag{8.30}
$$

From Eqs. (8.14), (8.25) and (8.30), it is clear that

$$
EIw_{xxxx}(x,t) = -k_{p4}w(x,t). \tag{8.31}
$$

It is worth noting that Eq. (8.31) is separable by using the technique of separation of variables. It is assumed that $w(x,t)$ can be separated as

$$
w(x,t) = \Phi(x)\cdot q(t) \tag{8.32}
$$

where $\Phi(x)$ and $q(t)$ are two undetermined functions related to space and time respectively. Introducing Eq. (8.32) into Eq. (8.31), we can get

$$\frac{\Phi_{xxxx}(x)}{\Phi(x)} = -\frac{k_{p4}}{EI} \tag{8.33}$$

which means

$$\Phi_{xxxx}(x) + \frac{k_{p4}}{EI}\Phi(x) = 0. \tag{8.34}$$

Setting $\alpha^4 = \frac{k_{p4}}{EI}$ and solving Eq. (8.34), we have

$$\begin{aligned}\Phi(x) = e^{\frac{\sqrt{2}}{2}\alpha x}\left(b_1 \cos\frac{\sqrt{2}}{2}\alpha x + b_2 \sin\frac{\sqrt{2}}{2}\alpha x\right) \\ + e^{-\frac{\sqrt{2}}{2}\alpha x}\left(b_3 \cos\frac{\sqrt{2}}{2}\alpha x + b_4 \sin\frac{\sqrt{2}}{2}\alpha x\right)\end{aligned} \tag{8.35}$$

where $b_i \in R$, $i = 1, 2, 3, 4$, are four constants to be determined.

From Eqs. (8.4) and (8.28), we know that

$$\begin{aligned}EIw_{xxx}(L,t) &= k_{p2}w(L,t) + \frac{e_2(t)}{k_{b2}^2 - e_2^2(t)} \\ &= k_{p2}w(L,t) + \frac{w(L,t)}{k_{b2}^2 - w^2(L,t)}.\end{aligned} \tag{8.36}$$

Hence

$$\left(k_{b2}^2 - w^2(L,t)\right)EIw_{xxx}(L,t) = \left[k_{p2}\left(k_{b2}^2 - w^2(L,t)\right) + 1\right]w(L,t). \tag{8.37}$$

Substituting Eq. (8.31) to Eq. (8.37), we obtain

$$\left(k_{b2}^2 - \Phi^2(L)q^2(t)\right)EI\Phi_{xxx}(L)q(t) = \left[k_{p2}\left(k_{b2}^2 - \Phi^2(L)q^2(t)\right) + 1\right]\Phi(L)q(t) \tag{8.38}$$

which means for $q(t) \neq 0$ and $\Phi(L) \neq 0$

$$EI\frac{\Phi_{xxx}(L)}{\Phi(L)} = k_{p2} + \frac{1}{k_{b2}^2 - \Phi^2(L)q^2(t)}. \tag{8.39}$$

Since Eq. (8.39) holds for all time and its right side is changing over time, Eq. (8.39) is satisfied only when

$$\Phi_{xxx}(L) = \Phi(L) = 0. \tag{8.40}$$

From boundary conditions in Eqs. (8.5)–(8.7) and Eq. (8.40), we have $\Phi(0) = \Phi_x(0) = \Phi_{xx}(L) = \Phi_{xxx}(L) = \Phi(L) = 0$. Then from Eq. (8.40), $b_i = 0$, $i = 1, 2, 3, 4$, can be easily obtained as the unique solution of Eq. (8.35) and thus $\Phi(x) = 0$, $w(x, t) = 0$, $w(L, t) = 0$ hold. Substituting control laws in Eqs. (8.11) and (8.13) into Eqs. (8.27) and (8.29) respectively, we can get

$$-k_{p1}e_1(t) - \frac{e_1(t)}{k_{b1}^2 - e_1^2(t)} = 0 \tag{8.41}$$

$$-k_{p3}e_3(t) - \frac{e_3(t)}{k_{b3}^2 - e_3^2(t)} = 0. \tag{8.42}$$

Therefore, for $|e_1(t)| < k_{b1}$ and $|e_3(t)| < k_{b3}$, we have

$$e_1(t) = \theta(t) - \theta_d = 0 \tag{8.43}$$

$$e_2(t) = w(L, t) = 0 \tag{8.44}$$

$$e_3(t) = r(t) - r_d = 0 \tag{8.45}$$

$$w(x, t) = 0. \tag{8.46}$$

As a result, $\theta(t) \to \theta_d, \dot{\theta}(t) \to 0, r(t) \to r_d, \dot{r}(t) \to 0, w(L, t) \to 0, \dot{w}(L, t) \to 0$, $w(x, t) \to 0$, $\dot{w}(x, t) \to 0$ can be guaranteed by applying extended LaSalle's invariance principle when $t \to \infty$. Under the output feedback control proposed in Eqs. (8.11)–(8.14), the asymptotical stability of the fault-free closed-loop system is verified. Besides, $\theta(t)$, $r(t)$, $w(L, t)$ all remain in the given constrained space for bounded initial conditions based on Lemma 2.3. That is to say, when initial conditions $|e_1(0)| < k_{b1}$, $|e_2(0)| < k_{b2}$, $|e_3(0)| < k_{b3}$ hold, output constraints of $|e_1(t)| < k_{b1}$, $|e_2(t)| < k_{b2}$, $|e_3(t)| < k_{b3}$ can be guaranteed. Therefore, Theorem 8.1 has been proved.

8.3.2 *Monitoring Functions and Switching Strategy Design*

In order to maintain performance levels and eliminate the negative impacts of actuator faults, we propose monitoring functions to monitor the system so that the occurrence of actuator failure can be instantly found out.

For accommodating actuator faults and minimizing their negative effect on the system, there exists actuation redundancy in the following proposed control scheme. Assume that there are total q actuators on the joint, end of the flexible link, and vehicle, whose output signals are defined as $\tau_i(t)$, $u_{Li}(t)$, $u_{mi}(t)$, $i = 1, 2 \cdots q$, respectively. Among the q actuators for each control object, only one healthy actuator is working until the fault occurs on it and other healthy ones are used as backups. Once the failure of the current actuator is detected by one of the proposed monitoring functions at some point, the present actuator will be switched to the next healthy one to preserve system stability even under the occurrence of failure.

Assumption 8.2 At $t_0 = 0$, actuators of $\tau_1(t)$, $u_{L1}(t)$, $u_{m1}(t)$ are failure free with normal operation. Once fault occurs, actuators of $\tau_i(t), u_{Li}(t), u_{mi}(t), i = 1, 2 \cdots q$, will be switched to next ones in sequence. There only exists one output signal of one fault-free actuator on the joint, end of the flexible link, and vehicle respectively at a time.

The number of actuator failures for each control object is set as n. Consequently, during the operation of the actuators of $\tau_i(t), u_{Li}(t), u_{mi}(t), i = 1, 2 \cdots q$, there are n respective actuator switchings on $t \in (0, \infty)$ due to actuator failures. We assume the switching instants of the actuators of $\tau_i(t), u_{Li}(t), u_{mi}(t)$ to be t_{1k}, t_{2k}, t_{3k} respectively for $k = 1, 2 \cdots n$, which meets $t_{j0} < t_{j1} < \cdots < t_{jk} < t_{j(k+1)} < \cdots < t_{jn} < t_{j(n+1)}$, $j = 1, 2, 3$, where $t_{j0} := 0$, $t_{j(n+1)} := +\infty$, $q \geq n + 1$. It is noteworthy that there is no switching at time $t_{j0} := 0$.

Integrating two sides of inequality (8.24) from 0 to t, we obtain

$$
\begin{aligned}
&V(t) - V(0) \\
&= \int_0^t \left(-k_{d1}\dot{\theta}^2(t) - k_{d2}\dot{w}^2(L, t) - k_{d3}\dot{r}^2(t) - k_{d4} \int_0^L \dot{w}^2(x, t)\mathrm{d}x \right) \mathrm{d}t \leq 0.
\end{aligned}
\tag{8.47}
$$

Therefore, we have

$$
\begin{aligned}
V(t) &\leq V(0) \\
&= \left(\frac{1}{2} I_h \dot{\theta}^2(0) + \frac{1}{2}\rho \int_0^L \dot{\mathbf{P}}^{\mathrm{T}}(x, 0)\dot{\mathbf{P}}(x, 0)\mathrm{d}x + \frac{1}{2} M \dot{\mathbf{P}}^{\mathrm{T}}(L, 0)\dot{\mathbf{P}}(L, 0) \right. \\
&\quad + \frac{1}{2} m \dot{r}^2(0) + \frac{EI}{2} \int_0^L w_{xx}^2(x, 0)\mathrm{d}x + \frac{1}{2} k_{p1} e_1^2(0) + \frac{1}{2} k_{p2} e_2^2(0) \\
&\quad + \frac{1}{2} k_{p3} e_3^2(0) + \frac{1}{2} k_{p4} \int_0^L w^2(x, 0)\mathrm{d}x + \frac{1}{2} \ln \frac{k_{b1}^2}{k_{b1}^2 - e_1^2(0)} \\
&\quad \left. + \frac{1}{2} \ln \frac{k_{b2}^2}{k_{b2}^2 - e_2^2(0)} + \frac{1}{2} \ln \frac{k_{b3}^2}{k_{b3}^2 - e_3^2(0)} \right) \\
&= \frac{1}{2} \left(\mu_0^2 - \varepsilon_0 \right)
\end{aligned}
\tag{8.48}
$$

where ε_0 is a small positive constant and

$$\mu_0 = \left(I_h\dot{\theta}^2(0) + \rho \int_0^L \dot{\mathbf{P}}^{\mathrm{T}}(x,0)\dot{\mathbf{P}}(x,0)\mathrm{d}x + M\dot{\mathbf{P}}^T(L,0)\dot{\mathbf{P}}(L,0) + m\dot{r}^2(0) \right.$$
$$+ EI\int_0^L w_{xx}^2(x,0)\mathrm{d}x + k_{p1}e_1^2(0) + k_{p2}e_2^2(0) + k_{p3}e_3^2(0)$$
$$+ k_{p4}\int_0^L w^2(x,0)\mathrm{d}x + \ln\frac{k_{b1}^2}{k_{b1}^2 - e_1^2(0)} + \ln\frac{k_{b2}^2}{k_{b2}^2 - e_2^2(0)}$$
$$\left. + \ln\frac{k_{b3}^2}{k_{b3}^2 - e_3^2(0)} + \varepsilon_0 \right)^{\frac{1}{2}}. \tag{8.49}$$

From inequality (8.48), we have

$$V(t) \le V(0) < \frac{1}{2}\mu_0^2. \tag{8.50}$$

Hence

$$\frac{1}{2}\ln\frac{k_{bj}^2}{k_{bj}^2 - e_j^2(t)} < \frac{1}{2}\mu_0^2 \tag{8.51}$$

which is equivalent to

$$e_j^2(t) < k_{bj}^2\left(1 - e^{-\mu_0^2}\right) \tag{8.52}$$

where $j = 1, 2, 3$.

Based on Assumption 8.2, for each instant t_{jk}, $k = 0, 1, 2 \cdots n$, the monitoring functions are proposed as

$$\left|e_j(t)\right| < k_{bj}\gamma_{jk} \tag{8.53}$$

where $t \in \left[t_{jk}, t_{j(k+1)}\right)$ and

$$\gamma_{jk} = \sqrt{\left(1 - e^{-\mu_{jk}^2}\right)} \tag{8.54}$$

$$\mu_{jk} = \left(I_h \dot{\theta}^2(t_{jk}) + \rho \int_0^L \dot{\mathbf{P}}^{\mathrm{T}}(x, t_{jk}) \dot{\mathbf{P}}(x, t_{jk}) \mathrm{d}x + M \dot{\mathbf{P}}^{\mathrm{T}}(L, t_{jk}) \dot{\mathbf{P}}(L, t_{jk}) \right.$$
$$+ m \dot{r}^2(t_{jk}) + EI \int_0^L w_{xx}^2(x, t_{jk}) \mathrm{d}x + k_{p1} e_1^2(t_{jk}) + k_{p2} e_2^2(t_{jk})$$
$$+ k_{p3} e_3^2(t_{jk}) + k_{p4} \int_0^L w^2(x, t_{jk}) \mathrm{d}x + \ln \frac{k_{b1}^2}{k_{b1}^2 - e_1^2(t_{jk})}$$
$$\left. + \ln \frac{k_{b2}^2}{k_{b2}^2 - e_2^2(t_{jk})} + \ln \frac{k_{b3}^2}{k_{b3}^2 - e_3^2(t_{jk})} + \varepsilon_j \right)^{\frac{1}{2}} \tag{8.55}$$

where ε_j is a positive constant for $j = 1, 2, 3$.

The switching time of each actuator is defined as

$$t_{j(k+1)} = \min\{t : |e_j(t)| = k_{bj} \gamma_{jk}\}, \ j = 1, 2, 3 \tag{8.56}$$

at which the current actuator fault is detected by the monitoring functions and a switching from the current actuator to the next healthy one is triggered.

A schematic diagram of the closed-loop system with the proposed switching fault-tolerant control is as shown in Fig. 8.2.

Then with monitoring functions in Eqs. (8.53)–(8.56), Theorem 8.2 is presented as follows.

Theorem 8.2 *If Assumptions 8.1–8.2 and initial conditions in Eqs.* (8.8)–(8.10) *are satisfied, then with the input signals of actuators in Eqs.* (8.11)–(8.14)*, the switching strategy, and the proposed monitoring functions given by Eqs.* (8.53)–(8.56)*, output errors of the closed-loop system still remain in* $|e_1(t)| < k_{b1}$, $|e_2(t)| < k_{b2}$, $|e_3(t)| < k_{b3}$ *even under actuator failures.*

Proof Let the $(k+1)$th actuator failures of $\tau_{k+1}(t)$, $u_{L(k+1)}(t)$, $u_{m(k+1)}(t)$ occur at $t'_{1(k+1)}$, $t'_{2(k+1)}$, $t'_{3(k+1)}$ respectively where $t'_{j(k+1)} \in (t_{jk}, t_{j(k+1)})$, $j = 1, 2, 3$, $k = 0, 1, 2 \ldots n$, then the output of the actuators can be expressed as

$$\tau_C(t) = \tau_{k+1}(t) = \begin{cases} \tau(t), \ t \in \left[t_{1k}, t'_{1(k+1)}\right) \\ \tau'_{k+1}(t), \ t \in \left[t'_{1(k+1)}, t_{1(k+1)}\right) \end{cases} \tag{8.57}$$

$$u_{LC}(t) = u_{L(k+1)}(t) = \begin{cases} u_L(t), \ t \in \left[t_{2k}, t'_{2(k+1)}\right) \\ u'_{L(k+1)}(t), \ t \in \left[t'_{2(k+1)}, t_{2(k+1)}\right) \end{cases} \tag{8.58}$$

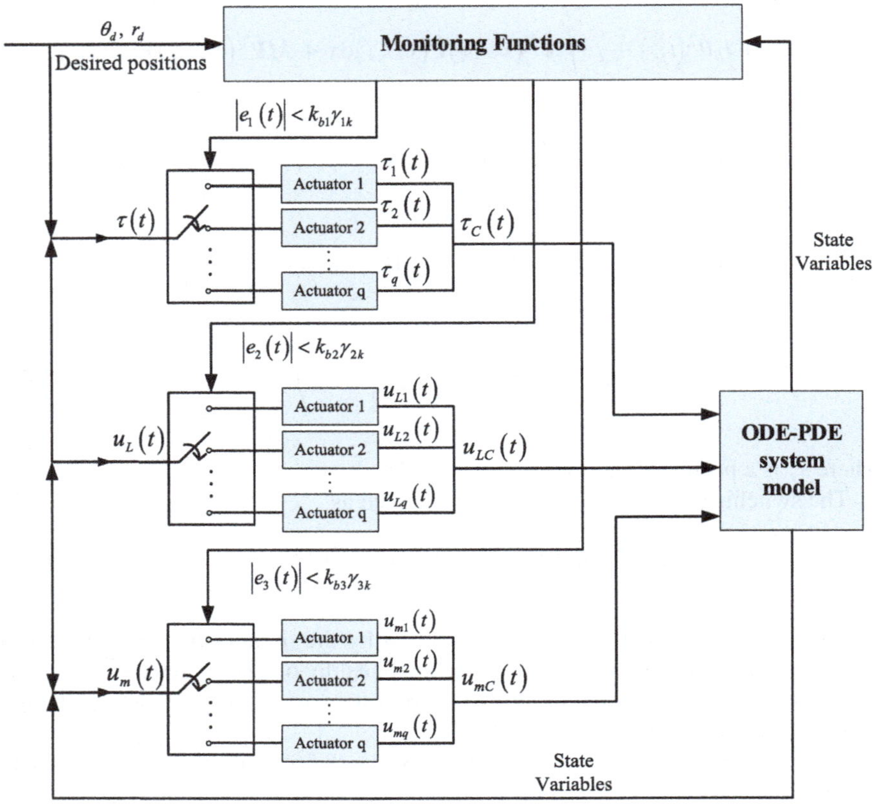

Fig. 8.2 A schematic diagram of the closed-loop system with switching fault-tolerant control

$$u_{mC}(t) = u_{m(k+1)}(t) = \begin{cases} u_m(t),\ t \in \left[t_{3k}, t'_{3(k+1)}\right) \\ u'_{m(k+1)}(t),\ t \in \left[t'_{3(k+1)}, t_{3(k+1)}\right) \end{cases} \tag{8.59}$$

where $\tau(t)$, $u_L(t)$, $u_m(t)$ are given in Eqs. (8.11)–(8.13). $\tau_C(t) = \tau(t)$, $u_{LC}(t) = u_L(t)$, $u_{mC}(t) = u_m(t)$ mean that the actuators are healthy. In Eqs. (8.57)–(8.59), $\tau'_{k+1}(t), u'_{L(k+1)}(t), u'_{m(k+1)}(t)$ are uncertain functions, which means that the actuators suffer an unknown failure.

For $\tau_C(t) = \tau(t), u_{LC}(t) = u_L(t), u_{mC}(t) = u_m(t)$ in $t \in \left[t_{jk}, t'_{j(k+1)}\right)$, $|e_1(t)| < k_{b1}$, $|e_2(t)| < k_{b2}$, $|e_3(t)| < k_{b3}$ are always ensured and the proof procedure is similar with Theorem 8.1 so that it is not described here to avoid repetition.

For $t \in \left[t'_{j(k+1)}, t_{j(k+1)}\right)$, although $\tau'_{k+1}(t), u'_{L(k+1)}(t), u'_{m(k+1)}(t)$ are uncertain, the proposed monitoring functions in Eqs. (8.53)–(8.56) can detect the actuator failure in time and ensure that $\left|e_j(t)\right|$ is bounded for $j = 1, 2, 3$.

In Eqs. (8.55) and (8.15), let $t = t_{j(k+1)}$ and we can obtain that

$$\frac{1}{2}\mu_{j(k+1)}^2 > V\left(t_{j(k+1)}\right) > \frac{1}{2}\ln\frac{k_{bj}^2}{k_{bj}^2 - e_j^2\left(t_{j(k+1)}\right)}. \tag{8.60}$$

According to the switching law, it is obvious that $\left|e_j\left(t_{j(k+1)}\right)\right| = k_{bj}\gamma_{jk}$ holds for $k = 0, 1, 2 \cdots n$. As a result, we get

$$\begin{aligned}\frac{1}{2}\mu_{j(k+1)}^2 &> V\left(t_{j(k+1)}\right) > \frac{1}{2}\ln\frac{k_{bj}^2}{k_{bj}^2 - e_j^2\left(t_{j(k+1)}\right)}\\ &= \frac{1}{2}\ln\frac{k_{bj}^2}{k_{bj}^2 - k_{bj}^2\left(1 - e^{-\mu_{jk}^2}\right)} = \frac{1}{2}\ln\frac{k_{bj}^2}{k_{bj}^2 e^{-\mu_{jk}^2}} = \frac{1}{2}\mu_{jk}^2.\end{aligned} \tag{8.61}$$

Hence, from the definition of γ_{jk}, $k = 0, 1, 2 \ldots n$, in Eq. (8.54), we know that the following inequalities are hold

$$0 < \gamma_{j0} < \gamma_{j1} < \cdots < \gamma_{jn} < 1 \tag{8.62}$$

$$\left|e_j(t)\right| < k_{bj}\gamma_{jk} < k_{bj},\ j = 1, 2, 3. \tag{8.63}$$

Then Theorem 8.2 has been completely proved.

As aforementioned, if one of the present actuators of $\tau_i(t)$, $u_{Li}(t)$, $u_{mi}(t)$, $i = 1, 2 \cdots q$, in Eqs. (8.11)–(8.13) is detected to be fault, a new actuator switching will be triggered at time $t_{j(k+1)}$, $j = 1, 2, 3$, $k = 0, 1, 2 \ldots n$, to continue functioning. Since the switching only occurs for $\left|e_j(t)\right| = k_{bj}\gamma_{jk}$, the number of switchings has been effectively reduced. Accordingly, under Assumptions 8.1–8.2 and initial conditions in Eqs. (8.8)–(8.10), the fault-free closed-loop system can be proved to be asymptotically stable. The desired output limitations are never violated with the developed switching fault-tolerant control by using a BLF. Further, the tracking error constraints can be guaranteed without needing to know the failure patterns and the switching time.

8.4 Simulations

In this section, a numerical simulation study is conducted to assess the performance of the proposed switching fault-tolerant control as-applied to the vehicle-mounted flexible manipulator with system output constraints. Without actuator failure, the signals of the system are expected to be $\theta(t) \to \theta_d$, $\dot{\theta}(t) \to 0$, $r(t) \to r_d$, $\dot{r}(t) \to 0$,

Table 8.1 Parameters of a moving vehicle-mounted flexible manipulator

Parameter	Description	Value
L	Length of the flexible link	0.6 m
ρ	Uniform mass per unit length of the flexible link	0.8 kg/m
M	Mass of the tip payload	0.3 kg
EI	Bending rigidity of the flexible link	12 Nm^2
I_h	Moment of inertia of the flexible link	2 kgm^2
m	Mass of the vehicle	1.5 kg

$w(L,t) \to 0$, $\dot{w}(L,t) \to 0$, $w(x,t) \to 0$, $\dot{w}(x,t) \to 0$ when $t \to \infty$. $|e_1(t)| < k_{b1}$, $|e_2(t)| < k_{b2}$, $|e_3(t)| < k_{b3}$ can be still guaranteed when some of the actuators fail.

The system parameters are given in Table 8.1.

Set $\varepsilon_1(t) = \varepsilon_2(t) = \varepsilon_3(t) = 0.001$. The desired positions are given as $\theta_d = 0.2$rad, $r_d = 0.2$m and we assume that the actuators of $\tau(t)$, $u_L(t)$, $u_m(t)$ suffer an abrupt failure at $t = 4$s, $t = 2.5$s, $t = 0.25$s respectively, where output signals of these three actuators drop to zero.

The dynamic responses of the system are simulated in three cases to validate the effect of the proposed control scheme.

Case 1: Without control inputs: $\tau_C(t)$, $u_{LC}(t)$, $u_{mC}(t)$, $F(x,t)$.

Case 2: With PD control under actuator failure: the boundary and distributed input signals of actuators are given as

$$\tau(t) = -k_{p1}e_1(t) - k_{d1}\dot{e}_1(t) \tag{8.64}$$

$$u_L(t) = -k_{p2}e_2(t) - k_{d2}\dot{e}_2(t) \tag{8.65}$$

$$u_m(t) = -k_{p3}e_3(t) - k_{d3}\dot{e}_3(t) \tag{8.66}$$

$$F(x,t) = -k_{p4}w(x,t) - k_{d4}\dot{w}(x,t) \tag{8.67}$$

where control gains are given as $k_{p1} = 5$, $k_{d1} = 40$, $k_{p2} = 40$, $k_{d2} = 60$, $k_{p3} = 25$, $k_{d3} = 20$, $k_{p4} = 40$, $k_{d4} = 20$, $k_{b1} = 0.52$, $k_{b2} = 0.015$, and $k_{b3} = 0.22$.

Case 3: With the proposed switching fault-tolerant control under actuator failure: the system is simulated with the proposed actuator inputs in Eqs. (8.11)–(8.14) and with switching law and monitoring functions in Eqs. (8.53)–(8.56).

In case 1, the tracking error $e_1(t)$ and speed of the joint are as shown in Fig. 8.3. The tracking error $e_2(t)$ and speed of the vehicle in the open-loop system are shown in Fig. 8.4. Without control, the end-point deflection of the flexible manipulator and its speed are shown in Fig. 8.5. To compare the effect of the designed control scheme more clearly, the deformation and speed of the vibration of the flexible manipulator are also depicted in Figs. 8.6 and 8.7. As shown in Figs. 8.3, 8.4, 8.5, 8.6, 8.7, it

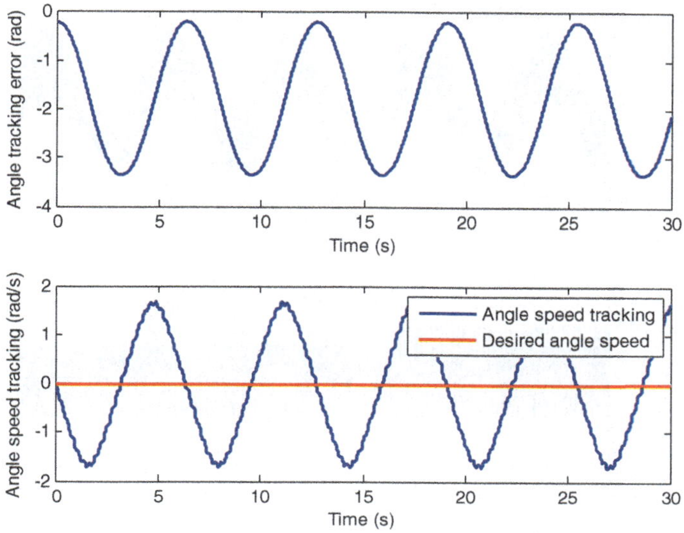

Fig. 8.3 Angular position error and speed of the joint without control

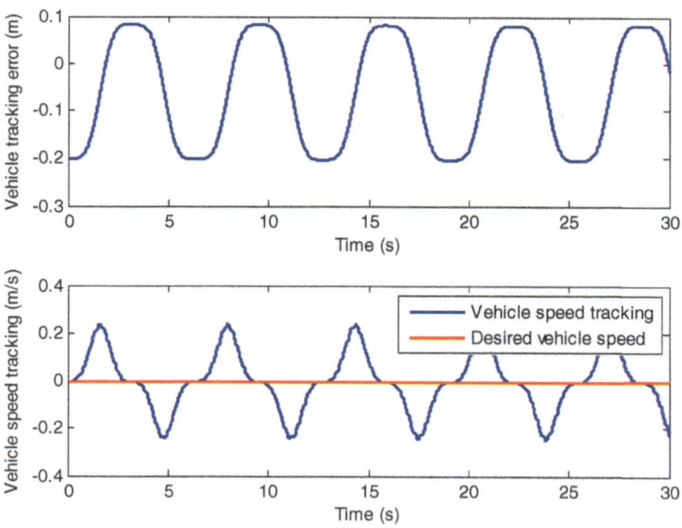

Fig. 8.4 Position error and speed of the vehicle without control

is obvious that the signals of the system are not convergent and the system forms undesired continuous oscillation in case 1. Therefore, in order to limit the system outputs within the specified ranges, developing proper control laws is quite necessary.

The simulation results for case 2 and case 3 are shown in Figs. 8.8, 8.9, 8.11,

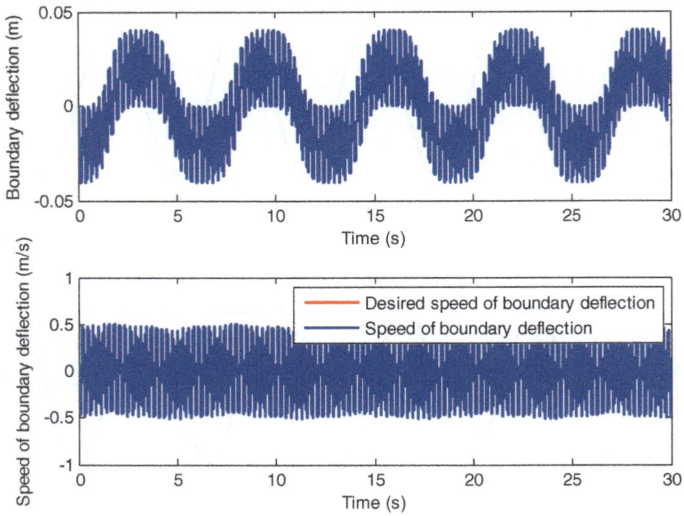

Fig. 8.5 Elastic deflection and speed of elastic deflection of the end of the flexible manipulator

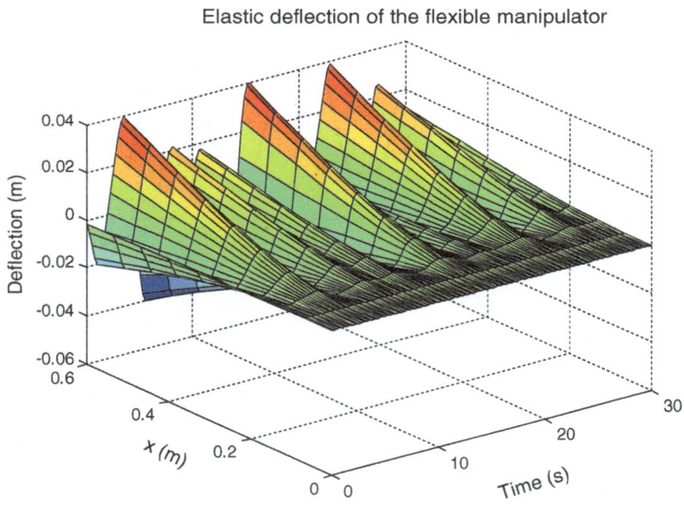

Fig. 8.6 Elastic deflection of the flexible manipulator without control

8.12. For simulating actuator faults, actuators of $\tau(t)$, $u_L(t)$, $u_m(t)$ are locked dead to be zero at $t = 4$s, $t = 2.5$s, $t = 0.25$s respectively. Under actuator failures, Fig. 8.8 gives the tracking error and speed of the joint. Figure 8.9 shows the tracking error of the vehicle and its speed. $w(L, t)$ and $\dot{w}(L, t)$ of the flexible manipulator

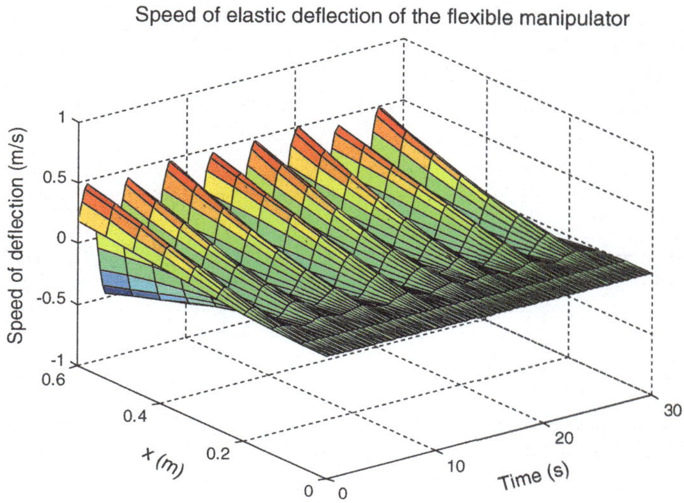

Fig. 8.7 Elastic deflection speed of the flexible manipulator without control

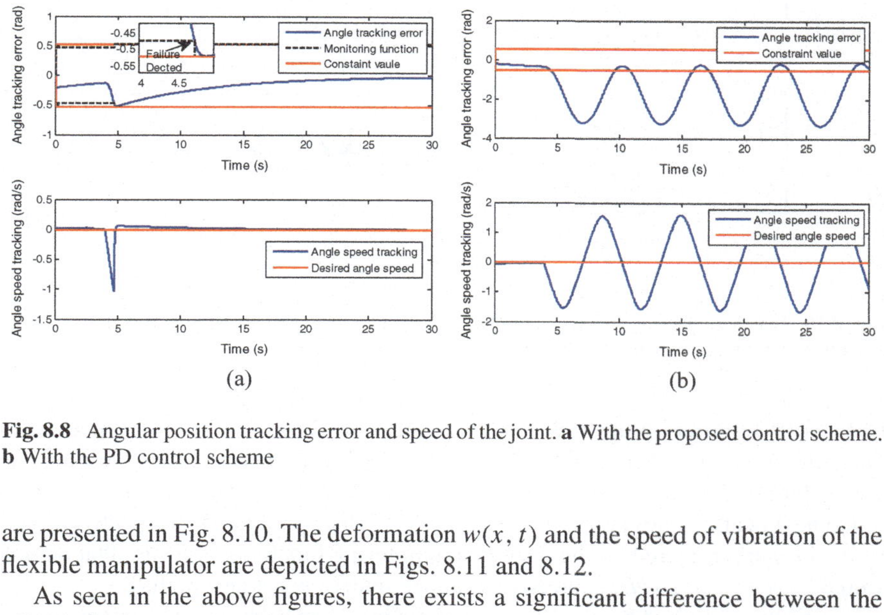

Fig. 8.8 Angular position tracking error and speed of the joint. **a** With the proposed control scheme. **b** With the PD control scheme

are presented in Fig. 8.10. The deformation $w(x, t)$ and the speed of vibration of the flexible manipulator are depicted in Figs. 8.11 and 8.12.

As seen in the above figures, there exists a significant difference between the effect of PD control and the proposed control scheme. It can be observed that if the actuators of the traditional PD control as described in Eqs. (8.64)–(8.66) fault, the output values of $\theta(t)$, $r(t)$, $w(L, t)$ will exceed the given constraints immediately. This comparison between the proposed control scheme and traditional PD control demonstrates the effectiveness of the proposed switching fault-tolerant control. As shown in Figs. 8.8, 8.9, 8.10, the actuators of $\tau(t)$, $u_L(t)$, $u_m(t)$ described in Eqs.

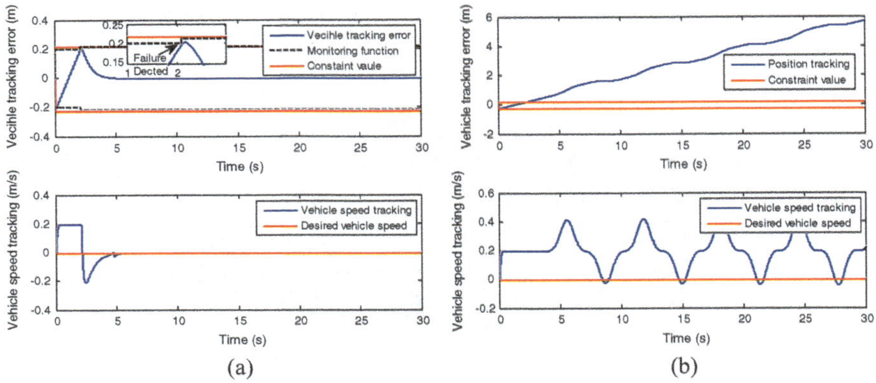

Fig. 8.9 Position tracking error and speed of the vehicle. **a** With the proposed control scheme. **b** With the PD control scheme

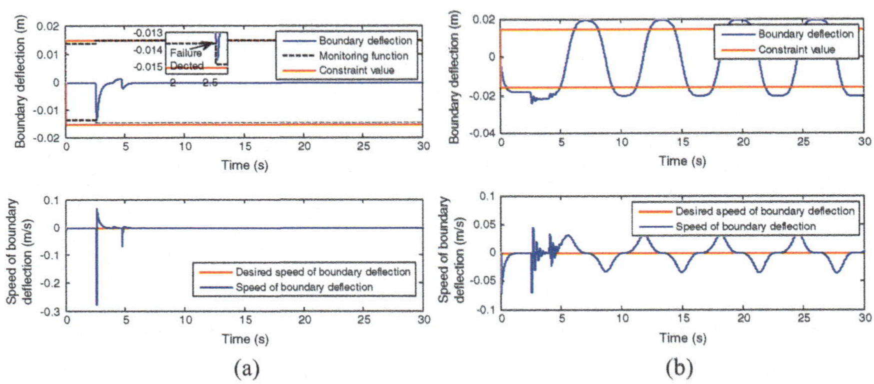

Fig. 8.10 Elastic deflection and speed of elastic deflection of the end of the flexible manipulator. **a** With the proposed control scheme. **b** With the PD control scheme

(8.11)–(8.13) suffer abrupt failures at $t = 4$s, $t = 2.5$s, $t = 0.25$s, respectively. Once the monitoring functions in Eqs. (8.53)–(8.56) are exceeded by the tracking errors, failure is detected and the faulty actuators of $\tau(t)$, $u_L(t)$, $u_m(t)$ are switched to the next healthy ones at $t = 4.7031$s, $t = 2.5705$s, $t = 2.0805$s, respectively. All switchings are triggered by the proposed monitoring functions to ensure that output valuables $\theta(t)$, $r(t)$, $w(L, t)$ are kept bounded under the actuator failure.

According to Figs. 8.8, 8.9, 8.10, actuator failures can be detected by the proposed monitoring functions in a timely manner and system output variables can remain in the given bounds even in the presence of actuator faults. After suffering the last failure at $t = 4$ s, it is easy to see that the proposed control scheme converges the speeds of the joint, vehicle and flexible manipulator to zero. The positions of the joint and vehicle can also be regulated to the desired values quickly. Moreover, vibration

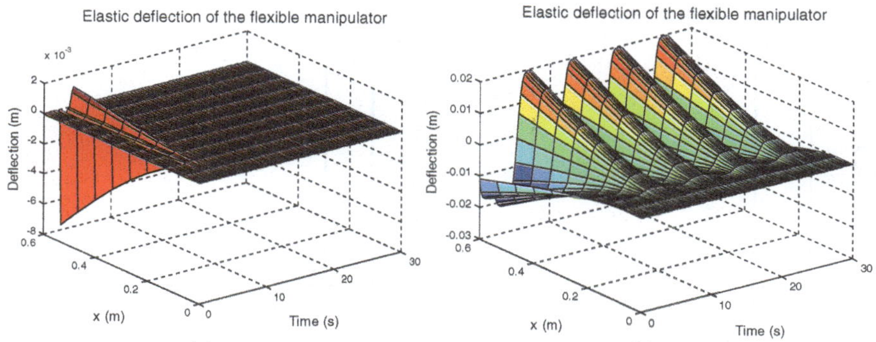

(a) (b)

Fig. 8.11 Elastic deflection of the flexible manipulator. **a** With the proposed control scheme. **b** With the PD control scheme

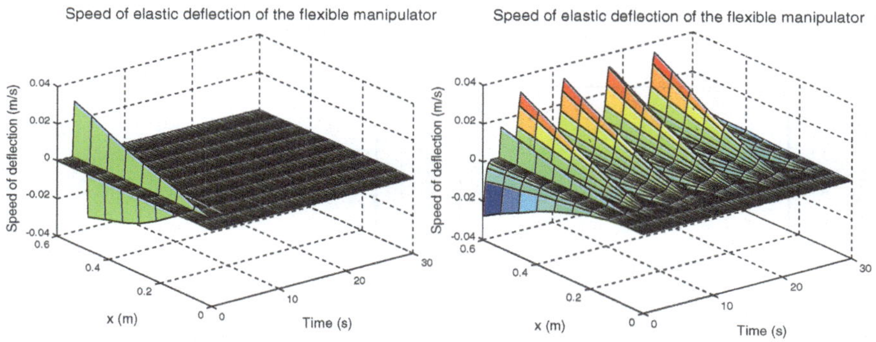

(a) (b)

Fig. 8.12 Speed of elastic deflection of the flexible manipulator. **a** With the proposed control scheme. **b** With the PD control scheme

in the flexible manipulator can also be eliminated in the closed-loop system with no fault occurring on the actuators.

Control inputs $\tau_C(t)$, $u_{LC}(t)$, $u_{mC}(t)$ and $F(x,t)$ are shown in Figs. 8.13, 8.14, 8.15, 8.16 as follows. The instants at which failure occur are shown in Figs. 8.13b, 8.14b, 8.15b. From Figs. 8.13a, 8.14a, 8.15a we can know that the proposed control scheme enforces actuator switchings of $\tau(t)$, $u_L(t)$, $u_m(t)$, at which point the control amplitude is abruptly enlarged to maintain performance levels under the proposed monitoring functions.

From the simulation results, we can find that the vibration of the flexible manipulator has been effectively suppressed and the actual positions can track the desired values with healthy actuators. $\theta(t) \to \theta_d$, $\dot{\theta}(t) \to 0$, $r(t) \to r_d$, $\dot{r}(t) \to 0$, $w(L,t) \to 0$, $\dot{w}(L,t) \to 0$, $w(x,t) \to 0$, $\dot{w}(x,t) \to 0$ can be realized when $t \to \infty$ without actuator failure. Once actuator failure does occur, actuator is switched with

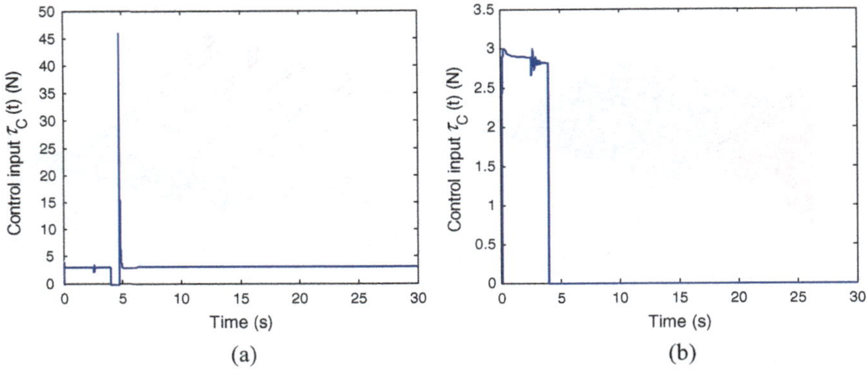

Fig. 8.13 Control input $\tau_C(t)$

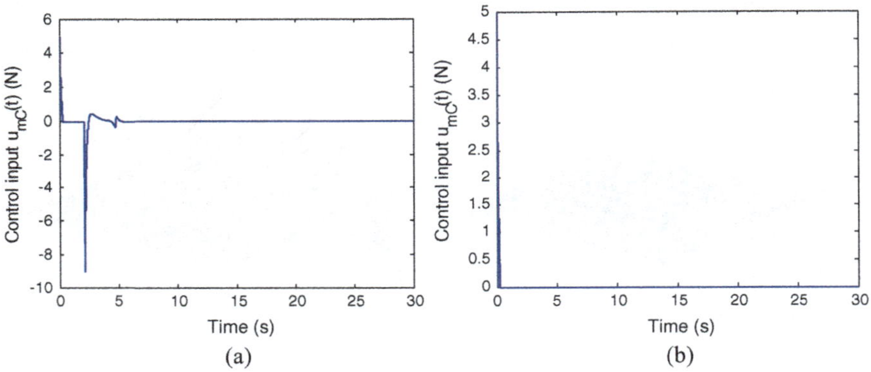

Fig. 8.14 Control input $u_{mC}(t)$

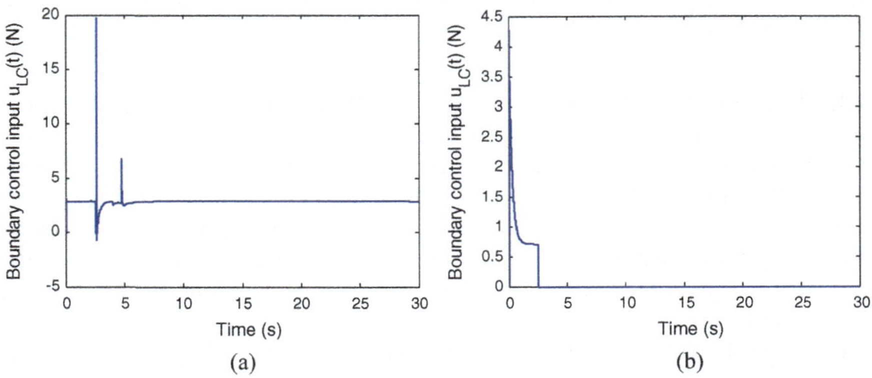

Fig. 8.15 Boundary control input $u_{LC}(t)$

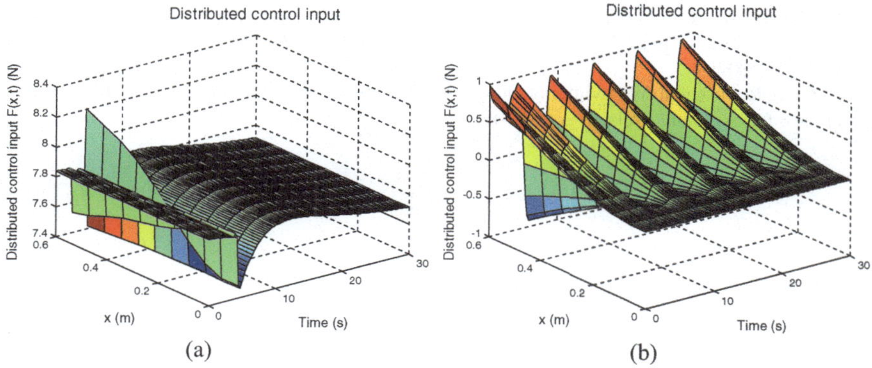

Fig. 8.16 Distributed control input $F(x, t)$

the supervision of the monitoring functions and $|e_1(t)| < k_{b1}$, $|e_2(t)| < k_{b2}$, $|e_3(t)| < k_{b3}$ can be consistently guaranteed. For all the cases, it is seen that the control objectives are able to be achieved by the proposed control scheme.

8.5 Conclusions

In this chapter, a switching fault-tolerant control strategy has been carried out for the vehicle-mounted flexible manipulator system with output constraints by using a BLF which grows to infinity whenever its arguments approach some limits. Vibration of the flexible link can be eliminated with the proposed control and healthy actuators. Vehicle and joint positions can be forced into the desired values without actuator failure. The proposed monitoring functions allow for effective fault detection and ensure that output constraints of the joint, vehicle, and end of the flexible manipulator are all satisfied under actuator failure with no need to know switching time and fault types. The asymptotical stability of the fault-free closed-loop system has been demonstrated by Lyapunov's direct method and extended LaSalle's invariance principle. Simulation examples verify the effectiveness and performance of the proposed control scheme.

References

1. C. Berbra, S. Lesecq, J.J. Martinez, A multi-observer switching strategy for fault-tolerant control of a quadrotor helicopter, in *16th Mediterranean Conference on Control and Automation Congress Centre* (Ajaccio, France, 2008), pp. 1094–1099
2. X. Wang, G.-H. Yang, Distributed fault-tolerant control for a class of cooperative uncertain systems with actuator failures and switching topologies. Inf. Sci. **370**, 650–666 (2016)

3. H.-J. Ma, G.-H. Yang, D. Ye, Switching fault-tolerant control for a class of nonlinear systems with actuator faults, in *25th Chinese Control and Decision Conference* (Guiyang, China, 2013), pp. 4797–4802
4. X.-Z. Jin, Robust adaptive switching fault-tolerant control of a class of uncertain systems against actuator faults. Math. Prob. Eng. (2013)

Chapter 9
Prescribed Performance Control of Vehicle-Mounted Flexible Manipulator

9.1 Introduction

Real control systems are rife with output constraints which, when violated, cause serious threats to the system. Thus, it is crucial to design proper controllers for systems with output constraints. Compared with the BLF-based control which restricts system output variables within a given range without considering the convergence rate, the prescribed performance control guarantees not only the convergence of the output tracking error to a pre-defined arbitrarily small residual set, but also the convergence rate no less than a pre-specified value [1–4].

In this chapter, control laws are designed for the vehicle-mounted flexible manipulator for position regulation and vibration suppression. Under the proposed prescribed performance control laws, both position regulation and vibration suppression can be achieved, and tracking errors and their convergence rates can be regulated within certain pre-specified values. Compared with Chap. 8, the convergence speed can also be restricted by the prescribed performance function.

9.2 Problem Formulation

In this chapter, the definition of system parameters and model are the same with ones in Chap. 6. We define the tracking errors of the joint angle and vehicle as $e_1(t) = \theta(t) - \theta_d$, $e_2(t) = r(t) - r_d$, respectively, where θ_d is the desired joint angle and r_d is the desired position of the vehicle. Similarly, $e_3(t) = w(L, t)$ denotes the error of the deflection at the end of the manipulator. As a result, we have $\dot{e}_1(t) = \dot{\theta}(t)$, $\dot{e}_2(t) = \dot{r}(t)$ and $\dot{e}_3(t) = \dot{w}(L, t)$.

The following control objectives will be achieved in this chapter

X. Xing and J. Liu, *Vibration Control Methods of Mechanical Distributed Parameter Systems*, Springer Tracts in Mechanical Engineering,
https://doi.org/10.1007/978-981-16-1532-0_9

(1) Regulate the positions and speeds of both the joint angle and vehicle. We also manage to suppress the elastic vibration while ensuring that the output signals of the closed-loop system are constrained within the given bounds. It means that $\theta(t) \to \theta_d$, $\dot{\theta}(t) \to 0$, $r(t) \to r_d$, $\dot{r}(t) \to 0$, $w(L,t) \to 0$, $\dot{w}(L,t) \to 0$, $w(x,t) \to 0$, $\dot{w}(x,t) \to 0$ are expected to be realized for $t \to \infty$.

(2) Prescribed performance bounds on the position tracking error and convergence rate of $e_i(t)$ are satisfied for $t > 0$ where $i = 1, 2, 3$.

9.3 Prescribed Performance Function

Strictly positive and decreasing function $\lambda_i(t), i = 1, 2, 3$, is given as a performance function as follows to guarantee that the error can converge to a small bounded neighborhood of zero with a constrained convergence rate. $\lambda_i(t)$ should have the following properties [5]:

(1) $\lambda_i(t) : R^+ \to R^+ - \{0\}$ is a smooth function;

(2) $\lambda_i(t)$ is decreasing;

(3) $\lim_{t\to\infty} \lambda_i(t) = \lambda_{i\infty} > 0$.

Based on the former description, $\lambda_i(t)$ is selected as an exponentially decaying performance function in the following math expression

$$\lambda_i(t) = (\lambda_{i0} - \lambda_{i\infty})e^{-l_i t} + \lambda_{i\infty} \tag{9.1}$$

where λ_{i0}, $\lambda_{i\infty}$, i_0, i_∞ and l_i are appropriate positive constants and $\lambda_{i0} > \lambda_{i\infty}$.

From the aforementioned equation, we can know that $\lambda_{i\infty}$ is the maximum value of the error $e_i(t)$ in the stable state, and the maximum overshoot of $e_i(t)$ is specified less than $\lambda_i(0)$. Moreover, the decline rate of $\lambda_i(t)$ also places a lower bound on the required convergence speed of $e_i(t)$.

As a result, the following prescribed constraint conditions in inequality (9.2) can guarantee the transient performances of tracking errors

$$-\lambda_i(t) < e_i(t) < \lambda_i(t), \ i = 1, 2, 3, \ \forall t > 0 \tag{9.2}$$

Assumption 9.1 Suppose that the initial tracking errors of joint and vehicle are respectively satisfied $|e_1(0)| < \lambda_1(0)$ and $|e_2(0)| < \lambda_2(0)$. Besides, the initial elastic deformation of the end point of the flexible arm meets $|e_3(0)| = |w(L,0)| < \lambda_3(0)$.

9.4 Control Design and Stability Analysis

In this section, the controller of the vehicle-mounted flexible manipulator with described performance is developed. To achieve control goals mentioned above, the error transformation is introduced.

Define

$$e_i(t) = \lambda_i(t) S_i(\varepsilon_i) \tag{9.3}$$

where ε_i is the transformed error and $S_i(\varepsilon_i)$ is a smooth, strictly increased function. $S_i(\varepsilon_i)$ should have the following properties:

(1) $S_i(\varepsilon_i)$ is a monotonically increasing function;
(2) $-1 < S_i(\varepsilon_i) < 1$;
(3) $\lim_{\varepsilon_i \to -\infty} S_i(\varepsilon_i) = -1$ and $\lim_{\varepsilon_i \to \infty} S_i(\varepsilon_i) = 1$.

To satisfy the above three properties, $S_i(\varepsilon_i)$ is designed as

$$S_i(\varepsilon_i) = \tanh(\varepsilon_i) = \frac{e^{\varepsilon_i} - e^{-\varepsilon_i}}{e^{\varepsilon_i} + e^{-\varepsilon_i}}. \tag{9.4}$$

Therefore, from Eq. (9.3) and the properties of $S_i(\varepsilon_i)$, we can get

$$-\lambda_i(t) < e_i(t) < \lambda_i(t). \tag{9.5}$$

From Eq. (9.4), we can obtain the inverse function of $S_i(\varepsilon_i)$ as

$$\varepsilon_i(t) = S_i^{-1}\left(\frac{e_i(t)}{\lambda_i(t)}\right) = \frac{1}{2}\ln\frac{1+S_i}{1-S_i} = \frac{1}{2}\ln\frac{1+\frac{e_i(t)}{\lambda_i(t)}}{1-\frac{e_i(t)}{\lambda_i(t)}} = \frac{1}{2}\ln\frac{\lambda_i(t)+e_i(t)}{\lambda_i(t)-e_i(t)}. \tag{9.6}$$

Differentiating Eq. (9.6) with respect to time yields

$$\dot{\varepsilon}_i(t) = \frac{1}{2}\left(\frac{\dot{\lambda}_i(t)+\dot{e}_i(t)}{\lambda_i(t)+e_i(t)} - \frac{\dot{\lambda}_i(t)-\dot{e}_i(t)}{\lambda_i(t)-e_i(t)}\right) = \frac{\lambda_i(t)\dot{e}_i(t) - \dot{\lambda}_i(t)e_i(t)}{\lambda_i^2(t) - e_i^2(t)}. \tag{9.7}$$

If $\varepsilon_i(t)$ can be made bounded for $t \geq 0$, inequality (9.5) will hold under the properties of the error transformation in Eq. (9.3) so that the transient and steady performance of the system can be guaranteed.

To ensure that tracking errors $e_1(t)$, $e_2(t)$, and the deflection error at the end of the flexible manipulator $e_3(t)$ converge to zero with the prescribed performance, control laws of the system are proposed as

$$\begin{aligned}\tau(t) = &-k_{p1}e_1(t) - k_{d1}\dot{e}_1(t) + \frac{1}{2}g\rho L^2\cos\theta(t) + MgL\cos\theta(t)\\ &- Mgw(L,t)\sin\theta(t)\end{aligned}$$

$$-\rho\int_0^L gw(x,t)\sin\theta(t)\mathrm{d}x+\frac{3}{2}e_1^2(t)\varepsilon_1(t)-\frac{3}{2}\lambda_1(t)e_1(t)-\frac{1}{2}\lambda_1^2(t)\varepsilon_1(t) \tag{9.8}$$

$$u_m(t)=-k_{p2}e_2(t)-k_{d2}\dot{e}_2(t)+\frac{3}{2}e_2^2(t)\varepsilon_2(t)-\frac{3}{2}\lambda_2(t)e_2(t)-\frac{1}{2}\lambda_2^2(t)\varepsilon_2(t) \tag{9.9}$$

$$\begin{aligned}u_L(t)=&-k_{p3}e_3(t)-k_{d3}\dot{e}_3(t)+Mg\cos\theta(t)+\frac{3}{2}e_3^2(t)\varepsilon_3(t)\\&-\frac{3}{2}\lambda_3(t)e_3(t)-\frac{1}{2}\lambda_3^2(t)\varepsilon_3(t)\end{aligned} \tag{9.10}$$

$$F(x,t)=-k_{p4}w(x,t)-k_{d4}\dot{w}(x,t)+g\rho\cos\theta(t) \tag{9.11}$$

where k_{p1}, k_{d1}, k_{p2}, k_{d2}, k_{p3}, k_{d3}, k_{p4} and k_{d4} are positive control coefficients and $\tau(t)$, $u_L(t)$, $u_m(t)$, $F(x,t)$ are four control inputs which act on the joint, payload, vehicle, flexible manipulator, respectively.

Theorem 9.1 Under Assumption 9.1, the system described by Eqs. (6.15)–(6.21) with the proposed control laws in Eqs. (9.8)–(9.11) satisfies the following properties

(1) Both position regulation and vibration suppression can be realized under the proposed control laws, that is, $\theta(t)\to\theta_d$, $\dot{\theta}(t)\to 0$, $r(t)\to r_d$, $\dot{r}(t)\to 0$, $w(L,t)\to 0$, $\dot{w}(L,t)\to 0$, $w(x,t)\to 0$, $\dot{w}(x,t)\to 0$ are guaranteed for $t\to\infty$.

(2) If initial boundary conditions $|e_1(0)|<\lambda_1(0)$, $|e_2(0)|<\lambda_2(0)$ and $|e_3(0)|<\lambda_3(0)$ hold, the tracking errors remain in the region of $\Omega_{e_i}=\{e_i(t)\in R:|e_i(t)|<\lambda_i(0)\}$ where $i=1,2$ and the deformation error of the end of the flexible link is also within the set of $\Omega_{e_3}=\{e_3(t)\in R:|e_3(t)|<\lambda_3(0)\}$ with a bounded speed. That is to say, the prescribed performance of both tracking errors and deflection error can be ensured with the given control laws.

Proof To realize the control goals, a Lyapunov function is proposed to be

$$V(t)=V_1(t)+V_2(t)+V_3(t) \tag{9.12}$$

where

$$\begin{aligned}V_1(t)=&\frac{1}{2}I_h\dot{\theta}^2(t)+\frac{1}{2}\rho\int_0^L\dot{\mathbf{P}}^{\mathrm{T}}(x,t)\dot{\mathbf{P}}(x,t)\mathrm{d}x+\frac{1}{2}M\dot{\mathbf{P}}^{\mathrm{T}}(L,t)\dot{\mathbf{P}}(L,t)\\&+\frac{1}{2}m\dot{r}^2(t)+\frac{EI}{2}\int_0^L w_{xx}^2(x,t)\mathrm{d}x\end{aligned} \tag{9.13}$$

$$V_2(t) = \frac{1}{2}k_{p1}e_1^2(t) + \frac{1}{2}k_{p2}e_2^2(t) + \frac{1}{2}k_{p3}e_3^2(t) + \frac{1}{2}k_{p4}\int_0^L w^2(x,t)\mathrm{d}x \quad (9.14)$$

$$\begin{aligned} V_3(t) = &\frac{1}{2}\big(\lambda_1^2(t) - e_1^2(t)\big)e_1(t)\varepsilon_1(t) + \frac{1}{2}\big(\lambda_2^2(t) - e_2^2(t)\big)e_2(t)\varepsilon_2(t) \\ &+ \frac{1}{2}\big(\lambda_3^2(t) - e_3^2(t)\big)e_3(t)\varepsilon_3(t) + \frac{1}{2}e_1^2(t)\lambda_1(t) + \frac{1}{2}e_2^2(t)\lambda_2(t) + \frac{1}{2}e_3^2(t)\lambda_3(t) \end{aligned} \quad (9.15)$$

in which $V_1(t)$ is a term of system energy; $V_2(t)$ is a term of tracking errors; $V_3(t)$ is a term to guarantee the described performance of the system.

The derivative of Eq. (9.12) with respect to time is given as

$$\begin{aligned} \dot{V}(t) =& I_h\dot{\theta}(t)\ddot{\theta}(t) \\ &+ \frac{1}{2}\rho\int_0^L \big(2x^2\dot{\theta}(t)\ddot{\theta}(t) + 2w(x,t)\dot{w}(x,t)\dot{\theta}^2(t) + 2w^2(x,t)\dot{\theta}(t)\ddot{\theta}(t) \\ &+ 2\dot{w}(x,t)\ddot{w}(x,t) + 2\dot{r}(t)\ddot{r}(t) + 2x\ddot{\theta}(t)\dot{w}(x,t) + 2x\dot{\theta}(t)\ddot{w}(x,t) \\ &- 2x\cos\theta(t)\dot{\theta}^2(t)\dot{r}(t) - 2x\sin\theta(t)\ddot{\theta}(t)\dot{r}(t) \\ &- 2x\sin\theta(t)\dot{\theta}(t)\ddot{r}(t) + 2\sin\theta(t)\dot{\theta}^2(t)w(x,t)\dot{r}(t) \\ &- 2\cos\theta(t)\ddot{\theta}(t)w(x,t)\dot{r}(t) - 4\cos\theta(t)\dot{\theta}(t)\dot{w}(x,t)\dot{r}(t) \\ &- 2\cos\theta(t)\dot{\theta}(t)w(x,t)\ddot{r}(t) - 2\sin\theta(t)\ddot{w}(x,t)\dot{r}(t) \\ &-2\sin\theta(t)\dot{w}(x,t)\ddot{r}(t)\big) + \frac{1}{2}M\big(2L^2\dot{\theta}(t)\ddot{\theta}(t) + 2w(L,t)\dot{w}(L,t)\dot{\theta}^2(t) \\ &+ 2w^2(L,t)\dot{\theta}(t)\ddot{\theta}(t) + 2\dot{w}(L,t)\ddot{w}(L,t) + 2\dot{r}(t)\ddot{r}(t) + 2L\ddot{\theta}(t)\dot{w}(L,t) \\ &+ 2L\dot{\theta}(t)\ddot{w}(L,t) - 2L\cos\theta(t)\dot{\theta}^2(t)\dot{r}(t) - 2L\sin\theta(t)\ddot{\theta}(t)\dot{r}(t) \\ &- 2L\sin\theta(t)\dot{\theta}(t)\ddot{r}(t) + 2\sin\theta(t)\dot{\theta}^2(t)w(L,t)\dot{r}(t) \\ &- 2\cos\theta(t)\ddot{\theta}(t)w(L,t)\dot{r}(t) - 4\cos\theta(t)\dot{\theta}(t)\dot{w}(L,t)\dot{r}(t) \\ &- 2\cos\theta(t)\dot{\theta}(t)w(L,t)\ddot{r}(t) \\ &-2\sin\theta(t)\ddot{w}(L,t)\dot{r}(t) - 2\sin\theta(t)\dot{w}(L,t)\ddot{r}(t)\big) \\ &+ m\dot{r}(t)\ddot{r}(t) + EI\int_0^L w_{xx}(x,t)\dot{w}_{xx}(x,t)\mathrm{d}x + k_{p1}e_1(t)\dot{e}_1(t) \\ &+ k_{p2}e_2(t)\dot{e}_2(t) + k_{p3}e_3(t)\dot{e}_3(t) + k_{p4}\int_0^L w(x,t)\dot{w}(x,t)\mathrm{d}x \\ &+ \big(\lambda_1(t)\dot{\lambda}_1(t) - e_1(t)\dot{e}_1(t)\big)e_1(t)\varepsilon_1(t) + \frac{1}{2}\big(\lambda_1^2(t) - e_1^2(t)\big)\dot{e}_1(t)\varepsilon_1(t) \\ &+ \frac{1}{2}\big(\lambda_1^2(t) - e_1^2(t)\big)e_1(t)\dot{\varepsilon}_1(t) + \big(\lambda_2(t)\dot{\lambda}_2(t) - e_2(t)\dot{e}_2(t)\big)e_2(t)\varepsilon_2(t) \\ &+ \frac{1}{2}\big(\lambda_2^2(t) - e_2^2(t)\big)\dot{e}_2(t)\varepsilon_2(t) + \frac{1}{2}\big(\lambda_2^2(t) - e_2^2(t)\big)e_2(t)\dot{\varepsilon}_2(t) \end{aligned}$$

$$
\begin{aligned}
&+ \big(\lambda_3(t)\dot{\lambda}_3(t) - e_3(t)\dot{e}_3(t)\big)e_3(t)\varepsilon_3(t) + \frac{1}{2}\big(\lambda_3^2(t) - e_3^2(t)\big)\dot{e}_3(t)\varepsilon_3(t) \\
&+ \frac{1}{2}\big(\lambda_3^2(t) - e_3^2(t)\big)e_3(t)\dot{\varepsilon}_3(t) + e_1(t)\dot{e}_1(t)\lambda_1(t) \\
&+ \frac{1}{2}e_1^2(t)\dot{\lambda}_1(t) + e_2(t)\dot{e}_2(t)\lambda_2(t) \\
&+ \frac{1}{2}e_2^2(t)\dot{\lambda}_2(t) + e_3(t)\dot{e}_3(t)\lambda_3(t) + \frac{1}{2}e_3^2(t)\dot{\lambda}_3(t).
\end{aligned} \tag{9.16}
$$

Substituting Eqs. (9.8)–(9.11) into Eq. (9.16) gives

$$
\begin{aligned}
\dot{V}(t) = &\dot{\theta}(t)\tau(t) + \dot{r}(t)u_m(t) + \dot{w}(L,t)u_L(t) \\
&+ \int_0^L F(x,t)\dot{w}(x,t)\mathrm{d}x - \frac{1}{2}g\rho L^2\cos\theta(t)\dot{\theta}(t) \\
&- MgL\cos\theta(t)\dot{\theta}(t) + Mgw(L,t)\sin\theta(t)\dot{\theta}(t) \\
&+ \rho\int_0^L gw(x,t)\sin\theta(t)\dot{\theta}(t)\mathrm{d}x \\
&- Mg\cos\theta(t)\dot{w}(L,t) - \int_0^L g\rho\cos\theta(t)\dot{w}(x,t)\mathrm{d}x \\
&+ EIw_{xxx}(L,t)\dot{w}(L,t) \\
&+ EI\int_0^L w_{xx}(x,t)\dot{w}_{xx}(x,t)\mathrm{d}x \\
&- \int_0^L EIw_{xxxx}(x,t)\dot{w}(x,t)\mathrm{d}x + k_{p1}e_1(t)\dot{e}_1(t) + k_{p2}e_2(t)\dot{e}_2(t) + k_{p3}e_3(t)\dot{e}_3(t) \\
&+ k_{p4}\int_0^L w(x,t)\dot{w}(x,t)\mathrm{d}x + \big(\lambda_1(t)\dot{\lambda}_1(t) - e_1(t)\dot{e}_1(t)\big)e_1(t)\varepsilon_1(t) \\
&+ \frac{1}{2}\Big(\lambda_1^2(t) - e_1^2(t)\Big)\dot{e}_1(t)\varepsilon_1(t) \\
&+ \frac{1}{2}\Big(\lambda_1^2(t) - e_1^2(t)\Big)e_1(t)\dot{\varepsilon}_1(t) + \big(\lambda_2(t)\dot{\lambda}_2(t) - e_2(t)\dot{e}_2(t)\big)e_2(t)\varepsilon_2(t) \\
&+ \frac{1}{2}\Big(\lambda_2^2(t) - e_2^2(t)\Big)\dot{e}_2(t)\varepsilon_2(t) + \frac{1}{2}\Big(\lambda_2^2(t) - e_2^2(t)\Big)e_2(t)\dot{\varepsilon}_2(t) \\
&+ \big(\lambda_3(t)\dot{\lambda}_3(t) - e_3(t)\dot{e}_3(t)\big)e_3(t)\varepsilon_3(t) + \frac{1}{2}\Big(\lambda_3^2(t) - e_3^2(t)\Big)\dot{e}_3(t)\varepsilon_3(t) \\
&+ \frac{1}{2}\Big(\lambda_3^2(t) - e_3^2(t)\Big)e_3(t)\dot{\varepsilon}_3(t) + e_1(t)\dot{e}_1(t)\lambda_1(t) \\
&+ \frac{1}{2}e_1^2(t)\dot{\lambda}_1(t) + e_2(t)\dot{e}_2(t)\lambda_2(t) \\
&+ \frac{1}{2}e_2^2(t)\dot{\lambda}_2(t) + e_3(t)\dot{e}_3(t)\lambda_3(t) + \frac{1}{2}e_3^2(t)\dot{\lambda}_3(t).
\end{aligned} \tag{9.17}
$$

Using integration by parts of the middle terms in Eq. (9.17), we get

$$
\begin{aligned}
&- \int_0^L EIw_{xxxx}(x,t)\dot{w}(x,t)\mathrm{d}x + EIw_{xxx}(L,t)\dot{w}(L,t) \\
&+ EI\int_0^L w_{xx}(x,t)\dot{w}_{xx}(x,t)\mathrm{d}x = 0.
\end{aligned} \tag{9.18}
$$

Substituting Eq. (9.18) into Eq. (9.17), we can obtain

$$
\begin{aligned}
\dot{V}(t) = &\dot{\theta}(t)\tau(t) + \dot{r}(t)u_m(t) + \dot{w}(L,t)u_L(t) \\
&+ \int_0^L F(x,t)\dot{w}(x,t)\mathrm{d}x - \frac{1}{2}g\rho L^2\cos\theta(t)\dot{\theta}(t) \\
&- MgL\cos\theta(t)\dot{\theta}(t) + Mgw(L,t)\sin\theta(t)\dot{\theta}(t) \\
&+ \rho\int_0^L gw(x,t)\sin\theta(t)\dot{\theta}(t)\mathrm{d}x \\
&- Mg\cos\theta(t)\dot{w}(L,t) - \int_0^L g\rho\cos\theta(t)\dot{w}(x,t)\mathrm{d}x \\
&+ k_{p1}e_1(t)\dot{e}_1(t) + k_{p2}e_2(t)\dot{e}_2(t) + k_{p3}e_3(t)\dot{e}_3(t) \\
&+ k_{p4}\int_0^L w(x,t)\dot{w}(x,t)\mathrm{d}x + \big(\lambda_1(t)\dot{\lambda}_1(t) - e_1(t)\dot{e}_1(t)\big)e_1(t)\varepsilon_1(t) \\
&+ \frac{1}{2}\big(\lambda_1^2(t) - e_1^2(t)\big)\dot{e}_1(t)\varepsilon_1(t) + \frac{1}{2}\big(\lambda_1^2(t) - e_1^2(t)\big)e_1(t)\dot{\varepsilon}_1(t) \\
&+ \big(\lambda_2(t)\dot{\lambda}_2(t) - e_2(t)\dot{e}_2(t)\big)e_2(t)\varepsilon_2(t) + \frac{1}{2}\big(\lambda_2^2(t) - e_2^2(t)\big)\dot{e}_2(t)\varepsilon_2(t) \\
&+ \frac{1}{2}\big(\lambda_2^2(t) - e_2^2(t)\big)e_2(t)\dot{\varepsilon}_2(t) + \big(\lambda_3(t)\dot{\lambda}_3(t) - e_3(t)\dot{e}_3(t)\big)e_3(t)\varepsilon_3(t) \\
&+ \frac{1}{2}\big(\lambda_3^2(t) - e_3^2(t)\big)\dot{e}_3(t)\varepsilon_3(t) + \frac{1}{2}\big(\lambda_3^2(t) - e_3^2(t)\big)e_3(t)\dot{\varepsilon}_3(t) + e_1(t)\dot{e}_1(t)\lambda_1(t) \\
&+ \frac{1}{2}e_1^2(t)\dot{\lambda}_1(t) + e_2(t)\dot{e}_2(t)\lambda_2(t) + \frac{1}{2}e_2^2(t)\dot{\lambda}_2(t) \\
&+ e_3(t)\dot{e}_3(t)\lambda_3(t) + \frac{1}{2}e_3^2(t)\dot{\lambda}_3(t).
\end{aligned} \tag{9.19}
$$

Substituting control laws in Eqs. (9.8)–(9.11) into Eq. (9.19), we have

$$
\begin{aligned}
\dot{V}(t) = &\dot{\theta}(t)\Big(-k_{p1}e_1(t) - k_{d1}\dot{e}_1(t) + \frac{1}{2}g\rho L^2\cos\theta(t) \\
&+ MgL\cos\theta(t) - Mgw(L,t)\sin\theta(t) \\
&-\rho\int_0^L gw(x,t)\sin\theta(t)\mathrm{d}x + \frac{3}{2}e_1^2(t)\varepsilon_1(t) - \frac{3}{2}\lambda_1(t)e_1(t) - \frac{1}{2}\lambda_1^2(t)\varepsilon_1(t)\Big) \\
&+ \dot{r}(t)\Big(-k_{p2}e_2(t) - k_{d2}\dot{e}_2(t) + \frac{3}{2}e_2^2(t)\varepsilon_2(t) - \frac{3}{2}\lambda_2(t)e_2(t) - \frac{1}{2}\lambda_2^2(t)\varepsilon_2(t)\Big)
\end{aligned}
$$

$$
\begin{aligned}
&+\dot{w}(L,t)\Big(-k_{p3}e_3(t)-k_{d3}\dot{e}_3(t)+Mg\cos\theta(t)+\frac{3}{2}e_3^2(t)\varepsilon_3(t)-\frac{3}{2}\lambda_3(t)e_3(t)\\
&-\frac{1}{2}\lambda_3^2(t)\varepsilon_3(t)\Big)+\int_0^L\big(-k_{p4}w(x,t)-k_{d4}\dot{w}(x,t)+g\rho\cos\theta(t)\big)\dot{w}(x,t)\mathrm{d}x\\
&-\frac{1}{2}g\rho L^2\cos\theta(t)\dot{\theta}(t)-MgL\cos\theta(t)\dot{\theta}(t)+Mgw(L,t)\sin\theta(t)\dot{\theta}(t)\\
&+\rho\int_0^L gw(x,t)\sin\theta(t)\dot{\theta}(t)\mathrm{d}x-Mg\cos\theta(t)\dot{w}(L,t)\\
&-\int_0^L g\rho\cos\theta(t)\dot{w}(x,t)\mathrm{d}x\\
&+k_{p1}e_1(t)\dot{e}_1(t)+k_{p2}e_2(t)\dot{e}_2(t)+k_{p3}e_3(t)\dot{e}_3(t)\\
&+k_{p4}\int_0^L w(x,t)\dot{w}(x,t)\mathrm{d}x+\big(\lambda_1(t)\dot{\lambda}_1(t)-e_1(t)\dot{e}_1(t)\big)e_1(t)\varepsilon_1(t)\\
&+\frac{1}{2}\Big(\lambda_1^2(t)-e_1^2(t)\Big)\dot{e}_1(t)\varepsilon_1(t)+\frac{1}{2}\Big(\lambda_1^2(t)-e_1^2(t)\Big)e_1(t)\dot{\varepsilon}_1(t)\\
&+\big(\lambda_2(t)\dot{\lambda}_2(t)-e_2(t)\dot{e}_2(t)\big)e_2(t)\varepsilon_2(t)+\frac{1}{2}\Big(\lambda_2^2(t)-e_2^2(t)\Big)\dot{e}_2(t)\varepsilon_2(t)\\
&+\frac{1}{2}\Big(\lambda_2^2(t)-e_2^2(t)\Big)e_2(t)\dot{\varepsilon}_2(t)+\big(\lambda_3(t)\dot{\lambda}_3(t)-e_3(t)\dot{e}_3(t)\big)e_3(t)\varepsilon_3(t)\\
&+\frac{1}{2}\Big(\lambda_3^2(t)-e_3^2(t)\Big)\dot{e}_3(t)\varepsilon_3(t)+\frac{1}{2}\Big(\lambda_3^2(t)-e_3^2(t)\Big)e_3(t)\dot{\varepsilon}_3(t)+e_1(t)\dot{e}_1(t)\lambda_1(t)\\
&+\frac{1}{2}e_1^2(t)\dot{\lambda}_1(t)+e_2(t)\dot{e}_2(t)\lambda_2(t)\\
&+\frac{1}{2}e_2^2(t)\dot{\lambda}_2(t)+e_3(t)\dot{e}_3(t)\lambda_3(t)+\frac{1}{2}e_3^2(t)\dot{\lambda}_3(t)\\
&=\dot{e}_1(t)\Big(-k_{d1}\dot{e}_1(t)+\frac{3}{2}e_1^2(t)\varepsilon_1(t)-\frac{3}{2}\lambda_1(t)e_1(t)-\frac{1}{2}\lambda_1^2(t)\varepsilon_1(t)\Big)\\
&+\dot{e}_2(t)\Big(-k_{d2}\dot{e}_2(t)+\frac{3}{2}e_2^2(t)\varepsilon_2(t)-\frac{3}{2}\lambda_2(t)e_2(t)-\frac{1}{2}\lambda_2^2(t)\varepsilon_2(t)\Big)\\
&+\dot{e}_3(t)\Big(-k_{d3}\dot{w}(L,t)+\frac{3}{2}e_3^2(t)\varepsilon_3(t)-\frac{3}{2}\lambda_3(t)e_3(t)-\frac{1}{2}\lambda_3^2(t)\varepsilon_3(t)\Big)\\
&-\int_0^L k_{d4}\dot{w}^2(x,t)\mathrm{d}x+\big(\lambda_1(t)\dot{\lambda}_1(t)-e_1(t)\dot{e}_1(t)\big)e_1(t)\varepsilon_1(t)\\
&+\frac{1}{2}\Big(\lambda_1^2(t)-e_1^2(t)\Big)\dot{e}_1(t)\varepsilon_1(t)+\frac{1}{2}\Big(\lambda_1^2(t)-e_1^2(t)\Big)e_1(t)\dot{\varepsilon}_1(t)\\
&+\big(\lambda_2(t)\dot{\lambda}_2(t)-e_2(t)\dot{e}_2(t)\big)e_2(t)\varepsilon_2(t)+\frac{1}{2}\Big(\lambda_2^2(t)-e_2^2(t)\Big)\dot{e}_2(t)\varepsilon_2(t)\\
&+\frac{1}{2}\Big(\lambda_2^2(t)-e_2^2(t)\Big)e_2(t)\dot{\varepsilon}_2(t)+\big(\lambda_3(t)\dot{\lambda}_3(t)-e_3(t)\dot{e}_3(t)\big)e_3(t)\varepsilon_3(t)\\
&+\frac{1}{2}\Big(\lambda_3^2(t)-e_3^2(t)\Big)\dot{e}_3(t)\varepsilon_3(t)+\frac{1}{2}\Big(\lambda_3^2(t)-e_3^2(t)\Big)e_3(t)\dot{\varepsilon}_3(t)\\
&+e_1(t)\dot{e}_1(t)\lambda_1(t)+\frac{1}{2}e_1^2(t)\dot{\lambda}_1(t)\\
&+e_2(t)\dot{e}_2(t)\lambda_2(t)+\frac{1}{2}e_2^2(t)\dot{\lambda}_2(t)+e_3(t)\dot{e}_3(t)\lambda_3(t)+\frac{1}{2}e_3^2(t)\dot{\lambda}_3(t).
\end{aligned}
\tag{9.20}
$$

Now, it follows from Eqs. (9.7) and (9.20) that

$$
\begin{aligned}
\dot{V}(t) = & -k_{d1}\dot{e}_1^2(t) + \frac{3}{2}e_1^2(t)\varepsilon_1(t)\dot{e}_1(t) - \frac{3}{2}\lambda_1(t)e_1(t)\dot{e}_1(t) - \frac{1}{2}\lambda_1^2(t)\varepsilon_1(t)\dot{e}_1(t) \\
& - k_{d2}\dot{e}_2^2(t) + \frac{3}{2}e_2^2(t)\varepsilon_2(t)\dot{e}_2(t) - \frac{3}{2}\lambda_2(t)e_2(t)\dot{e}_2(t) - \frac{1}{2}\lambda_2^2(t)\varepsilon_2(t)\dot{e}_2(t) \\
& - k_{d3}\dot{e}_3^2(t) + \frac{3}{2}e_3^2(t)\varepsilon_3(t)\dot{e}_3(t) - \frac{3}{2}\lambda_3(t)e_3(t)\dot{e}_3(t) - \frac{1}{2}\lambda_3^2(t)\varepsilon_3(t)\dot{e}_3(t) \\
& - k_{d4}\int_0^L \dot{w}^2(x,t)\mathrm{d}x + \left(\lambda_1(t)\dot{\lambda}_1(t) - e_1(t)\dot{e}_1(t)\right)e_1(t)\varepsilon_1(t) \\
& + \frac{1}{2}\left(\lambda_1^2(t) - e_1^2(t)\right)\dot{e}_1(t)\varepsilon_1(t) \\
& + \frac{1}{2}\left(\lambda_1^2(t) - e_1^2(t)\right)e_1(t)\frac{\lambda_1(t)\dot{e}_1(t) - \dot{\lambda}_1(t)e_1(t)}{\lambda_1^2(t) - e_1^2(t)} \\
& + \left(\lambda_2(t)\dot{\lambda}_2(t) - e_2(t)\dot{e}_2(t)\right)e_2(t)\varepsilon_2(t) \\
& + \frac{1}{2}\left(\lambda_2^2(t) - e_2^2(t)\right)\dot{e}_2(t)\varepsilon_2(t) + \frac{1}{2}\left(\lambda_2^2(t) - e_2^2(t)\right)e_2(t)\frac{\lambda_2(t)\dot{e}_2(t) - \dot{\lambda}_2(t)e_2(t)}{\lambda_2^2(t) - e_2^2(t)} \\
& + \left(\lambda_3(t)\dot{\lambda}_3(t) - e_3(t)\dot{e}_3(t)\right)e_3(t)\varepsilon_3(t) + \frac{1}{2}\left(\lambda_3^2(t) - e_3^2(t)\right)\dot{e}_3(t)\varepsilon_3(t) \\
& + \frac{1}{2}\left(\lambda_3^2(t) - e_3^2(t)\right)e_3(t)\frac{\lambda_3(t)\dot{e}_3(t) - \dot{\lambda}_3(t)e_3(t)}{\lambda_3^2(t) - e_3^2(t)} \\
& + e_1(t)\dot{e}_1(t)\lambda_1(t) + \frac{1}{2}e_1^2(t)\dot{\lambda}_1(t) \\
& + e_2(t)\dot{e}_2(t)\lambda_2(t) + \frac{1}{2}e_2^2(t)\dot{\lambda}_2(t) + e_3(t)\dot{e}_3(t)\lambda_3(t) + \frac{1}{2}e_3^2(t)\dot{\lambda}_3(t) \\
= & -k_{d1}\dot{e}_1^2(t) - k_{d2}\dot{e}_2^2(t) - k_{d3}\dot{e}_3^2(t) - k_{d4}\int_0^L \dot{w}^2(x,t)\mathrm{d}x \\
& + \lambda_1(t)\dot{\lambda}_1(t)e_1(t)\varepsilon_1(t) \\
& + \lambda_2(t)\dot{\lambda}_2(t)e_2(t)\varepsilon_2(t) + \lambda_3(t)\dot{\lambda}_3(t)\dot{e}_3(t)\varepsilon_3(t).
\end{aligned} \tag{9.21}
$$

Since

$$\lambda_i(t)\dot{\lambda}_i(t)e_i(t)\varepsilon_i(t) \le 0,\ i = 1, 2, 3 \tag{9.22}$$

can be obtained from Eqs. (9.1) and (9.6), we can further obtain $\dot{V}(t) \le 0$ on the basis of Eq. (9.21), which means that $V(t)$ is negative semi-definite.

In this chapter, the extended LaSalle's invariance principle is applied to prove the asymptotic stability of the system.

If $\dot{V}(t) \equiv 0$ then from Eq. (9.21), we get

$$\dot{\theta}(t) = \dot{r}(t) = \dot{w}(L,t) = \dot{w}(x,t) = e_1(t)\varepsilon_1(t) = e_2(t)\varepsilon_2(t) = e_3(t)\varepsilon_3(t) \equiv 0 \tag{9.23}$$

which means

$$\ddot{\theta}(t) = \ddot{r}(t) = \ddot{w}(L, t) = \ddot{w}(x, t) \equiv 0. \tag{9.24}$$

Applying Eqs. (9.23) and (9.24) into the model of the system yields

$$\frac{1}{2} g\rho L^2 \cos\theta(t) + MgL\cos\theta(t) - Mgw(L, t)\sin\theta(t) \\ - \rho \int_0^L gw(x, t)\sin\theta(t)\mathrm{d}x - \tau(t) = 0 \tag{9.25}$$

$$-EIw_{xxx}(L, t) + Mg\cos\theta(t) - u_L(t) = 0 \tag{9.26}$$

$$u_m(t) = 0 \tag{9.27}$$

$$EIw_{xxxx}(x, t) + \rho g\cos\theta(t) - F(x, t) = 0. \tag{9.28}$$

From Eqs. (9.23) and (9.28), it is clear that

$$EIw_{xxxx}(x, t) = -\rho g\cos\theta(t) + F(x, t) = -k_{p4}w(x, t). \tag{9.29}$$

By using the technique of separation of variables, we separate $w(x, t)$ as

$$w(x, t) = \Phi(x) \cdot q(t) \tag{9.30}$$

where $\Phi(x)$ and $q(t)$ are two undetermined functions with respect to space and time respective.

Substituting Eq. (9.30) into Eq. (9.29), we can get

$$\frac{\Phi_{xxxx}(x)}{\Phi(x)} = -\frac{k_{p4}}{EI}. \tag{9.31}$$

Setting $\alpha^4 = \frac{k_{p4}}{EI}$ and solving Eq. (9.31), we obtain

$$\Phi(x) = e^{\frac{\sqrt{2}}{2}\alpha x}\left(b_1 \cos\frac{\sqrt{2}}{2}\alpha x + b_2 \sin\frac{\sqrt{2}}{2}\alpha x\right) \\ + e^{-\frac{\sqrt{2}}{2}\alpha x}\left(b_3 \cos\frac{\sqrt{2}}{2}\alpha x + b_4 \sin\frac{\sqrt{2}}{2}\alpha x\right) \tag{9.32}$$

where $b_i \in R,\ i = 1, 2, 3, 4$ are four constants to be determined.

From Eqs. (9.10), (9.23) and (9.26), we know that

$$k_{p3}w(L,t) + \frac{3}{2}\lambda_3(t)e_3(t) + \frac{1}{2}\lambda_3^2(t)\varepsilon_3(t) = EIw_{xxx}(L,t). \tag{9.33}$$

Hence, we have

$$w(L,t) = \frac{k_{p3}}{EI}w_{xxx}(L,t) = 0. \tag{9.34}$$

Using boundary conditions in Eqs. (6.19)–(6.21) and $w_{xxx}(L,t) = 0$, we have $\Phi(0) = \Phi_x(0) = \Phi_{xx}(L) = \Phi_{xxx}(L) = 0$. By resolving Eqs. (9.32), we can derive that $b_i = 0,\ i = 1, 2, 3, 4$, is the unique solution. Therefore, $\Phi(x) = 0$, $w(x,t) = 0$, $w_{xxx}(L,t) = 0$, $w_{xxxx}(x,t) = 0$ can be obtained.

Substituting control laws in Eqs. (9.8)–(9.11) into Eqs. (9.25)–(9.28), we can get

$$e_1(t) = \theta(t) - \theta_d = 0 \tag{9.35}$$

$$e_2(t) = r(t) - r_d = 0 \tag{9.36}$$

$$e_3(t) = w(L,t) = 0 \tag{9.37}$$

$$w(x,t) = 0. \tag{9.38}$$

On the basis of the above analysis, the asymptotical stability of the closed-loop system is verified and $\theta(t) \to \theta_d$, $\dot{\theta}(t) \to 0$, $r(t) \to r_d$, $\dot{r}(t) \to 0$, $w(L,t) \to 0$, $\dot{w}(L,t) \to 0$, $w(x,t) \to 0$, $\dot{w}(x,t) \to 0$ when $t \to \infty$ can be proved by applying extended LaSalle's invariance principle. Furthermore, under the Assumption 9.1 and control inputs proposed in Eqs. (9.8)–(9.11), the transient performance of the system is adjustable and output errors $e_1(t)$, $e_2(t)$, $e_3(t)$ can be constrained within prescribed performance bounds as $-\lambda_i(t) < e_i(t) < \lambda_i(t), i = 1, 2, 3$. Therefore, Theorem 9.1 has been proved.

9.5 Simulations

In this part, we run several simulations on the proposed controller of the vehicle-mounted flexible manipulator system to validate the control performance of both position regulation and vibration suppression, and the transient and steady prescribed performance. In order to obtain the approximate solution of the nonlinear PDE system, we choose the temporal step as $\Delta t = 4 \times 10^{-4}$ s and spatial step size as $\Delta x = 6 \times 10^{-2}$ m. The initial conditions of the system are set as $w(x,0) = 0$, $\dot{w}(x,0) = 0$, $\theta(0) = 0$, $\dot{\theta}(0) = 0$, $r(0) = 0$, $\dot{r}(0) = 0$ in order to satisfy the Assumption 9.1.

The desired positions of the joint and vehicle are given as $\theta_d = 0.2$ rad and $r_d = 1$ m. The final signals of the closed-loop system are expected to be $\theta(t) \to \theta_d$, $\dot{\theta}(t) \to 0$, $r(t) \to r_d$, $\dot{r}(t) \to 0$, $w(L,t) \to 0$, $\dot{w}(L,t) \to 0$, $w(x,t) \to 0$, $\dot{w}(x,t) \to 0$ for $t \to \infty$. The parameters of the system can be found in Table 9.1 and $\lambda_i(t)$ in Eq. (9.1) are given as $k_{p1} = 40$, $k_{d1} = 10$, $k_{p2} = 40$, $k_{d2} = 20$, $k_{p3} = 40$, $k_{d3} = 15$, $k_{p4} = 10$, $k_{d4} = 20$, $\lambda_{10} = 0.4$, $\lambda_{1\infty} = 0.002$, $l_1 = 1.6$, $\lambda_{20} = 1.2$, $\lambda_{2\infty} = 0.003$, $l_2 = 1.8$, $\lambda_{30} = 0.002$, $\lambda_{3\infty} = 2 \times 10^{-5}$, $l_3 = 2$, for $i = 1, 2, 3$.

To validate the control performance and demonstrate the advantages of the proposed control over the traditional control scheme, the dynamic responses of the system are simulated with the above parameters in the following three cases:

Case 1. With an improved PID control with gravity compensation:

$$\tau(t) = -k_1 e_1(t) - k_2 \dot{e}_1(t) - k_3 \int_0^t e_1(\mu)\mathrm{d}\mu + \frac{1}{2} g\rho L^2 \cos\theta(t) + MgL\cos\theta(t) - Mgw(L,t)\sin\theta(t) - \rho \int_0^L gw(x,t)\sin\theta(t)\mathrm{d}x \quad (9.39)$$

$$u_m(t) = -k_4 e_2(t) - k_5 \dot{e}_2(t) - k_6 \int_0^t e_2(\mu)\mathrm{d}\mu \quad (9.40)$$

$$u_L(t) = -k_7 e_3(t) - k_8 \dot{e}_3(t) - k_9 \int_0^t e_3(\mu)\mathrm{d}\mu + Mg\cos\theta(t) \quad (9.41)$$

$$F(x,t) = -k_{10} w(x,t) - k_{11} \dot{w}(x,t) - k_{12} \int_0^t w(x,\mu)\mathrm{d}\mu + g\rho\cos\theta(t) \quad (9.42)$$

where $k_1 = 20$, $k_2 = 10$, $k_3 = 10$, $k_4 = 20$, $k_5 = 20$, $k_6 = 2$, $k_7 = 20$, $k_8 = 15$, $k_9 = 10$, $k_{10} = 10$, $k_{11} = 20$, $k_{12} = 10$;

Case 2. With the proposed control in Eqs. (9.8)–(9.11).

The performance of the system with control input for cases 1 and 2 is shown in Figs. 9.1, 9.2, 9.3, 9.4 and 9.5. Figure 9.1 gives the position of the joint and its speed. Figure 9.2 shows both the position and speed of the vehicle. Figure 9.3 shows the

Table 9.1 Parameters of the vehicle-mounted flexible manipulator

Parameter	Description	Value
L	Length of the flexible link	0.6 m
ρ	Uniform mass per unit length of the flexible link	0.8 kg m^{-1}
M	Mass of the tip payload	0.3 kg
EI	Bending rigidity of the flexible link	20 N m^2
I_h	Moment of inertia of the flexible link	2 kg m^2
m	Mass of the vehicle	1.5 kg

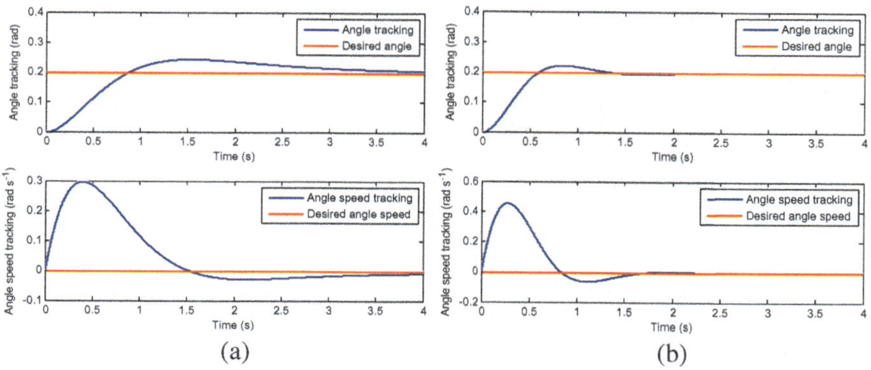

Fig. 9.1 Angular position and speed of the joint. **a** With the PID control for case 1. **b** With the proposed control for case 2

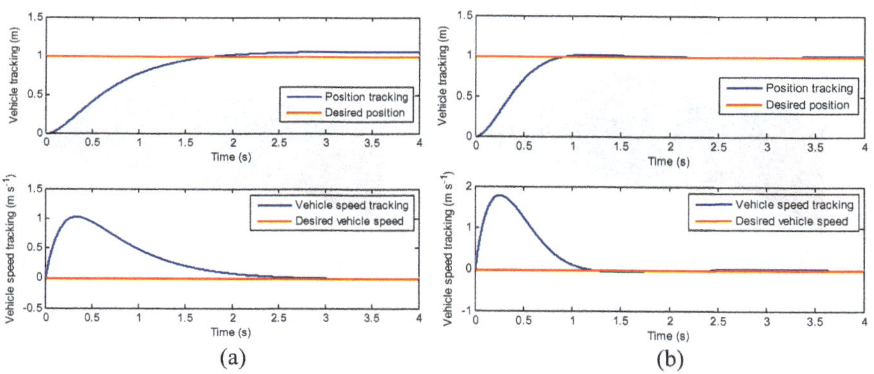

Fig. 9.2 Position and speed of the vehicle. **a** With the PID control for case 1. **b** With the proposed control for case 2

elastic deflection and speed of the elastic deflection at the free end of the flexible manipulator. The elastic vibration of the flexible manipulator with control inputs is shown in Fig. 9.4, and the speed of the elastic deflection in the closed-loop system is shown in Fig. 9.5. Figures 9.1b, 9.2b, 9.3b, 9.4b and 9.5b indicate that under the proposed control, the joint and vehicle are positioned to the desired set points and the speeds of the joint and vehicle converge to zero over time. The vibration of the flexible manipulator is also significantly suppressed under the proposed control within 4 s, which indicates that the proposed controller has very good control effect. Compared with Figs. 9.1b, 9.2b, 9.3b, 9.4b and 9.5b, Figs. 9.1a, 9.2a, 9.3a, 9.4a, 9.5a show that the control performance is poorer and the convergence rate is slower with the PID control.

Tracking errors of the joint and vehicle are shown in Figs. 9.6 and 9.7, respectively, with the proposed described performance function curves. The error of boundary

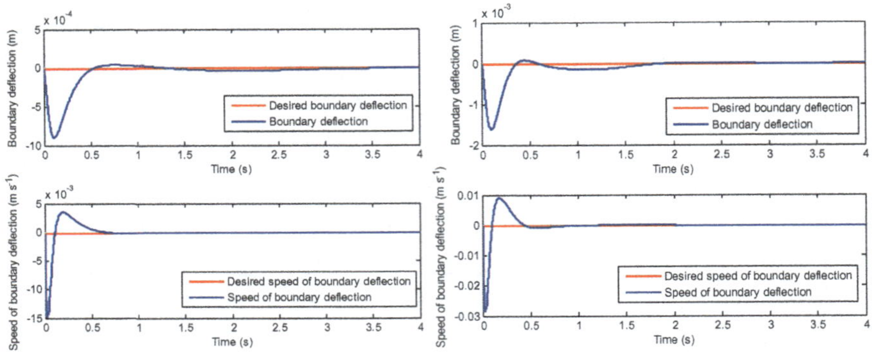

(a) (b)

Fig. 9.3 Elastic deflection and speed of the elastic deflection at the end of the flexible manipulator. **a** With the PID control for case 1. **b** With the proposed control for case 2

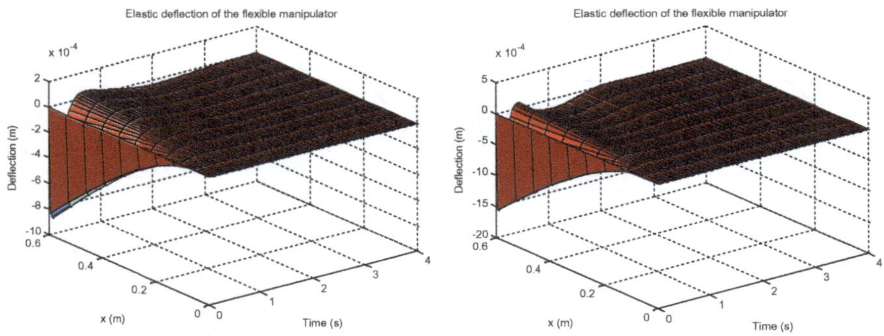

(a) (b)

Fig. 9.4 Elastic deflection of the flexible manipulator. **a** With the PID control for case 1. **b** With the proposed control for case 2

deflection of the flexible manipulator is shown in Fig. 9.8. As per our theoretical analysis, in Figs. 9.6b, 9.7b and 9.8b, system error $e_i(t)$, $i = 1, 2, 3$, can converge to the prescribed performance bounds without transgression of the prescribed speed under the proposed control. As compared with the PID control in Figs. 9.6a, 9.7a and 9.8a, it is obvious that the prescribed performance can always be satisfied under the proposed control.

Four control inputs $\tau(t)$, $u_m(t)$, $u_L(t)$, and $F(x, t)$ are shown in Figs. 9.9, 9.10, 9.11 and 9.12 as follows.

The proposed control laws are indeed effective according to the simulation results shown in Figs. 9.1, 9.2, 9.3, 9.4, 9.5, 9.6, 9.7, 9.8, 9.9, 9.10, 9.11 and 9.12. Under the proposed control scheme, positions of the joint and vehicle can be regulated to the desired values while the vibration of the flexible manipulator is suppressed. When

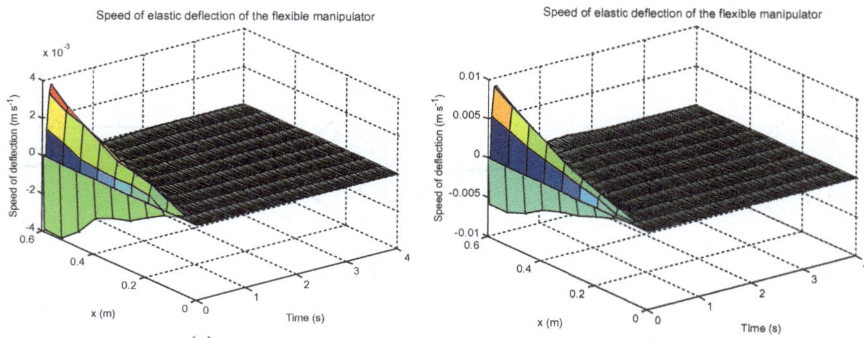

(a) (b)

Fig. 9.5 Speed of the elastic deflection of the flexible manipulator. **a** With the PID control for case 1. **b** With the proposed control for case 2

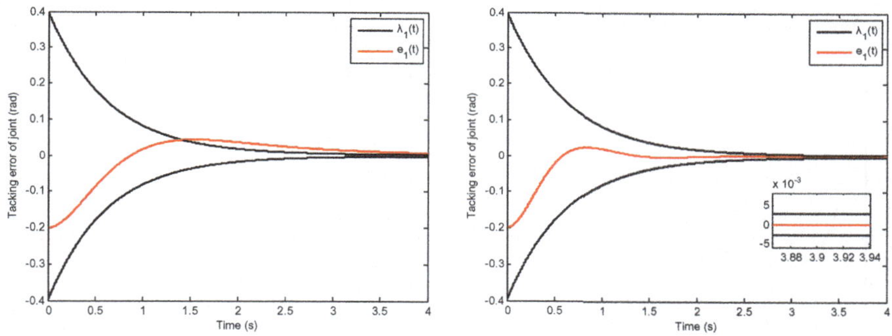

(a) (b)

Fig. 9.6 Tracking error of the joint. **a** With the PID control for case 1. **b** With the proposed control for case 2

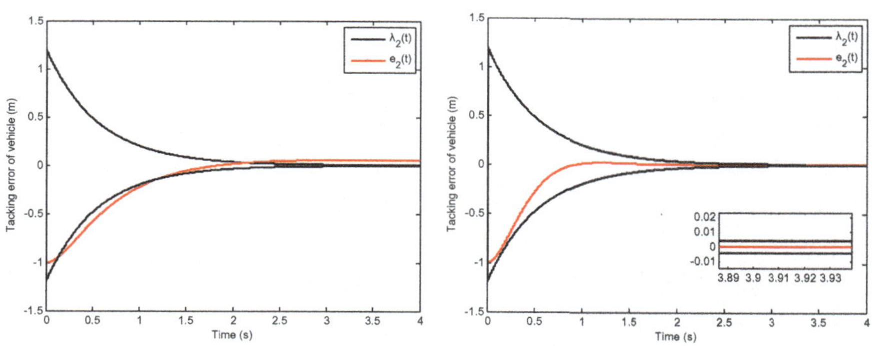

(a) (b)

Fig. 9.7 Tracking error of the vehicle. **a** With the PID control for case 1. **b** With the proposed control for case 2

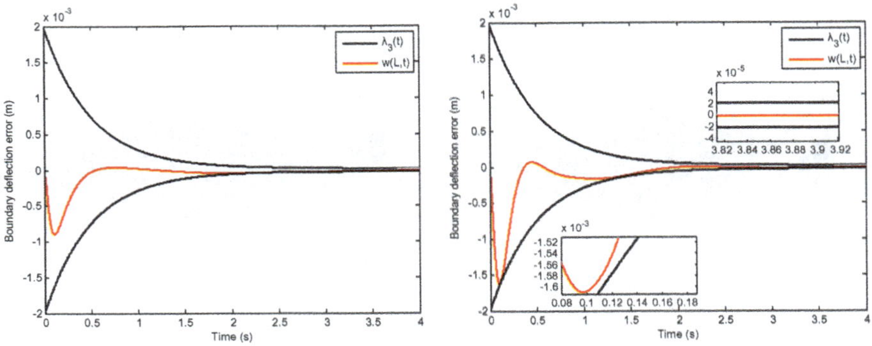

(a) (b)

Fig. 9.8 Error of the boundary deflection of the flexible manipulator. **a** With the PID control for case 1. **b** With the proposed control for case 2

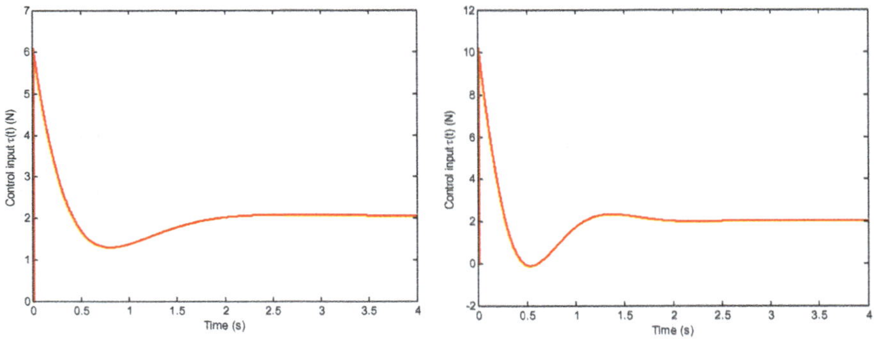

(a) (b)

Fig. 9.9 Control input $\tau(t)$. **a** PID control for case 1. **b** Proposed control for case 2

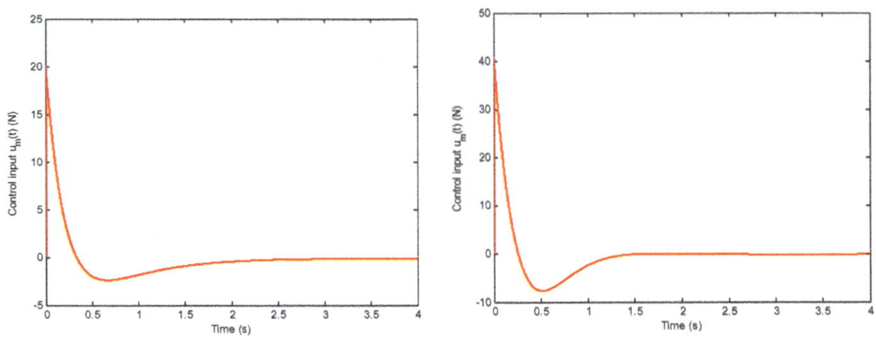

(a) (b)

Fig. 9.10 Control input $u_m(t)$. **a** PID control for case 1. **b** Proposed control for case 2

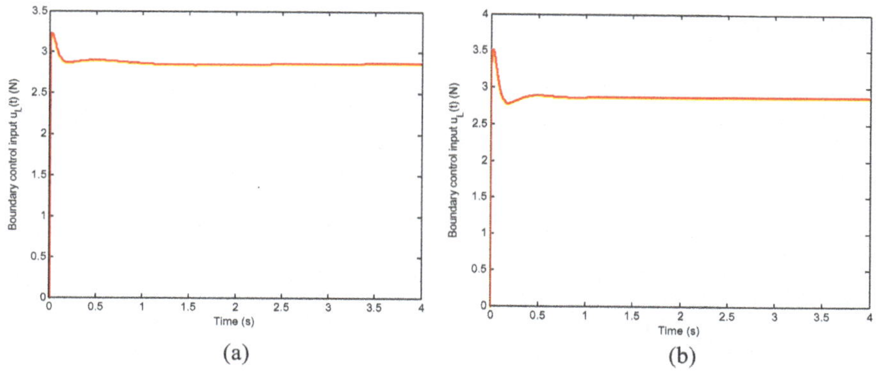

Fig. 9.11 Boundary control input $u_L(t)$. **a** PID control for case 1. **b** Proposed control for case 2

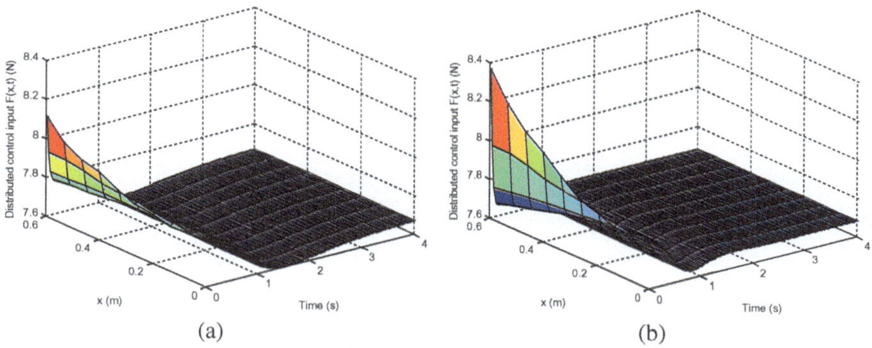

Fig. 9.12 Distributed control input $F(x, t)$. **a** PID control for case 1. **b** Proposed control for case 2

$t \to \infty$, $\theta(t) \to \theta_d$, $\dot{\theta}(t) \to 0$, $r(t) \to r_d$, $\dot{r}(t) \to 0$, $w(L, t) \to 0$, $\dot{w}(L, t) \to 0$, $w(x, t) \to 0$, $\dot{w}(x, t) \to 0$ can be realized in the closed-loop system. Furthermore, prescribed performance bounds are guaranteed under the proposed controller.

9.6 Conclusions

This chapter addresses a control problem of the vehicle-mounted flexible manipulator with the prescribed performance of system outputs. Based on the PDE model, state-feedback control laws have been proposed to not only regulate joint and vehicle positions, but also eliminate flexible link vibration with the prescribed performance. With the proposed control scheme, the output errors converge to predefined arbitrarily small residual sets and the convergence rates are no less than pre-specified values. Simulations have been conducted to validate the control scheme.

References

1. A.K. Kostarigka, Z. Doulgeri, G.A. Rovithakis, Prescribed performance tracking for flexible joint robots with unknown dynamics and elasticity, in *IEEE International Conference on Robotics and Automation*, Saint Paul, Minnesota (USA, 2012), pp. 5365–5370
2. Y. Karayiannidis, Z. Doulgeri, Model-free robot joint position regulation and tracking with prescribed performance guarantees. Robot. Auton. Syst. **60**, 214–226 (2012)
3. C. Bechlioulis, Z. Doulgeri, G. Rovithakis, Prescribed performance adaptive control for robot force/position tracking. Control Appl. (CCA) Intell. Control 920–925 (2009)
4. Z. Doulgeri, Y. Karayiannidis, PID type robot joint position regulation with prescribed performance guaranties, in *IEEE International Conference on Robotics and Automation Anchorage Convention District* (Anchorage, Alaska, USA, 2010), pp. 4137–4142
5. C.P. Bechlioulis, G.A. Rovithakis, Adaptive control with guaranteed transient and steady state tracking error bounds for strict feedback systems. Automatica **45**(2), 532–538 (2009)

Chapter 10
PDE Modeling and Basic Vibration Control of Overhead Crane Bridge

10.1 Introduction

Due to requirements of convenient operation, easy assembly and cost saving, the overhead crane system is highly desirable with the development of automation, and it also plays a significant role in the process of material transportation. However, most of the researches on the control design of overhead crane systems focus on the cable stabilization and trolley position regulation [1–4], which ignore the modeling and vibration control of the overhead crane bridge, and that will lead to a lower control accuracy.

In this chapter, a PDE model is developed for an overhead crane bridge, which is regarded as an infinite-dimensional Timoshenko beam with an attached rigid body. By preserving all the modal information, the PDE model in this chapter can accurately delineate the nonlinear dynamic characteristics of the system to significantly avoid the control spillover problems caused by using traditional truncated ODE models. The consideration of the effect of the rotational inertia, shear deformation, and strong nonlinearity coupling between the bridge and the trolley greatly complicate the control design. Under the proposed control method, states of the overhead crane bridge are globally bounded and finally converge to zero.

10.2 Problem Formulation

The simplified diagram of the overhead crane is presented in Fig. 10.1. In this chapter, an overhead crane bridge is considered. As is shown in Fig. 10.2, the overhead crane bridge model is composed of an overhead crane trolley and a main Timoshenko beam. We consider the trolley as a rigid body. The length of the crane bridge is defined as L. The distance of the rigid body to the left and right wall are defined as L_1 and L_2, respectively ($L = L_1 + L_2$). Denote the beam deflection on the left side of the

X. Xing and J. Liu, *Vibration Control Methods of Mechanical Distributed Parameter Systems*, Springer Tracts in Mechanical Engineering,
https://doi.org/10.1007/978-981-16-1532-0_10

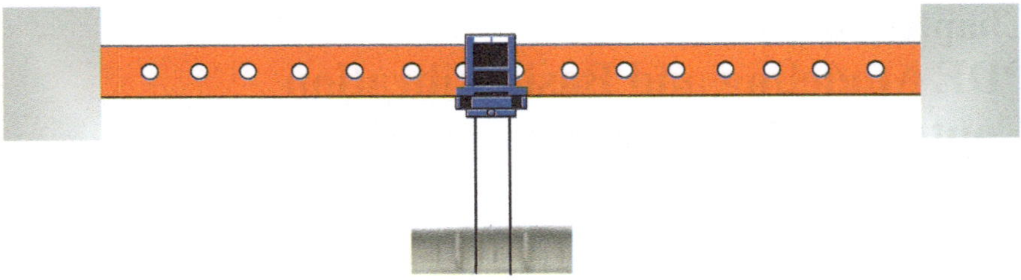

Fig. 10.1 A typical overhead crane

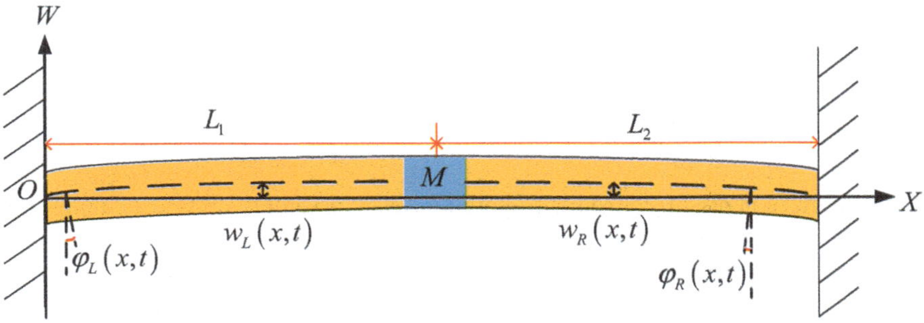

Fig. 10.2 Configuration of an overhead crane bridge with consideration of the trolley dynamics

trolley as $w_L(x,t)$ at position x and time t. Similarly, the beam deflection on the right side of the trolley is expressed by $w_R(x,t)$. In addition, M is the central mass of the trolley, and $w_0(t)$ represents the displacement of the middle-lumped mass. On the left of the trolley, the rotation of the crane bridge beam's cross-section is defined as $\varphi_L(x,t)$ at the position x for time t. The right rotation of the crane bridge beam's cross-section is denoted by $\varphi_R(x,t)$. $\varphi_0(t)$ is the angular deflection of the middle rigid body. Other system parameters are defined as: I_p is the uniform mass moment of inertia of the crane bridge beam's cross-section; ρ is the density of the crane bridge beam; EI is the bending stiffness of the crane bridge beam; J is the moment of inertia of the middle lumped mass; k is a positive constant fixed by the shape of the crane bridge beam's cross-section; A is the cross-sectional area of the crane bridge beam; γ_1 and γ_2 are the friction coefficients of the crane bridge beam caused by bending and rotation, respectively; G denotes the modulus of elasticity of the crane bridge beam in shear; $u(t)$ is the control force and $\tau(t)$ is the control torque at the trolley for suppressing the system vibrations.

Remark 10.1 For simplification, $w(x,t)$ and $\varphi(x,t)$ are denoted as $w(x,t) = \begin{cases} w_L(x,t),\ 0 \le x < L_1 \\ w_0(t),\quad x = L_1 \\ w_R(x,t),\ L_1 < x \le L \end{cases}$ and $\varphi(x,t) = \begin{cases} \varphi_L(x,t),\ 0 \le x < L_1 \\ \varphi_0(t),\quad x = L_1 \\ \varphi_R(x,t),\ L_1 < x \le L \end{cases}$, and we also define following notations: $(\cdot)_x = \frac{\partial(\cdot)}{\partial x}$, $(\cdot)_{xx} = \frac{\partial^2(\cdot)}{\partial^2 x}$, $(\dot{\cdot}) = \frac{\partial(\cdot)}{\partial t}$, $(\ddot{\cdot}) = \frac{\partial^2(\cdot)}{\partial^2 t}$.

The kinetic energy of the overhead crane bridge system is calculated as

$$E_k(t) = \frac{1}{2}\rho \int_0^{L_1} \dot{w}_L^2(x,t)\mathrm{d}x + \frac{1}{2}\rho \int_{L_1}^{L} \dot{w}_R^2(x,t)\mathrm{d}x + \frac{1}{2}M\dot{w}_0^2(t) + \frac{1}{2}I_p \int_0^{L_1} \dot{\varphi}_L^2(x,t)\mathrm{d}x + \frac{1}{2}I_p \int_{L_1}^{L} \dot{\varphi}_R^2(x,t)\mathrm{d}x + \frac{1}{2}J\dot{\varphi}_0^2(t). \tag{10.1}$$

The potential energy can be attained from

$$E_p(t) = \frac{1}{2}K \int_0^{L_1} \left(w_{Lx}(x,t) - \varphi_L(x,t)\right)^2 \mathrm{d}x + \frac{1}{2}K \int_{L_1}^{L} \left(w_{Rx}(x,t) - \varphi_R(x,t)\right)^2 \mathrm{d}x + \frac{1}{2}EI \int_0^{L_1} \varphi_{Lx}^2(x,t)\mathrm{d}x + \frac{1}{2}EI \int_{L_1}^{L} \varphi_{Rx}^2(x,t)\mathrm{d}x \tag{10.2}$$

where $K = kGA$.

The virtual work done by control signals and frictions is expressed by

$$W(t) = u(t)w_0(t) + \tau(t)\varphi_0(t) - \gamma_1 \int_0^{L_1} \dot{w}_L(x,t)w_L(x,t)\mathrm{d}x - \gamma_1 \int_{L_1}^{L} \dot{w}_R(x,t)w_R(x,t)\mathrm{d}x - \gamma_2 \int_0^{L_1} \varphi_L(x,t)\dot{\varphi}_L(x,t)\mathrm{d}x - \gamma_2 \int_{L_1}^{L} \varphi_R(x,t)\dot{\varphi}_R(x,t)\mathrm{d}x. \tag{10.3}$$

According to Eq. (10.1), the variation of kinetic energy of the system can be obtained as follows

$$\int_{t_1}^{t_2} \delta E_k(t)\mathrm{d}t = \rho \int_{t_1}^{t_2}\int_0^{L_1} \dot{w}_L(x,t)\delta\dot{w}_L(x,t)\mathrm{d}x\mathrm{d}t + \rho \int_{t_1}^{t_2}\int_{L_1}^{L} \dot{w}_R(x,t)\delta\dot{w}_R(x,t)\mathrm{d}x\mathrm{d}t$$

$$+ M\int_{t_1}^{t_2}\dot{w}_0(t)\delta\dot{w}_0(t)\mathrm{d}t + I_p\int_{t_1}^{t_2}\int_0^{L_1}\dot{\varphi}_L(x,t)\delta\dot{\varphi}_L(x,t)\mathrm{d}x\mathrm{d}t$$

$$+ I_p\int_{t_1}^{t_2}\int_{L_1}^{L}\dot{\varphi}_R(x,t)\delta\dot{\varphi}_R(x,t)\mathrm{d}x\mathrm{d}t + J\int_{t_1}^{t_2}\dot{\varphi}_0(t)\delta\dot{\varphi}_0(t)\mathrm{d}t \tag{10.4}$$

where δ is the variational operator; t_1 and t_2 are time constants, which satisfy $t_1 < t < t_2$.

Using integration by parts, Eq. (10.4) can be recalculated as

$$\begin{aligned}\int_{t_1}^{t_2}\delta E_k(t)\mathrm{d}t = &-\rho\int_{t_1}^{t_2}\int_0^{L_1}\ddot{w}_L(x,t)\delta w_L(x,t)\mathrm{d}x\mathrm{d}t\\ &-\rho\int_{t_1}^{t_2}\int_{L_1}^{L}\ddot{w}_R(x,t)\delta w_R(x,t)\mathrm{d}x\mathrm{d}t\\ &- M\int_{t_1}^{t_2}\ddot{w}_0(t)\delta w_0(t)\mathrm{d}t - I_p\int_{t_1}^{t_2}\int_0^{L_1}\ddot{\varphi}_L(x,t)\delta\varphi_L(x,t)\mathrm{d}x\mathrm{d}t\\ &- I_p\int_{t_1}^{t_2}\int_{L_1}^{L}\ddot{\varphi}_R(x,t)\delta\varphi_R(x,t)\mathrm{d}x\mathrm{d}t - J\int_{t_1}^{t_2}\ddot{\varphi}_0(t)\delta\varphi_0(t)\mathrm{d}t.\end{aligned} \tag{10.5}$$

The variation of $E_p(t)$ is given as

$$\begin{aligned}\int_{t_1}^{t_2}\delta E_p(t)\mathrm{d}t = &K\int_{t_1}^{t_2}\int_0^{L_1}\big(w_{Lx}(x,t)-\varphi_L(x,t)\big)\big(\delta w_{Lx}(x,t)-\delta\varphi_L(x,t)\big)\mathrm{d}x\mathrm{d}t\\ &+ K\int_{t_1}^{t_2}\int_{L_1}^{L}\big(w_{Rx}(x,t)-\varphi_R(x,t)\big)\big(\delta w_{Rx}(x,t)-\delta\varphi_R(x,t)\big)\mathrm{d}x\mathrm{d}t\\ &+ EI\int_{t_1}^{t_2}\int_0^{L_1}\varphi_{Lx}(x,t)\delta\varphi_{Lx}(x,t)\mathrm{d}x\mathrm{d}t\\ &+ EI\int_{t_1}^{t_2}\int_{L_1}^{L}\varphi_{Rx}(x,t)\delta\varphi_{Rx}(x,t)\mathrm{d}x\mathrm{d}t\end{aligned}$$

$$
\begin{aligned}
&= K\int_{t_1}^{t_2}\int_{0}^{L_1}\big(w_{Lx}(x,t)\delta w_{Lx}(x,t) - w_{Lx}(x,t)\delta\varphi_L(x,t)\\
&- \varphi_L(x,t)\delta w_{Lx}(x,t)\\
&+\varphi_L(x,t)\delta\varphi_L(x,t)\big)\mathrm{d}x\mathrm{d}t\\
&+ K\int_{t_1}^{t_2}\int_{L_1}^{L}\big(w_{Rx}(x,t)\delta w_{Rx}(x,t)\\
&- w_{Rx}(x,t)\delta\varphi_R(x,t)\\
&- \varphi_R(x,t)\delta w_{Rx}(x,t)+\varphi_R(x,t)\delta\varphi_R(x,t)\big)\mathrm{d}x\mathrm{d}t\\
&+ EI\int_{t_1}^{t_2}\int_{0}^{L_1}\varphi_{Lx}(x,t)\delta\varphi_{Lx}(x,t)\mathrm{d}x\mathrm{d}t\\
&+ EI\int_{t_1}^{t_2}\int_{L_1}^{L}\varphi_{Rx}(x,t)\delta\varphi_{Rx}(x,t)\mathrm{d}x\mathrm{d}t
\end{aligned}
\tag{10.6}
$$

which can be rewritten as

$$
\begin{aligned}
\int_{t_1}^{t_2}\delta E_p(t)\mathrm{d}t = &\int_{t_1}^{t_2}\big(EI\varphi_{Lx}(L_1,t)\delta\varphi_L(L_1,t) - EI\varphi_{Lx}(0,t)\delta\varphi_L(0,t)\\
&- EI\varphi_{Rx}(L_1,t)\delta\varphi_R(L_1,t)+EI\varphi_{Rx}(L,t)\delta\varphi_R(L,t)\\
&+Kw_{Lx}(L_1,t)\delta w_L(L_1,t) - K\varphi_L(L_1,t)\delta w_L(L_1,t)\\
&- Kw_{Lx}(0,t)\delta w_L(0,t)+K\varphi_L(0,t)\delta w_L(0,t)\\
&- Kw_{Rx}(L_1,t)\delta w_R(L_1,t)\\
&+ K\varphi_R(L_1,t)\delta w_R(L_1,t)+Kw_{Rx}(L,t)\delta w_R(L,t)\\
&-K\varphi_R(L,t)\delta w_R(L,t)\big)\mathrm{d}t\\
&+\int_{t_1}^{t_2}\int_{0}^{L_1}\big(K\varphi_{Lx}(x,t)\delta w_L(x,t) + K\varphi_L(x,t)\delta\varphi_L(x,t)\\
&- EI\varphi_{Lxx}(x,t)\delta\varphi_L(x,t) - Kw_{Lxx}(x,t)\delta w_L(x,t)\\
&-Kw_{Lx}(x,t)\delta\varphi_L(x,t)\big)\mathrm{d}x\mathrm{d}t\\
&+\int_{t_1}^{t_2}\int_{L_1}^{L}\big(-EI\varphi_{Rxx}(x,t)\delta\varphi_R(x,t)\\
&- Kw_{Rxx}(x,t)\delta w_R(x,t) - Kw_{Rx}(x,t)\delta\varphi_R(x,t)
\end{aligned}
$$

$$+K\varphi_{Rx}(x,t)\delta w_R(x,t)+K\varphi_R(x,t)\delta\varphi_R(x,t)\big)\mathrm{d}x\mathrm{d}t \tag{10.7}$$

The variation of $W(t)$ in Eq. (10.3) is expressed as

$$\begin{aligned}\int_{t_1}^{t_2}\delta W(t)\mathrm{d}t=&\int_{t_1}^{t_2}u(t)\delta v_0(t)\mathrm{d}t+\int_{t_1}^{t_2}\tau(t)\delta\varphi_0(t)\mathrm{d}t\\&-\gamma_1\int_{t_1}^{t_2}\int_0^{L_1}\dot{w}_L(x,t)\delta w_L(x,t)\mathrm{d}x\mathrm{d}t\\&-\gamma_1\int_{t_1}^{t_2}\int_{L_1}^{L}\dot{w}_R(x,t)\delta w_R(x,t)\mathrm{d}x\mathrm{d}t\\&-\gamma_2\int_{t_1}^{t_2}\int_0^{L_1}\varphi_L(x,t)\delta\dot{\varphi}_L(x,t)\mathrm{d}x\mathrm{d}t\\&-\gamma_2\int_{t_1}^{t_2}\int_{L_1}^{L}\varphi_R(x,t)\delta\dot{\varphi}_R(x,t)\mathrm{d}x\mathrm{d}t.\end{aligned} \tag{10.8}$$

Natural boundary conditions of the overhead crane bridge system are

$$w_L(0,t)=w_R(L,t)=\varphi_L(0,t)=\varphi_R(L,t)=0 \tag{10.9}$$

$$w_L(L_1,t)=w_R(L_1,t)=w_0(t),\ \varphi_L(L_1,t)=\varphi_R(L_1,t)=\varphi_0(t). \tag{10.10}$$

According to the following Hamilton principle in Eq. (2.5)

$$\int_{t_1}^{t_2}\big(\delta E_k(t)-\delta E_p(t)+\delta W(t)\big)\mathrm{d}t=0 \tag{10.11}$$

and substituting Eqs. (10.5), (10.7)–(10.10) into Eq. (10.11) and applying integration by parts, we obtain

$$\begin{aligned}&-\int_{t_1}^{t_2}\int_0^{L_1}\bar{A}\delta w_L(x,t)\mathrm{d}x\mathrm{d}t-\int_{t_1}^{t_2}\int_{L_1}^{L}\bar{B}\delta w_R(x,t)\mathrm{d}x\mathrm{d}t\\&-\int_{t_1}^{t_2}\bar{C}\delta w_0(t)\mathrm{d}t-\int_{t_1}^{t_2}\int_0^{L_1}\bar{D}\delta\varphi_L(x,t)\mathrm{d}x\mathrm{d}t\end{aligned}$$

$$-\int_{t_1}^{t_2}\int_{L_1}^{L} \bar{E}\delta\varphi_R(x,t)\mathrm{d}x\mathrm{d}t - \int_{t_1}^{t_2} \bar{F}\delta\varphi_0(t)\mathrm{d}t = 0 \tag{10.12}$$

where

$$\bar{A} = \rho\ddot{w}_L(x,t) - Kw_{Lxx}(x,t) + K\varphi_{Lx}(x,t) + \gamma_1\dot{w}_L(x,t)$$
$$\bar{B} = \rho\ddot{w}_R(x,t) - Kw_{Rxx}(x,t) + K\varphi_{Rx}(x,t) + \gamma_1\dot{w}_R(x,t)$$
$$\bar{C} = M\ddot{w}_0(t) + Kw_{Lx}(L_1,t) - Kw_{Rx}(L_1,t) - u(t)$$
$$\bar{D} = I_p\ddot{\varphi}_L(x,t) - EI\varphi_{Lxx}(x,t) - Kw_{Lx}(x,t) + K\varphi_L(x,t) + \gamma_2\dot{\varphi}_L(x,t)$$
$$\bar{E} = I_p\ddot{\varphi}_R(x,t) - EI\varphi_{Rxx}(x,t) - Kw_{Rx}(x,t) + K\varphi_R(x,t) + \gamma_2\dot{\varphi}_R(x,t)$$
$$\bar{F} = J\ddot{\varphi}_0(t) + EI\varphi_{Lx}(L_1,t) - EI\varphi_{Rx}(L_1,t) - \tau(t)$$

Since $\delta w_L(x,t)$, $\delta w_R(x,t)$, $\delta w_0(t)$, $\delta\varphi_L(x,t)$, $\delta\varphi_R(x,t)$, and $\delta\varphi_0(t)$ are independent variations, we have $\bar{A} = \bar{B} = \bar{C} = \bar{D} = \bar{E} = \bar{F} = 0$. Combing the natural boundary conditions in Eqs. (10.9) and (10.10), the PDE model of the overhead crane bridge system can be obtained as

$$\rho\ddot{w}_L(x,t) - Kw_{Lxx}(x,t) + K\varphi_{Lx}(x,t) + \gamma_1\dot{w}_L(x,t) = 0 \tag{10.13}$$

$$I_p\ddot{\varphi}_L(x,t) - EI\varphi_{Lxx}(x,t) - Kw_{Lx}(x,t) + K\varphi_L(x,t) + \gamma_2\dot{\varphi}_L(x,t) = 0 \tag{10.14}$$

for $\forall(x,t) \in (0, L_1) \times [0, \infty)$, and

$$\rho\ddot{w}_R(x,t) - Kw_{Rxx}(x,t) + K\varphi_{Rx}(x,t) + \gamma_1\dot{w}_R(x,t) = 0 \tag{10.15}$$

$$I_p\ddot{\varphi}_R(x,t) - EI\varphi_{Rxx}(x,t) - Kw_{Rx}(x,t) + K\varphi_R(x,t) + \gamma_2\dot{\varphi}_R(x,t) = 0 \tag{10.16}$$

for $\forall(x,t) \in (L_1, L)\times[0, \infty)$, and boundary conditions of the system can be received as

$$M\ddot{w}_0(t) + Kw_{Lx}(L_1,t) - Kw_{Rx}(L_1,t) - u(t) = 0 \tag{10.17}$$

$$J\ddot{\varphi}_0(t) + EI\varphi_{Lx}(L_1,t) - EI\varphi_{Rx}(L_1,t) - \tau(t) = 0 \tag{10.18}$$

$$w_L(0,t) = w_R(L,t) = \varphi_L(0,t) = \varphi_R(L,t) = 0 \tag{10.19}$$

$$w_L(L_1,t) = w_R(L_1,t) = w_0(t),\ \varphi_L(L_1,t) = \varphi_R(L_1,t) = \varphi_0(t) \tag{10.20}$$

for $\forall t \in [0, \infty)$.

Remark 10.2 It should be pointed out that the strong coupling effect between the central rigid mass and both the left and right beam greatly enhances the nonlinearity of the system and complicates the PDE model established in this chapter, which further increases the difficulty of the following control design. In practical engineering, if the bridge beam is characterized with large slender ratio, the Euler-Bernoulli beam can be used to approximately describe beam's dynamics with ignoring the rotational inertia and shear deformation effects.

Assumption 10.1 It is reasonable to assume that $L_1 > 0$ and $L_2 > 0$, and all system parameters are constants and bounded.

10.3 Control Design and Stability Analysis

In this section, to eliminate the deflection and rotation of the crane bridge beam with a lumped mass in the middle, a control strategy is proposed. The control inputs are all applied on the middle trolley.

Four auxiliary variables are defined as

$$z_1(t) = w_0(t) \tag{10.21}$$

$$z_2(t) = \varphi_0(t) \tag{10.22}$$

$$z_3(t) = \dot{z}_1(t) + \eta_1(t) \tag{10.23}$$

$$z_4(t) = \dot{z}_2(t) + \eta_2(t) \tag{10.24}$$

where $\eta_1(t)$ and $\eta_2(t)$ are virtual control laws needing to be designed.

Design

$$\eta_1(t) = \frac{\alpha}{\beta} z_1(t) = \frac{\alpha}{\beta} w_0(t) \tag{10.25}$$

$$\eta_2(t) = \frac{\alpha}{\beta} z_2(t) = \frac{\alpha}{\beta} \varphi_0(t) \tag{10.26}$$

where $\alpha > 0$ and $\beta > 0$.

A Lyapunov function is selected as

$$V_{11}(t) = \frac{\beta \bar{k}_1}{2} z_1^2(t) + \frac{\beta \bar{k}_2}{2} z_2^2(t) \tag{10.27}$$

where $\bar{k}_1 > 0$ and $\bar{k}_2 > 0$.

Differentiating Eq. (10.27) and substituting Eqs. (10.23) and (10.24) into Eq. (10.27), we have

$$\begin{aligned}\dot{V}_{11}(t) &= \beta\bar{k}_1 z_1(t)\dot{z}_1(t) + \beta\bar{k}_2 z_2(t)\dot{z}_2(t)\\ &= \beta\bar{k}_1 z_1(t)(z_3(t) - \eta_1(t)) + \beta\bar{k}_2 z_2(t)(z_4(t) - \eta_2(t))\\ &= \beta\bar{k}_1 z_1(t)\left(z_3(t) - \frac{\alpha}{\beta}z_1(t)\right) + \beta\bar{k}_2 z_2(t)\left(z_4(t) - \frac{\alpha}{\beta}z_2(t)\right)\\ &= -\alpha\bar{k}_1 z_1^2(t) - \alpha\bar{k}_2 z_2^2(t) + \beta\bar{k}_1 z_1(t)z_3(t) + \beta\bar{k}_2 z_2(t)z_4(t). \end{aligned} \tag{10.28}$$

Then a new Lyapunov function is chosen on the basis of Eq. (10.27) as

$$V_1(t) = V_{11}(t) + \frac{\beta}{2}Mz_3^2(t) + \frac{\beta}{2}Jz_4^2(t). \tag{10.29}$$

The derivative of Eq. (10.29) can be obtained from Eqs. (10.23)–(10.26) and (10.28)

$$\begin{aligned}\dot{V}_1(t) =& -\alpha\bar{k}_1 z_1^2(t) - \alpha\bar{k}_2 z_2^2(t) + \beta\bar{k}_1 z_1(t)z_3(t) + \beta\bar{k}_2 z_2(t)z_4(t)\\ &+ \beta M z_3(t)\dot{z}_3(t) + \beta J z_4(t)\dot{z}_4(t)\\ =& -\alpha\bar{k}_1 z_1^2(t) - \alpha\bar{k}_2 z_2^2(t) + \beta\bar{k}_1 z_1(t)z_3(t) + \beta\bar{k}_2 z_2(t)z_4(t)\\ &+ \beta z_3(t)(M\ddot{w}_0(t) + M\dot{\eta}_1(t)) + \beta z_4(t)(J\ddot{\varphi}_0(t) + J\dot{\eta}_2(t))\\ =& -\alpha\bar{k}_1 z_1^2(t) - \alpha\bar{k}_2 z_2^2(t) + \beta\bar{k}_1 z_1(t)z_3(t) + \beta\bar{k}_2 z_2(t)z_4(t)\\ &+ \beta z_3(t)(-Kw_{Lx}(L_1,t) + Kw_{Rx}(L_1,t) + u(t) + M\dot{\eta}_1(t))\\ &+ \beta z_4(t)(-EI\varphi_{Lx}(L_1,t) + EI\varphi_{Rx}(L_1,t) + \tau(t) + J\dot{\eta}_2(t)). \end{aligned} \tag{10.30}$$

Then control laws are developed as follows

$$u(t) = -\kappa_1 z_3(t) - M\dot{\eta}_1(t) - \bar{k}_1 z_1(t) \tag{10.31}$$

$$\tau(t) = -\kappa_2 z_4(t) - J\dot{\eta}_2(t) - \bar{k}_2 z_2(t). \tag{10.32}$$

Theorem 10.1 *If the PDE model of overhead crane bridge in Eqs.* (10.13)–(10.20) *satisfies Assumption* 10.1 *and system initial conditions are bounded, then the closed-loop system is globally stable, and following properties of the crane bridge system with control laws in Eqs.* (10.31) *and* (10.32) *hold:*

(1) *The states of the closed-loop system are respectively uniformly bounded for* $\forall(x,t) \in [0, L] \times [0, \infty)$;
(2) *The undesirable deflection and rotation of the system can eventually converge to the given compact set which is defined as* $\Omega_1 =$

$\left\{w(x,t), \varphi(x,t) \in R \mid \lim_{t\to\infty}|w(x,t)|, \lim_{t\to\infty}|\varphi(x,t)| = 0, \forall x \in [0,L]\right\}$ *when* $t \to \infty$.

Proof The following Lyapunov function is constructed as

$$V(t) = V_1(t) + V_2(t) + V_3(t) + V_4(t) \tag{10.33}$$

where

$$\begin{aligned}V_2(t) = &\frac{\beta}{2}\rho\int_0^{L_1}\dot{w}_L^2(x,t)\mathrm{d}x + \frac{\beta}{2}I_p\int_0^{L_1}\dot{\varphi}_L^2(x,t)\mathrm{d}x + \frac{\beta}{2}EI\int_0^{L_1}\varphi_{Lx}^2(x,t)\mathrm{d}x\\&+\frac{\beta}{2}K\int_0^{L_1}\left(\varphi_L(x,t)-w_{Lx}(x,t)\right)^2\mathrm{d}x\\&+\frac{\beta}{2}\rho\int_{L_1}^{L}\dot{w}_R^2(x,t)\mathrm{d}x + \frac{\beta}{2}I_p\int_{L_1}^{L}\dot{\varphi}_R^2(x,t)\mathrm{d}x\\&+\frac{\beta}{2}EI\int_{L_1}^{L}\varphi_{Rx}^2(x,t)\mathrm{d}x + \frac{\beta}{2}K\int_{L_1}^{L}\left(\varphi_R(x,t)-w_{Rx}(x,t)\right)^2\mathrm{d}x\end{aligned} \tag{10.34}$$

$$\begin{aligned}V_3(t) = &\alpha\rho\int_0^{L_1}\dot{w}_L(x,t)w_L(x,t)\mathrm{d}x + \alpha I_p\int_0^{L_1}\dot{\varphi}_L(x,t)\varphi_L(x,t)\mathrm{d}x\\&+\alpha\rho\int_{L_1}^{L}\dot{w}_R(x,t)w_R(x,t)\mathrm{d}x + \alpha I_p\int_{L_1}^{L}\dot{\varphi}_R(x,t)\varphi_R(x,t)\mathrm{d}x\end{aligned} \tag{10.35}$$

$$\begin{aligned}V_4(t) = &\frac{\alpha\gamma_1}{2}\int_0^{L_1}w_L^2(x,t)\mathrm{d}x + \frac{\alpha\gamma_2}{2}\int_0^{L_1}\varphi_L^2(x,t)\mathrm{d}x\\&+\frac{\alpha\gamma_1}{2}\int_{L_1}^{L}w_R^2(x,t)\mathrm{d}x + \frac{\alpha\gamma_2}{2}\int_{L_1}^{L}\varphi_R^2(x,t)\mathrm{d}x.\end{aligned} \tag{10.36}$$

Define a positive function as

$$V_0(t) = \int_0^{L_1}\dot{w}_L^2(x,t)\mathrm{d}x + \int_0^{L_1}\dot{\varphi}_L^2(x,t)\mathrm{d}x + \int_0^{L_1}\varphi_{Lx}^2(x,t)\mathrm{d}x + \int_0^{L_1}w_{Lx}^2(x,t)\mathrm{d}x$$

$$+\int_0^{L_1}\varphi_L^2(x,t)\mathrm{d}x+\int_0^{L_1}w_L^2(x,t)\mathrm{d}x+\int_{L_1}^{L}\dot{w}_R^2(x,t)\mathrm{d}x+\int_{L_1}^{L}\dot{\varphi}_R^2(x,t)\mathrm{d}x$$
$$+\int_{L_1}^{L}\varphi_{Rx}^2(x,t)\mathrm{d}x+\int_{L_1}^{L}w_{Rx}^2(x,t)\mathrm{d}x+\int_{L_1}^{L}\varphi_R^2(x,t)\mathrm{d}x+\int_{L_1}^{L}w_R^2(x,t)\mathrm{d}x. \tag{10.37}$$

Remark 10.3 According to Lemma 2.6, we have

$$\int_0^{L_1}w_L^2(x,t)\mathrm{d}x\le 4L_1^2\int_0^{L_1}w_{Lx}^2(x,t)\mathrm{d}x,\quad \int_0^{L_1}\varphi_L^2(x,t)\mathrm{d}x\le 4L_1^2\int_0^{L_1}\varphi_{Lx}^2(x,t)\mathrm{d}x \tag{10.38}$$

$$\int_{L_1}^{L}w_R^2(x,t)\mathrm{d}x\le 4L_2^2\int_{L_1}^{L}w_{Rx}^2(x,t)\mathrm{d}x,\quad \int_{L_1}^{L}\varphi_R^2(x,t)\mathrm{d}x\le 4L_2^2\int_{L_1}^{L}\varphi_{Rx}^2(x,t)\mathrm{d}x \tag{10.39}$$

Using Lemmas 2.2, 2.4, 2.5 and inequalities (10.38) and (10.39), the following inequality can be obtained

$$V_2(t)+V_4(t)\ge\frac{\beta}{2}\rho\int_0^{L_1}\dot{w}_L^2(x,t)\mathrm{d}x+\frac{\beta}{2}I_p\int_0^{L_1}\dot{\varphi}_L^2(x,t)\mathrm{d}x+\frac{\beta}{4}EI\int_0^{L_1}\varphi_{Lx}^2(x,t)\mathrm{d}x$$
$$+\frac{\beta}{16L_1^2}EI\int_0^{L_1}\varphi_L^2(x,t)\mathrm{d}x+\frac{\beta}{2}K\int_0^{L_1}\varphi_L^2(x,t)\mathrm{d}x+\frac{\beta}{2}K\int_0^{L_1}w_{Lx}^2(x,t)\mathrm{d}x$$
$$-\beta K\int_0^{L_1}\varphi_L(x,t)w_{Lx}(x,t)\mathrm{d}x+\frac{\beta}{2}\rho\int_{L_1}^{L}\dot{w}_R^2(x,t)\mathrm{d}x+\frac{\beta}{2}I_p\int_{L_1}^{L}\dot{\varphi}_R^2(x,t)\mathrm{d}x$$
$$+\frac{\beta}{4}EI\int_{L_1}^{L}\varphi_{Rx}^2(x,t)\mathrm{d}x+\frac{\beta}{16L_2^2}EI\int_{L_1}^{L}\varphi_R^2(x,t)\mathrm{d}x+\frac{\beta}{2}K\int_{L_1}^{L}\varphi_R^2(x,t)\mathrm{d}x$$
$$+\frac{\beta}{2}K\int_{L_1}^{L}w_{Rx}^2(x,t)\mathrm{d}x-\beta K\int_{L_1}^{L}\varphi_R(x,t)w_{Rx}(x,t)\mathrm{d}x+\frac{\alpha\gamma_1}{2}\int_0^{L_1}w_L^2(x,t)\mathrm{d}x$$
$$+\frac{\alpha\gamma_2}{2}\int_0^{L_1}\varphi_L^2(x,t)\mathrm{d}x+\frac{\alpha\gamma_1}{2}\int_{L_1}^{L}w_R^2(x,t)\mathrm{d}x+\frac{\alpha\gamma_2}{2}\int_{L_1}^{L}\varphi_R^2(x,t)\mathrm{d}x$$

$$
\begin{aligned}
&\geq \frac{\beta}{2}\rho\int_0^{L_1}\dot{w}_L^2(x,t)\mathrm{d}x+\frac{\beta}{2}I_p\int_0^{L_1}\dot{\varphi}_L^2(x,t)\mathrm{d}x+\frac{\beta}{4}EI\int_0^{L_1}\varphi_{Lx}^2(x,t)\mathrm{d}x\\
&+\int_0^{L_1}\mathbf{\Phi}^{\mathrm{T}}(x,t)\mathbf{\Delta}_1\mathbf{\Phi}(x,t)\mathrm{d}x+\frac{\beta}{2}\rho\int_{L_1}^{L}\dot{w}_R^2(x,t)\mathrm{d}x+\frac{\beta}{2}I_p\int_{L_1}^{L}\dot{\varphi}_R^2(x,t)\mathrm{d}x\\
&+\frac{\beta}{4}EI\int_{L_1}^{L}\varphi_{Rx}^2(x,t)\mathrm{d}x+\int_{L_1}^{L}\mathbf{\Psi}^{\mathrm{T}}(x,t)\mathbf{\Delta}_2\mathbf{\Psi}(x,t)\mathrm{d}x\\
&+\frac{\alpha\gamma_1}{2}\int_0^{L_1}w_L^2(x,t)\mathrm{d}x+\frac{\alpha\gamma_1}{2}\int_{L_1}^{L}w_R^2(x,t)\mathrm{d}x
\end{aligned}
\tag{10.40}
$$

where $\mathbf{\Phi}(x,t)=\left[\varphi_L(x,t)\ w_{Lx}(x,t)\right]^{\mathrm{T}}$, $\mathbf{\Psi}(x,t)=\left[\varphi_R(x,t)\ w_{Rx}(x,t)\right]^{\mathrm{T}}$, and $\mathbf{\Delta}_i\in R^{2\times 2}$, $i=1,2$, is a positive matrix defined as $\mathbf{\Delta}_i=\begin{bmatrix}\frac{\beta EI}{16L_i^2}+\frac{\beta}{2}K+\frac{\alpha\gamma_2}{2} & -\frac{\beta}{2}K\\ -\frac{\beta}{2}K & \frac{\beta}{2}K\end{bmatrix}>0$.

From Eq. (10.37) and inequality (10.40), it is obvious that

$$
V_2(t)+V_4(t)\geq\lambda_1V_0(t)
\tag{10.41}
$$

where $\lambda_1=\min\left\{\frac{\beta}{2}\rho,\frac{\beta}{2}I_p,\frac{\beta}{4}EI,\sigma_{\min}(\mathbf{\Delta}_1),\sigma_{\min}(\mathbf{\Delta}_2),\frac{\alpha\gamma_1}{2}\right\}>0$, and $\sigma_{\min}(\mathbf{\Delta}_i)$ represents the minimal eigenvalue of $\mathbf{\Delta}_i$, $i=1,2$.

Similarly, we have

$$
\begin{aligned}
V_2(t)+V_4(t)&\leq\frac{\beta}{2}\rho\int_0^{L_1}\dot{w}_L^2(x,t)\mathrm{d}x+\frac{\beta}{2}I_p\int_0^{L_1}\dot{\varphi}_L^2(x,t)\mathrm{d}x+\frac{\beta}{2}EI\int_0^{L_1}\varphi_{Lx}^2(x,t)\mathrm{d}x\\
&+\left(\beta K+\frac{\alpha\gamma_2}{2}\right)\int_0^{L_1}\varphi_L^2(x,t)\mathrm{d}x+\beta K\int_0^{L_1}w_{Lx}^2(x,t)\mathrm{d}x+\frac{\beta}{2}\rho\int_{L_1}^{L}\dot{w}_R^2(x,t)\mathrm{d}x\\
&+\frac{\beta}{2}I_p\int_{L_1}^{L}\dot{\varphi}_R^2(x,t)\mathrm{d}x+\frac{\beta}{2}EI\int_{L_1}^{L}\varphi_{Rx}^2(x,t)\mathrm{d}x+\left(\beta K+\frac{\alpha\gamma_2}{2}\right)\int_{L_1}^{L}\varphi_R^2(x,t)\mathrm{d}x\\
&+\beta K\int_{L_1}^{L}w_{Rx}^2(x,t)\mathrm{d}x+\frac{\alpha\gamma_1}{2}\int_0^{L_1}w_L^2(x,t)\mathrm{d}x+\frac{\alpha\gamma_1}{2}\int_{L_1}^{L}w_R^2(x,t)\mathrm{d}x\\
&\leq\lambda_2V_0(t)
\end{aligned}
\tag{10.42}
$$

where $\lambda_2 = \max\left\{\frac{1}{2}\beta\rho, \frac{\beta}{2}I_p, \frac{\beta}{2}EI, \beta K, \beta K + \frac{\alpha\gamma_2}{2}, \frac{\alpha\gamma_1}{2}\right\} > 0$.

According to Eq. (10.35), we know

$$\begin{aligned}|V_3(t)| \le\ & \frac{\alpha\rho}{2}\int_0^{L_1} \dot{w}_L^2(x,t)\mathrm{d}x + \frac{\alpha\rho}{2}\int_0^{L_1} w_L^2(x,t)\mathrm{d}x \\ & + \frac{\alpha I_p}{2}\int_0^{L_1} \dot{\varphi}_L^2(x,t)\mathrm{d}x + \frac{\alpha I_p}{2}\int_0^{L_1} \varphi_L^2(x,t)\mathrm{d}x \\ & + \frac{\alpha\rho}{2}\int_{L_1}^{L} \dot{w}_R^2(x,t)\mathrm{d}x + \frac{\alpha\rho}{2}\int_{L_1}^{L} w_R^2(x,t)\mathrm{d}x \\ & + \frac{\alpha I_p}{2}\int_{L_1}^{L} \dot{\varphi}_R^2(x,t)\mathrm{d}x + \frac{\alpha I_p}{2}\int_{L_1}^{L} \varphi_R^2(x,t)\mathrm{d}x \\ \le\ & \lambda_3 V_0(t) \end{aligned} \tag{10.43}$$

in which $\lambda_3 = \max\left\{\frac{1}{2}\alpha I_p, \frac{1}{2}\alpha\rho\right\} > 0$.

Let $\lambda_1 > \lambda_3$ by choosing proper control coefficients α and β, and then we have

$$0 < (\lambda_1 - \lambda_3)V_0(t) \le V_2(t) + V_3(t) + V_4(t) \le (\lambda_2 + \lambda_3)V_0(t). \tag{10.44}$$

Define $a_1 = \lambda_1 - \lambda_3 > 0$ and $a_2 = \lambda_2 + \lambda_3 > 0$

$$\lambda_5(V_0(t) + V_1(t)) \le V(t) \le \lambda_4(V_0(t) + V_1(t)) \tag{10.45}$$

where $\lambda_4 = \max\{a_2, 1\}$ and $\lambda_5 = \min\{a_1, 1\}$.

With Eqs. (10.31) and (10.32), the time derivative of $V_1(t)$ can be written as

$$\begin{aligned}\dot{V}_1(t) =& -\alpha\bar{k}_1 z_1^2(t) - \alpha\bar{k}_2 z_2^2(t) + \beta\bar{k}_1 z_1(t)z_3(t) + \beta\bar{k}_2 z_2(t)z_4(t) \\ & +\beta z_3(t)\big(-Kw_{Lx}(L_1,t) + Kw_{Rx}(L_1,t) - \kappa_1 z_3(t) - \bar{k}_1 z_1(t)\big) \\ & +\beta z_4(t)\big(-EI\varphi_{Lx}(L_1,t) + EI\varphi_{Rx}(L_1,t) - \kappa_2 z_4(t) - \bar{k}_2 z_2(t)\big)\end{aligned} \tag{10.46}$$

which equals

$$\begin{aligned}\dot{V}_1(t) =\ & \beta z_3(t)(-Kw_{Lx}(L_1,t) + Kw_{Rx}(L_1,t)) \\ & + \beta z_4(t)(-EI\varphi_{Lx}(L_1,t) + EI\varphi_{Rx}(L_1,t)) \\ & - \alpha\bar{k}_1 z_1^2(t) - \beta\kappa_1 z_3^2(t) - \alpha\bar{k}_2 z_2^2(t) - \beta\kappa_2 z_4^2(t).\end{aligned} \tag{10.47}$$

Differentiating Eq. (10.34) and substituting Eqs. (10.13)–(10.16), we have

$$
\begin{aligned}
\dot{V}_2(t) =&\beta \int_0^{L_1} \dot{w}_L(x,t)(Kw_{Lxx}(x,t) - K\varphi_{Lx}(x,t) - \gamma_1 \dot{w}_L(x,t))\mathrm{d}x \\
&+ \beta \int_0^{L_1} \dot{\varphi}_L(x,t)(EI\varphi_{Lxx}(x,t) - \gamma_2 \dot{\varphi}_L(x,t))\mathrm{d}x \\
&+ \beta EI \int_0^{L_1} \varphi_{Lx}(x,t)\dot{\varphi}_{Lx}(x,t)\mathrm{d}x \\
&+ \beta K \int_0^{L_1} \big(-\varphi_L(x,t)\dot{w}_{Lx}(x,t) + w_{Lx}(x,t)\dot{w}_{Lx}(x,t)\big)\mathrm{d}x \\
&+ \beta \int_{L_1}^{L} \dot{w}_R(x,t)(Kw_{Rxx}(x,t) - K\varphi_{Rx}(x,t) - \gamma_1 \dot{w}_R(x,t))\mathrm{d}x \\
&+ \beta \int_{L_1}^{L} \dot{\varphi}_R(x,t)(EI\varphi_{Rxx}(x,t) - \gamma_2 \dot{\varphi}_R(x,t))\mathrm{d}x \\
&+ \beta EI \int_{L_1}^{L} \varphi_{Rx}(x,t)\dot{\varphi}_{Rx}(x,t)\mathrm{d}x \\
&+ \beta K \int_{L_1}^{L} \big(-\varphi_R(x,t)\dot{w}_{Rx}(x,t) + w_{Rx}(x,t)\dot{w}_{Rx}(x,t)\big)\mathrm{d}x \\
&= \beta K\dot{w}_L(L_1,t)(w_{Lx}(L_1,t) - \varphi_L(L_1,t)) + \beta EI\dot{\varphi}_L(L_1,t)\varphi_{Lx}(L_1,t) \\
&- \beta K\dot{w}_R(L_1,t)(w_{Rx}(L_1,t) - \varphi_R(L_1,t)) - \beta EI\dot{\varphi}_R(L_1,t)\varphi_{Rx}(L_1,t) \\
&- \beta\gamma_1 \int_0^{L_1} \dot{w}_L^2(x,t)\mathrm{d}x - \beta\gamma_2 \int_0^{L_1} \dot{\varphi}_L^2(x,t)\mathrm{d}x \\
&- \beta\gamma_1 \int_{L_1}^{L} \dot{w}_R^2(x,t)\mathrm{d}x - \beta\gamma_2 \int_{L_1}^{L} \dot{\varphi}_R^2(x,t)\mathrm{d}x.
\end{aligned} \tag{10.48}
$$

By using Eq. (10.35), $\dot{V}_3(t)$ is obtained as

$$
\dot{V}_3(t) = \alpha \int_0^{L_1} (Kw_{Lxx}(x,t) - K\varphi_{Lx}(x,t) - \gamma_1 \dot{w}_L(x,t))w_L(x,t)\mathrm{d}x
$$

$$
\begin{aligned}
&+\alpha\rho\int_0^{L_1}\dot{w}_L^2(x,t)\mathrm{d}x\\
&+\alpha\int_0^{L_1}(EI\varphi_{Lxx}(x,t)+Kw_{Lx}(x,t)\\
&-K\varphi_L(x,t)-\gamma_2\dot{\varphi}_L(x,t))\varphi_L(x,t)\mathrm{d}x\\
&+\alpha I_p\int_0^{L_1}\dot{\varphi}_L^2(x,t)\mathrm{d}x\\
&+\alpha\int_{L_1}^{L}(Kw_{Rxx}(x,t)-K\varphi_{Rx}(x,t)\\
&-\gamma_1\dot{w}_R(x,t))w_R(x,t)\mathrm{d}x\\
&+\alpha\int_{L_1}^{L}(EI\varphi_{Rxx}(x,t)+Kw_{Rx}(x,t)\\
&-K\varphi_R(x,t)-\gamma_2\dot{\varphi}_R(x,t))\varphi_R(x,t)\mathrm{d}x\\
&+\alpha I_p\int_{L_1}^{L}\dot{\varphi}_R^2(x,t)\mathrm{d}x+\alpha\rho\int_{L_1}^{L}\dot{w}_R^2(x,t)\mathrm{d}x. \qquad (10.49)
\end{aligned}
$$

Integrating Eq. (10.49) by parts leads to

$$
\begin{aligned}
\dot{V}_3(t)=&\ \alpha EI\varphi_L(L_1,t)\varphi_{Lx}(L_1,t)-\alpha EI\int_0^{L_1}\varphi_{Lx}^2(x,t)\mathrm{d}x\\
&+2\alpha K\int_0^{L_1}\varphi_L(x,t)w_{Lx}(x,t)\mathrm{d}x\\
&-\alpha K\int_0^{L_1}\varphi_L^2(x,t)\mathrm{d}x+\alpha I_p\int_0^{L_1}\dot{\varphi}_L^2(x,t)\mathrm{d}x-\alpha\gamma_2\int_0^{L_1}\varphi_L(x,t)\dot{\varphi}_L(x,t)\mathrm{d}x\\
&-\alpha EI\varphi_R(L_1,t)\varphi_{Rx}(L_1,t)-\alpha EI\int_{L_1}^{L}\varphi_{Rx}^2(x,t)\mathrm{d}x\\
&+2\alpha K\int_{L_1}^{L}\varphi_R(x,t)w_{Rx}(x,t)\mathrm{d}x
\end{aligned}
$$

$$
\begin{aligned}
&-\alpha K\int_{L_1}^{L}\varphi_R^2(x,t)\mathrm{d}x+\alpha I_p\int_{L_1}^{L}\dot{\varphi}_R^2(x,t)\mathrm{d}x-\alpha\gamma_2\int_{L_1}^{L}\varphi_R(x,t)\dot{\varphi}_R(x,t)\mathrm{d}x\\
&+\alpha K w_L(L_1,t)w_{Lx}(L_1,t)-\alpha K\int_0^{L_1}w_{Lx}^2(x,t)\mathrm{d}x-\alpha K w_L(L_1,t)\varphi_L(L_1,t)\\
&+\alpha\rho\int_0^{L_1}\dot{w}_L^2(x,t)\mathrm{d}x-\alpha\gamma_1\int_0^{L_1}w_L(x,t)\dot{w}_L(x,t)\mathrm{d}x\\
&-\alpha K w_R(L_1,t)w_{Rx}(L_1,t)\\
&-\alpha K\int_{L_1}^{L}w_{Rx}^2(x,t)\mathrm{d}x+\alpha K w_R(L_1,t)\varphi_R(L_1,t)\\
&+\alpha\rho\int_{L_1}^{L}\dot{w}_R^2(x,t)\mathrm{d}x-\alpha\gamma_1\int_{L_1}^{L}w_R(x,t)\dot{w}_R(x,t)\mathrm{d}x. \qquad (10.50)
\end{aligned}
$$

The derivative of Eq. (10.36) is attained as

$$
\begin{aligned}
\dot{V}_4(t)=&\alpha\gamma_1\int_0^{L_1}w_L(x,t)\dot{w}_L(x,t)\mathrm{d}x+\alpha\gamma_2\int_0^{L_1}\varphi_L(x,t)\dot{\varphi}_L(x,t)\mathrm{d}x\\
&+\alpha\gamma_1\int_{L_1}^{L}w_R(x,t)\dot{w}_R(x,t)\mathrm{d}x+\alpha\gamma_2\int_{L_1}^{L}\varphi_R(x,t)\dot{\varphi}_R(x,t)\mathrm{d}x. \qquad (10.51)
\end{aligned}
$$

Combing Eqs. (10.50), (10.51), (10.47), and (10.48), we have

$$
\begin{aligned}
\dot{V}(t)\leq&-\beta\gamma_1\int_0^{L_1}\dot{w}_L^2(x,t)\mathrm{d}x-\beta\gamma_2\int_0^{L_1}\dot{\varphi}_L^2(x,t)\mathrm{d}x\\
&-\beta\gamma_1\int_{L_1}^{L}\dot{w}_R^2(x,t)\mathrm{d}x-\beta\gamma_2\int_{L_1}^{L}\dot{\varphi}_R^2(x,t)\mathrm{d}x\\
&-\alpha EI\int_0^{L_1}\varphi_{Lx}^2(x,t)\mathrm{d}x+2\alpha K\int_0^{L_1}\varphi_L(x,t)w_{Lx}(x,t)\mathrm{d}x\\
&-\alpha K\int_0^{L_1}\varphi_L^2(x,t)\mathrm{d}x+\alpha I_p\int_0^{L_1}\dot{\varphi}_L^2(x,t)\mathrm{d}x
\end{aligned}
$$

$$
\begin{aligned}
& -\alpha EI \int_{L_1}^{L} \varphi_{Rx}^2(x,t)\mathrm{d}x + 2\alpha K \int_{L_1}^{L} \varphi_R(x,t) w_{Rx}(x,t)\mathrm{d}x \\
& -\alpha K \int_{L_1}^{L} \varphi_R^2(x,t)\mathrm{d}x + \alpha I_p \int_{L_1}^{L} \dot{\varphi}_R^2(x,t)\mathrm{d}x \\
& -\alpha K \int_{0}^{L_1} w_{Lx}^2(x,t)\mathrm{d}x + \alpha\rho \int_{0}^{L_1} \dot{w}_L^2(x,t)\mathrm{d}x \\
& -\alpha K \int_{L_1}^{L} w_{Rx}^2(x,t)\mathrm{d}x + \alpha\rho \int_{L_1}^{L} \dot{w}_R^2(x,t)\mathrm{d}x \\
& -\alpha\bar{k}_1 z_1^2(t) - \beta\kappa_1 z_3^2(t) - \alpha\bar{k}_2 z_2^2(t) - \beta\kappa_2 z_4^2(t) \\
= & -(\beta\gamma_1 - \alpha\rho) \int_{0}^{L_1} \dot{w}_L^2(x,t)\mathrm{d}x - \left(\beta\gamma_2 - \alpha I_p\right) \int_{0}^{L_1} \dot{\varphi}_L^2(x,t)\mathrm{d}x \\
& -(\beta\gamma_1 - \alpha\rho) \int_{L_1}^{L} \dot{w}_R^2(x,t)\mathrm{d}x - \left(\beta\gamma_2 - \alpha I_p\right) \int_{L_1}^{L} \dot{\varphi}_R^2(x,t)\mathrm{d}x \\
& -\alpha EI \int_{0}^{L_1} \varphi_{Lx}^2(x,t)\mathrm{d}x + 2\alpha K \int_{0}^{L_1} \varphi_L(x,t) w_{Lx}(x,t)\mathrm{d}x \\
& -\alpha K \int_{0}^{L_1} \varphi_L^2(x,t)\mathrm{d}x \\
& -\alpha EI \int_{L_1}^{L} \varphi_{Rx}^2(x,t)\mathrm{d}x + 2\alpha K \int_{L_1}^{L} \varphi_R(x,t) w_{Rx}(x,t)\mathrm{d}x \\
& -\alpha K \int_{L_1}^{L} \varphi_R^2(x,t)\mathrm{d}x \\
& -\alpha K \int_{0}^{L_1} w_{Lx}^2(x,t)\mathrm{d}x - \alpha K \int_{L_1}^{L} w_{Rx}^2(x,t)\mathrm{d}x - \alpha\bar{k}_1 z_1^2(t) \\
& -\beta\kappa_1 z_3^2(t) - \alpha\bar{k}_2 z_2^2(t) - \beta\kappa_2 z_4^2(t).
\end{aligned}
\tag{10.52}
$$

Then we have

$$
\begin{aligned}
\dot{V}(t) \le & -(\beta\gamma_1 - \alpha\rho)\int_0^{L_1} \dot{w}_L^2(x,t)\mathrm{d}x - \left(\beta\gamma_2 - \alpha I_p\right)\int_0^{L_1} \dot{\varphi}_L^2(x,t)\mathrm{d}x \\
& - (\beta\gamma_1 - \alpha\rho)\int_{L_1}^{L} \dot{w}_R^2(x,t)\mathrm{d}x - \left(\beta\gamma_2 - \alpha I_p\right)\int_{L_1}^{L} \dot{\varphi}_R^2(x,t)\mathrm{d}x \\
& - \alpha EI\int_0^{L_1} \varphi_{Lx}^2(x,t)\mathrm{d}x - \left(\alpha K - \frac{2\alpha K}{\varepsilon_1}\right)\int_0^{L_1} \varphi_L^2(x,t)\mathrm{d}x \\
& - \alpha EI\int_{L_1}^{L} \varphi_{Rx}^2(x,t)\mathrm{d}x \\
& - \left(\alpha K - \frac{2\alpha K}{\varepsilon_2}\right)\int_{L_1}^{L} \varphi_R^2(x,t)\mathrm{d}x - (\alpha K - 2\alpha K\varepsilon_1)\int_0^{L_1} w_{Lx}^2(x,t)\mathrm{d}x \\
& - (\alpha K - 2\alpha K\varepsilon_2)\int_{L_1}^{L} w_{Rx}^2(x,t)\mathrm{d}x - \alpha\bar{k}_1 z_1^2(t) \\
& - \beta\kappa_1 z_3^2(t) - \alpha\bar{k}_2 z_2^2(t) - \beta\kappa_2 z_4^2(t).
\end{aligned}
\tag{10.53}
$$

where $\varepsilon_1 > 0$ and $\varepsilon_2 > 0$.

Substituting inequalities (10.38) and (10.39) into inequality (10.53) and using Lemmas 2.2, 2.4, 2.5, the following inequality can be achieved

$$
\begin{aligned}
\dot{V}(t) \le & -(\beta\gamma_1 - \alpha\rho)\int_0^{L_1} \dot{w}_L^2(x,t)\mathrm{d}x - \left(\beta\gamma_2 - \alpha I_p\right)\int_0^{L_1} \dot{\varphi}_L^2(x,t)\mathrm{d}x \\
& - (\beta\gamma_1 - \alpha\rho)\int_{L_1}^{L} \dot{w}_R^2(x,t)\mathrm{d}x - \left(\beta\gamma_2 - \alpha I_p\right)\int_{L_1}^{L} \dot{\varphi}_R^2(x,t)\mathrm{d}x \\
& - \frac{\alpha}{2}EI\int_0^{L_1} \varphi_{Lx}^2(x,t)\mathrm{d}x - \left(\alpha K + \frac{\alpha}{8L_1^2}EI - \frac{2\alpha K}{\varepsilon_1}\right)\int_0^{L_1} \varphi_L^2(x,t)\mathrm{d}x \\
& - \frac{\alpha}{2}EI\int_{L_1}^{L} \varphi_{Rx}^2(x,t)\mathrm{d}x - \left(\alpha K + \frac{\alpha}{8L_2^2}EI - \frac{2\alpha K}{\varepsilon_2}\right)\int_{L_1}^{L} \varphi_R^2(x,t)\mathrm{d}x \\
& - \xi_1\int_0^{L_1} w_L^2(x,t)\mathrm{d}x
\end{aligned}
$$

$$
\begin{aligned}
&- \xi_2 \int_{L_1}^{L} w_R^2(x,t)\mathrm{d}x - \left(\alpha K - 2\alpha K\varepsilon_1 - 4L_1^2\xi_1\right)\int_0^{L_1} w_{Lx}^2(x,t)\mathrm{d}x \\
&- \left(\alpha K - 2\alpha K\varepsilon_2 - 4L_2^2\xi_2\right)\int_{L_1}^{L} w_{Rx}^2(x,t)\mathrm{d}x - \alpha\bar{k}_1 z_1^2(t) - \beta\kappa_1 z_3^2(t) \\
&- \alpha\bar{k}_2 z_2^2(t) - \beta\kappa_2 z_4^2(t)
\end{aligned}
\tag{10.54}
$$

where $\xi_1 > 0$, $\xi_2 > 0$, and α, β, ε_1, ε_2, ξ_1, ξ_2 are selected to satisfy $\beta\gamma_1 - \alpha\rho > 0$, $\beta\gamma_2 - \alpha I_p > 0$, $\alpha K + \frac{\alpha}{8L_1^2}EI - \frac{2\alpha K}{\varepsilon_1} > 0$, $\alpha K + \frac{\alpha}{8L_2^2}EI - \frac{2\alpha K}{\varepsilon_2} > 0$, $\alpha K - 2\alpha K\varepsilon_1 - 4L_1^2\xi_1 > 0$, $\alpha K - 2\alpha K\varepsilon_2 - 4L_2^2\xi_2 > 0$.

Then from inequality (10.54), we have

$$
\dot{V}(t) \le -\lambda_6(V_0(t) + V_1(t)) \tag{10.55}
$$

where

$$
\begin{aligned}
\lambda_6 =\min\Bigg\{&\beta\gamma_1 - \alpha\rho, \beta\gamma_2 - \alpha I_p, \frac{\alpha}{2}EI, \alpha K + \frac{\alpha}{8L_1^2}EI - \frac{2\alpha K}{\varepsilon_1}, \\
&\alpha K + \frac{\alpha}{8L_2^2}EI - \frac{2\alpha K}{\varepsilon_2}, \alpha K - 2\alpha K\varepsilon_1 - 4L_1^2\xi_1, \\
&\alpha K - 2\alpha K\varepsilon_2 - 4L_2^2\xi_2, \xi_1, \xi_2, \frac{2\kappa_1}{M}, \frac{2\kappa_2}{J}, \frac{2\alpha}{\beta}\Bigg\} > 0.
\end{aligned}
$$

According to inequality (10.55), we have

$$
\dot{V}(t) \le -\lambda V(t) \tag{10.56}
$$

where $\lambda = \frac{\lambda_6}{\lambda_4} > 0$.

Multiplying inequality (10.56) by $e^{\lambda t}$ and integrating yield

$$
V(t) \le V(0)e^{-\lambda t}. \tag{10.57}
$$

From inequality (10.57), we can know that $V(t)$ is bounded. Then using Lemmas 2.2 and 2.5, we have

$$
\frac{1}{L_1}\varphi_L^2(x,t) \le \int_0^{L_1} \varphi_{Lx}^2(x,t)\mathrm{d}x \le \frac{1}{\lambda_5}V(t) \le \frac{1}{\lambda_5}V(0)e^{-\lambda t} \tag{10.58}
$$

$$
\frac{\beta\bar{k}_2}{2}\varphi_0^2(t) \le \frac{1}{\lambda_5}V(t) \le \frac{1}{\lambda_5}V(0)e^{-\lambda t} \tag{10.59}
$$

$$\frac{1}{L_2}\varphi_R^2(x,t) \le \int_{L_1}^{L} \varphi_{Rx}^2(x,t)\mathrm{d}x \le \frac{1}{\lambda_5}V(t) \le \frac{1}{\lambda_5}V(0)e^{-\lambda t} \tag{10.60}$$

$$\frac{1}{L_1}w_L^2(x,t) \le \int_{0}^{L_1} w_{Lx}^2(x,t)\mathrm{d}x \le \frac{1}{\lambda_5}V(t) \le \frac{1}{\lambda_5}V(0)e^{-\lambda t} \tag{10.61}$$

$$\frac{\beta\bar{k}_1}{2}w_0^2(t) \le \frac{1}{\lambda_5}V(t) \le \frac{1}{\lambda_5}V(0)e^{-\lambda t} \tag{10.62}$$

$$\frac{1}{L_2}w_R^2(x,t) \le \int_{L_1}^{L} w_{Rx}^2(x,t)\mathrm{d}x \le \frac{1}{\lambda_5}V(t) \le \frac{1}{\lambda_5}V(0)e^{-\lambda t}. \tag{10.63}$$

Then according to inequalities (10.58)–(10.63), we get

$$\left|\varphi_L(x,t)\right|, \left|w_L(x,t)\right| \le \sqrt{\frac{L_1}{\lambda_5}V(0)e^{-\lambda t}} \tag{10.64}$$

$$\left|\varphi_R(x,t)\right|, \left|w_R(x,t)\right| \le \sqrt{\frac{L_2}{\lambda_5}V(0)e^{-\lambda t}} \tag{10.65}$$

$$|w_0(t)| \le \sqrt{\frac{2}{\beta\bar{k}_1\lambda_5}V(0)e^{-\lambda t}} \tag{10.66}$$

$$|\varphi_0(t)| \le \sqrt{\frac{2}{\beta\bar{k}_2\lambda_5}V(0)e^{-\lambda t}} \tag{10.67}$$

and

$$\lim_{t\to\infty}\left|\varphi_L(x,t)\right|, \lim_{t\to\infty}\left|w_L(x,t)\right| = 0 \tag{10.68}$$

$$\lim_{t\to\infty}\left|\varphi_R(x,t)\right|, \lim_{t\to\infty}\left|w_R(x,t)\right| = 0 \tag{10.69}$$

$$\lim_{t\to\infty}|w_0(t)| = 0, \ \lim_{t\to\infty}|\varphi_0(t)| = 0 \tag{10.70}$$

Therefore, if the boundedness of system initial conditions can be guaranteed, states of the closed-loop system are bounded subject to the developed control laws in Eqs. (10.31) and (10.32), and can eventually converge to small regions of zero by adjusting corresponding control parameters. Then the Theorem 10.1 has been completely proven.

10.4 Simulations

Numerical simulations are conducted to verify the effectiveness of the control strategy in Eqs. (10.31) and (10.32). The time and space steps of the finite difference method are set as $\Delta t = 0.02$ s and $\Delta x = 0.2$ m.

System parameters are given as $L = 4\,\text{m}$, $L_1 = 2\,\text{m}$, $L_2 = 2\,\text{m}$, $\rho = 2\,\text{kg/m}$, $\gamma_1 = 2.5\,\text{kg/m}$, $\gamma_2 = 2.5\,\text{kg/m}$, $M = 60\,\text{kg}$, $EI = 50\,\text{Nm}^2$, $I_p = 2\,\text{kgm}^2$, $K = 0.5\,\text{N}$, $J = 1\,\text{kgm}^2$, and other control parameters are selected as: $\alpha = 1$, $\beta = 10$, $\bar{k}_1 = 30$, $\bar{k}_2 = 0.5$, $\kappa_1 = 30$, $\kappa_2 = 0.5$. System initial states are given as $\dot{w}(x, 0) = 0$,
$\dot{\varphi}(x, 0) = 0$, $w(x, 0) = \begin{cases} 0.3x, & \text{if } 0 \le x < L_1 \\ 0.3L_1, & \text{if } x = L_1 \\ -0.3\frac{L_1}{L_2}(x - L), & \text{if } L_1 < x \le L \end{cases}$, $\varphi(x, 0) = 0$.

The vibration and rotation of the overhead crane bridge without control are given by Figs. 10.3, 10.4, 10.5, 10.6 and 10.7. Vibration displacement $w(x, t)$ and rotation angle $\varphi(x, t)$ of the open-loop overhead crane bridge system are presented in Figs. 10.3 and 10.4. Figures 10.5, 10.6 and 10.7 show $w_L\left(\frac{L_1}{2}, t\right)$, $\varphi_L\left(\frac{L_1}{2}, t\right)$, $w_0(t)$, $\varphi_0(t)$, $w_R\left(L_1 + \frac{L_2}{2}, t\right)$, $w_R\left(L_1 + \frac{L_2}{2}, t\right)$ of the crane bridge system respectively without control. It can be seen that the oscillatory of the system without control inputs is obvious and needs to be suppressed by designing control laws.

States of the closed-loop system under the control laws in Eqs. (10.31) and (10.32) are as shown in Figs. 10.8, 10.9, 10.10, 10.11 and 10.12. Bending displacement $w(x, t)$ and rotation $\varphi(x, t)$ of the system under the proposed control laws are given in Figs. 10.8 and 10.9. States $w_L\left(\frac{L_1}{2}, t\right)$, $\varphi_L\left(\frac{L_1}{2}, t\right)$, $w_0(t)$, $\varphi_0(t)$, $w_R\left(L_1 + \frac{L_2}{2}, t\right)$, $\varphi_R\left(L_1 + \frac{L_2}{2}, t\right)$ of the closed-loop system are presented in Figs. 10.10, 10.11 and

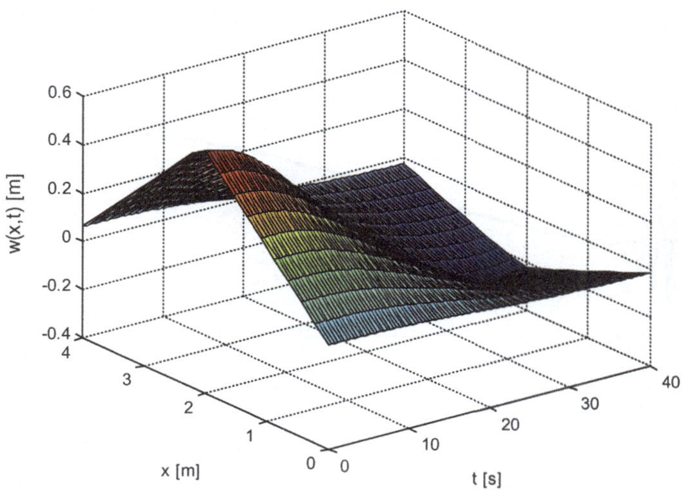

Fig. 10.3 Bending displacement $w(x, t)$ of the crane bridge without control

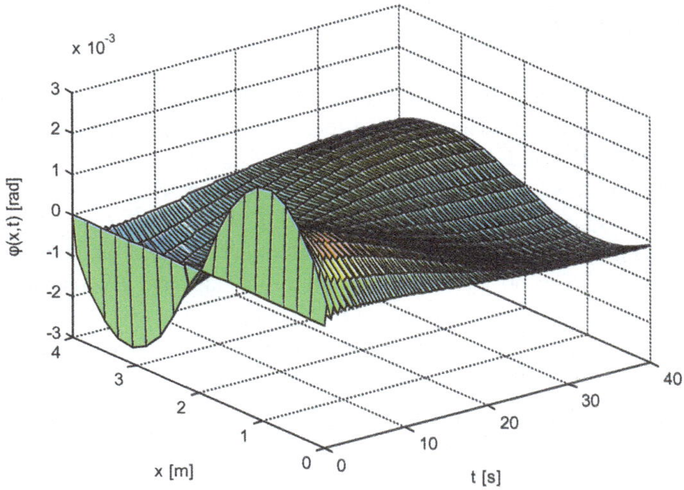

Fig. 10.4 Rotation $\varphi(x, t)$ of the crane bridge without control

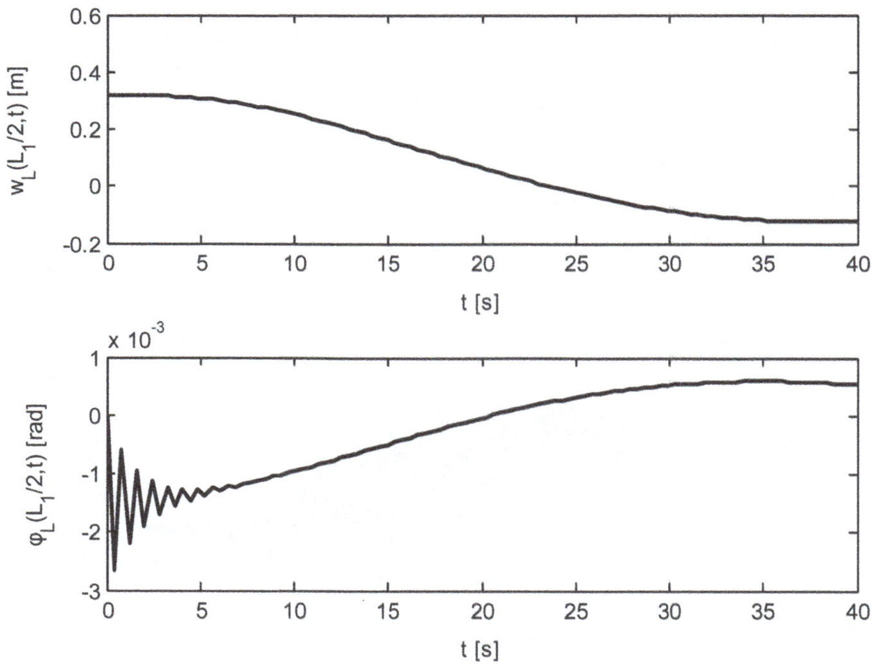

Fig. 10.5 Bending displacement $w_L\left(\frac{L_1}{2}, t\right)$ and rotation $\varphi_L\left(\frac{L_1}{2}, t\right)$ of the crane bridge without control

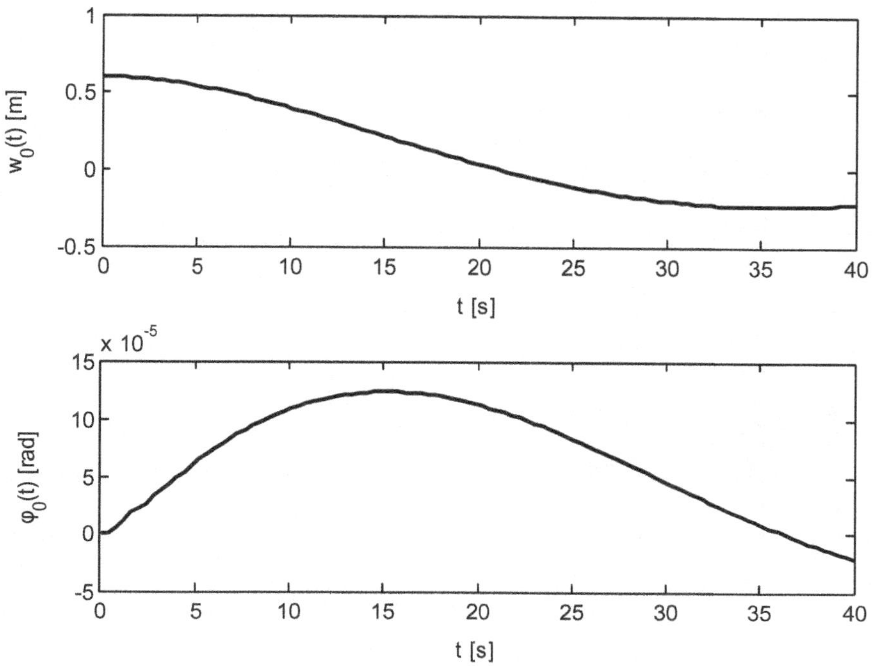

Fig. 10.6 Displacement $w_0(t)$ and rotation $\varphi_0(t)$ of the middle rigid body without control

10.12. By observing Figs. 10.10, 10.11 and 10.12, it is obvious that the vibration of the trolley can be suppressed under the designed control.

Control input force $u(t)$ and control torque $\tau(t)$ are shown in Figs. 10.13 and 10.14, respectively.

From Figs. 10.3, 10.4, 10.5, 10.6, 10.7, 10.8, 10.9, 10.10, 10.11, 10.12, 10.13 and 10.14, it is known that the states of the crane bridge system can converge to zero, and the undesirable vibration and rotation of the system are suppressed with the proposed control scheme such that the adaptive control laws in Eqs. (10.31) and (10.32) are effective.

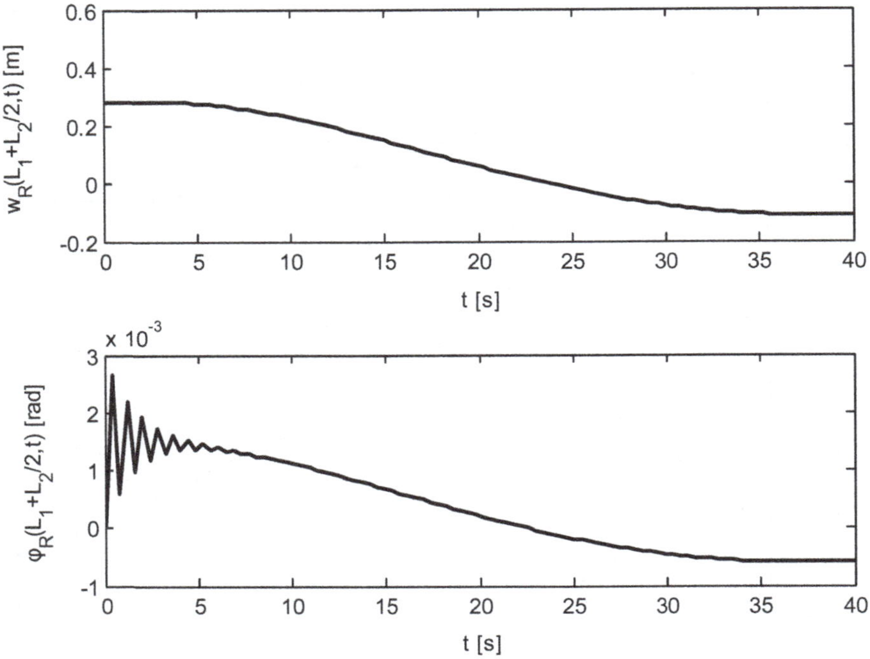

Fig. 10.7 Bending displacement $w_R\left(L_1 + \frac{L_2}{2}, t\right)$ and rotation $\varphi_R\left(L_1 + \frac{L_2}{2}, t\right)$ of the crane bridge without control

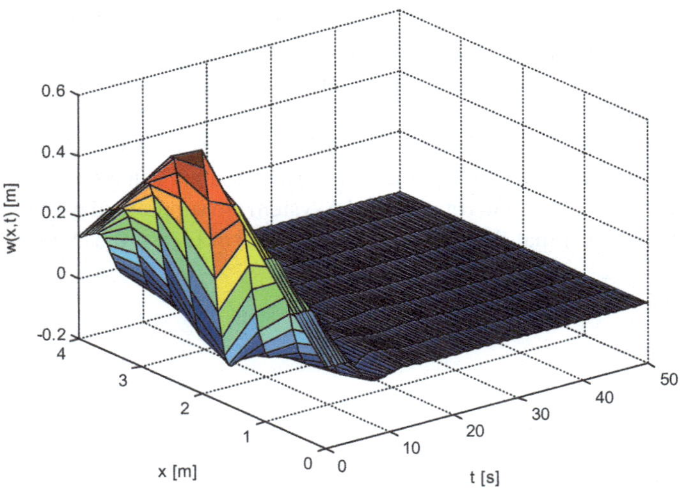

Fig. 10.8 Bending displacement $w(x, t)$ of the crane bridge with the proposed control

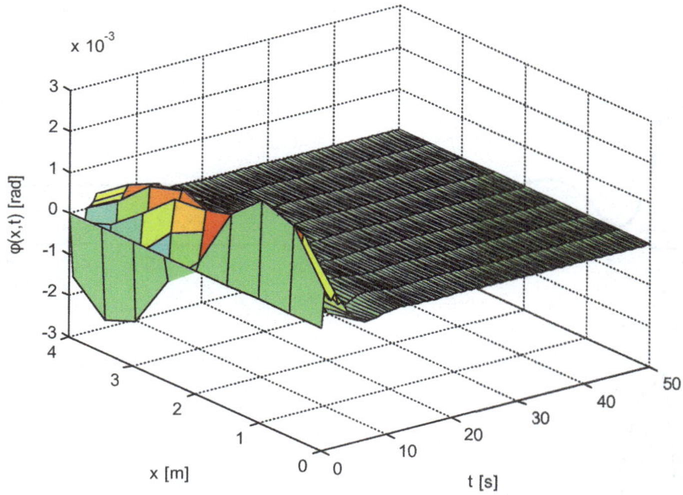

Fig. 10.9 Rotation $\varphi(x, t)$ of the crane bridge with the proposed control

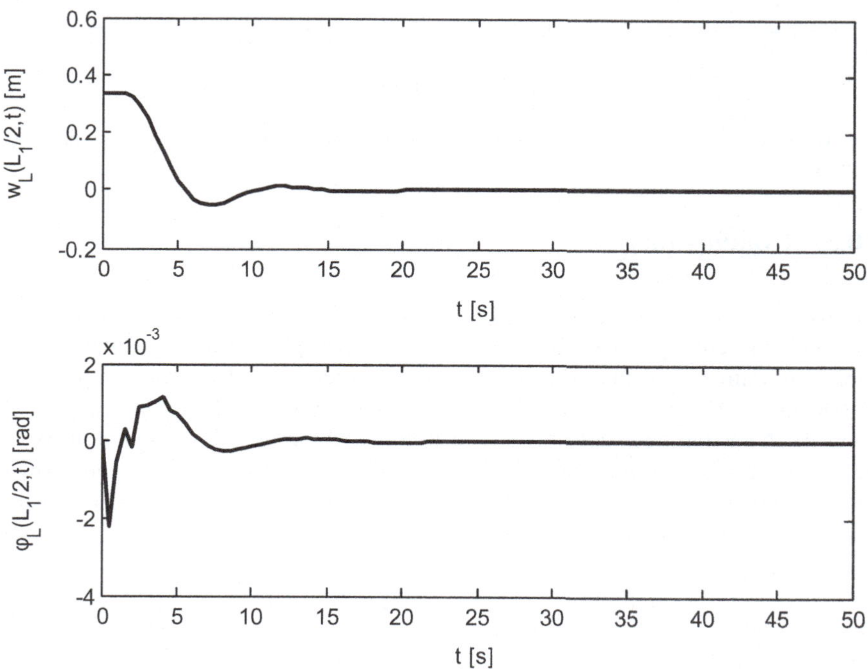

Fig. 10.10 Bending displacement $w_L\left(\frac{L_1}{2}, t\right)$ and rotation $\varphi_L\left(\frac{L_1}{2}, t\right)$ of the crane bridge with the proposed control

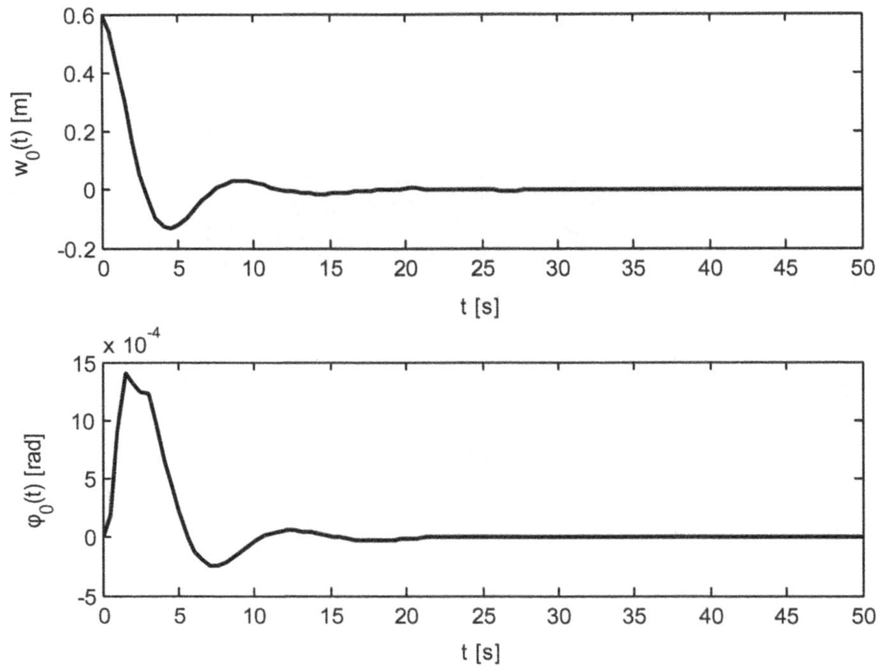

Fig. 10.11 Displacement $w_0(t)$ and rotation $\varphi_0(t)$ of the middle rigid body with the proposed control

10.5 Conclusions

In this chapter, the overhead crane bridge system has been considered as a Timoshenko beam equipped with a rigid body representing an attached trolley and its vibration control problem has been settled out. In order to avoid the spillover problem, a PDE model has been established for the system by using Hamilton principle. A control scheme has been developed to restrict the vibration and rotation of the overhead crane bridge system. The stability of the closed-loop system has been verified by Lyapunov's direct method and the effectiveness of the designed control laws has been verified by numerical simulations.

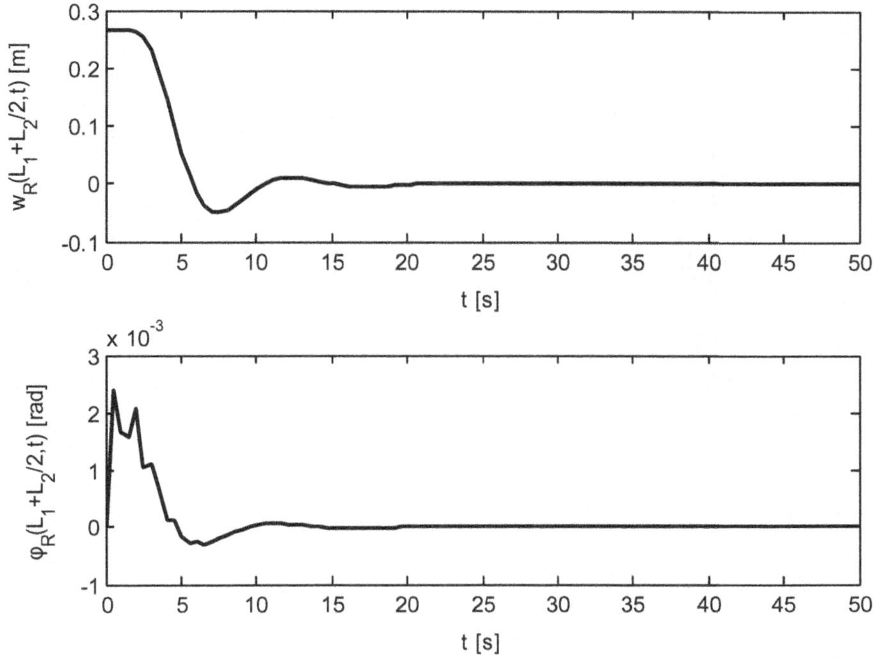

Fig. 10.12 Bending displacement $w_R\left(L_1 + \frac{L_2}{2}, t\right)$ and rotation $\varphi_R\left(L_1 + \frac{L_2}{2}, t\right)$ of the crane bridge with the proposed control

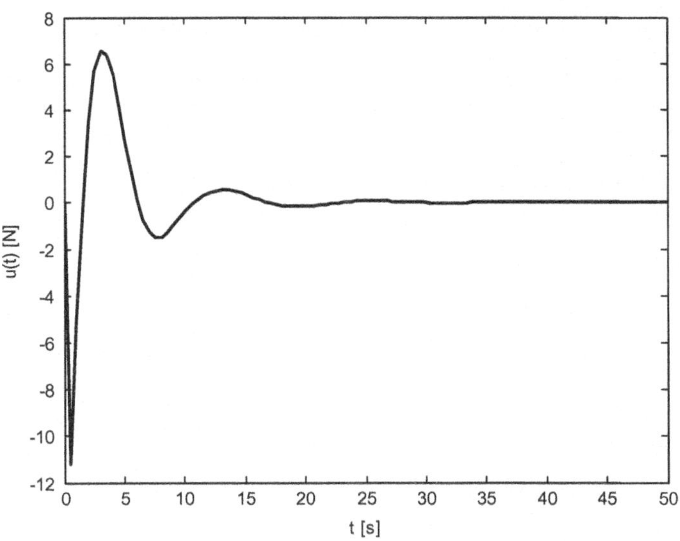

Fig. 10.13 Control input $u(t)$

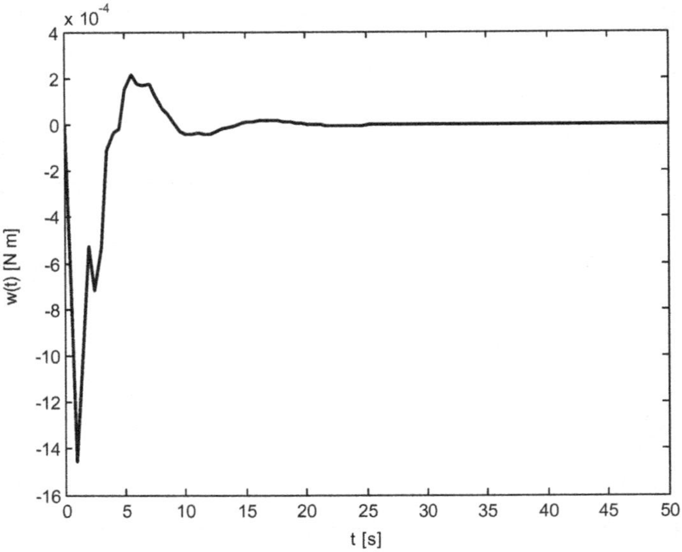

Fig. 10.14 Control input $\tau(t)$

References

1. W. He, S.S. Ge, Cooperative control of a nonuniform gantry crane with constrained tension. Automatica **66**(4), 146–154 (2009)
2. H.C. Cho, K.S. Lee, Adaptive control and stability analysis of nonlinear crane systems with perturbation. J. Mech. Sci. Technol. **22**(6), 1091–1098 (2008)
3. N. Sun, Y. Fang, H. Chen, B. He, Adaptive nonlinear crane control with load hoisting/lowering and unknown parameters: design and experiments. IEEE/ASME Trans. Mechatron. **20**(5), 2107–2119 (2015)
4. B. Ma, Y. Fang, Y. Zhang, Switching-based emergency braking control for an overhead crane system. IET Control Theory Appl. **4**(9), 1739–1747 (2010)

Chapter 11
Vibration Control for Overhead Crane Bridge with Output Constraints

11.1 Introduction

Output constraint is a major and widespread problem in engineering. Once the output constraints of control objects are exceeded, serious harm, such as sharp frictions, high-speed impact, and strong extrusion, may occur to control equipments. As a result, lots of related control strategies have been addressed to deal with output constraints [1–3].

Overhead crane systems universally exist in many industrial sites such as warehouses, workshop halls, and harbors to lift and transport the cargo to the desired position. In this chapter, a control scheme is developed for the crane bridge system based on the PDE model. Under the proposed control laws, the vibration and rotation can be suppressed, and output constraints of the middle trolley are effectively ensured with environmental disturbances via employing BLF. Numerical simulations are given to illustrate the performance of the proposed control method.

11.2 Problem Formulation

Different from the model given in Chap. 10, environmental disturbances $d_1(t)$ and $d_2(t)$ are considered in this chapter, which can be regarded as the effect of the sway of the payload and wind effect, such as, vortex shedding, flutter and buffeting, acting on the overhead crane bridge system, which may aggravate the vibration of the crane bridge and can be effectively resisted by the coming designed robust control laws. The vertical gravity effect of the payload mass can be equivalent to the mass of the trolley M in practice.

Then the virtual work is given by

X. Xing and J. Liu, *Vibration Control Methods of Mechanical Distributed Parameter Systems*, Springer Tracts in Mechanical Engineering,
https://doi.org/10.1007/978-981-16-1532-0_11

$$W(t) = (u(t) + d_1(t))w(L_1, t) + (\tau(t) + d_2(t))\varphi(L_1, t) - \gamma_1 \int_0^{L_1} \dot{w}_L(x, t)w_L(x, t)\mathrm{dx} - \gamma_1 \int_{L_1}^{L} \dot{w}_R(x, t)w_R(x, t)\mathrm{dx} - \gamma_2 \int_0^{L_1} \varphi_L(x, t)\dot{\varphi}_L(x, t)\mathrm{dx} - \gamma_2 \int_{L_1}^{L} \varphi_R(x, t)\dot{\varphi}_R(x, t)\mathrm{dx} \tag{11.1}$$

where $d_1(t)$ and $d_2(t)$ are disturbances acting on actuators of $u(t)$ and $\tau(t)$, respectively.

By applying Hamilton principle in Chap. 2, the governing equations can be achieved as

$$\rho\ddot{w}_L(x, t) - Kw_{Lxx}(x, t) + K\varphi_{Lx}(x, t) + \gamma_1\dot{\varphi}_L(x, t) = 0 \tag{11.2}$$

$$I_p\ddot{\varphi}_L(x, t) - EI\varphi_{Lxx}(x, t) - Kw_{Lx}(x, t) + K\varphi_L(x, t) + \gamma_2\dot{\varphi}_L(x, t) = 0 \tag{11.3}$$

for $\forall(x, t) \in (0, L_1) \times [0, \infty)$,

$$\rho\ddot{w}_R(x, t) - Kw_{Rxx}(x, t) + K\varphi_{Rx}(x, t) + \gamma_1\dot{\varphi}_R(x, t) = 0 \tag{11.4}$$

$$I_p\ddot{\varphi}_R(x, t) - EI\varphi_{Rxx}(x, t) - Kw_{Rx}(x, t) + K\varphi_R(x, t) + \gamma_2\dot{\varphi}_R(x, t) = 0 \tag{11.5}$$

for $\forall(x, t) \in (L_1, L) \times [0, \infty)$, and the boundary conditions are

$$M\ddot{w}(L_1, t) + Kw_{Lx}(L_1, t) - Kw_{Rx}(L_1, t) - u(t) - d_1(t) = 0 \tag{11.6}$$

$$J\ddot{\varphi}(L_1, t) + EI\varphi_{Lx}(L_1, t) - EI\varphi_{Rx}(L_1, t) - \tau(t) - d_2(t) = 0 \tag{11.7}$$

$$w_L(0, t) = w_R(L, t) = \varphi_L(0, t) = \varphi_R(L, t) = 0 \tag{11.8}$$

$$w_L(L_1, t) = w_R(L_1, t) = w_0(t), \varphi_L(L_1, t) = \varphi_R(L_1, t) = \varphi_0(t) \tag{11.9}$$

for $\forall t \in [0, \infty)$.

Assumption 11.1 The external disturbances acting on the system are bounded and their upper bounds are set as $|d_1(t)| \leq \bar{D}_1$, $|d_2(t)| \leq \bar{D}_2$, where $\bar{D}_1$ and $\bar{D}_2$ are positive constants.

Assumption 11.2 Initial conditions of the system are bounded, and the initial conditions of the trolley satisfy $|w_0(0)| < k_1$ and $|\varphi_0(0)| < k_2$, where k_1 and k_2 are positive constraints of $w_0(t)$ and $\varphi_0(t)$ for $\forall t \in [0, \infty)$, respectively.

11.3 Control Design and Stability Analysis

In this section, a BLF-based robust control scheme is designed using backstepping technology to handle disturbances and output constraints of the middle trolley with system stabilization.

First, the following transform of coordinate is made

$$\eta_1(t) = w_0(t), \eta_2(t) = \varphi_0(t) \tag{11.10}$$

$$\eta_3(t) = \dot{\eta}_1(t)+\varpi_1(t), \eta_4(t) = \dot{\eta}_2(t)+\varpi_2(t) \tag{11.11}$$

where $\varpi_1(t)$ and $\varpi_2(t)$ are the virtual control laws designed as

$$\varpi_1(t) = \frac{\alpha}{\beta}\eta_1(t) \tag{11.12}$$

$$\varpi_2(t) = \frac{\alpha}{\beta}\eta_2(t) \tag{11.13}$$

where $\alpha > 0$ and $\beta > 0$.

Then a Lyapunov function is selected as

$$V_a(t) = \frac{\beta}{2}k_1\eta_1^2(t) + \frac{\beta}{2}k_2\eta_2^2(t) \tag{11.14}$$

where $k_1 > 0$, $k_2 > 0$ and its time derivative can be calculated as

$$\dot{V}_a(t) = \beta k_1 w_0(t)\dot{w}_0(t) + \beta k_2\varphi_0(t)\dot{\varphi}_0(t). \tag{11.15}$$

To ensure output constraints, a BLF is introduced to the Lyapunov function, and let

$$V_1(t) = V_a(t) + \frac{\beta}{2}M\eta_3^2(t)a(t) + \frac{\beta}{2}J\eta_4^2(t)b(t) \tag{11.16}$$

where $a(t) = \ln\frac{2k_1^2}{k_1^2-w_0^2(t)}$ and $b(t) = \ln\frac{2k_2^2}{k_2^2-\varphi_0^2(t)}$.

Taking the derivative of $V_1(t)$ with respect to time, we have

$$
\begin{aligned}
\dot{V}_1(t) &= \dot{V}_a(t)+\beta M\eta_3(t)\dot{\eta}_3(t)a(t) + \beta J\eta_4(t)\dot{\eta}_4(t)b(t) \\
&\quad + \frac{\beta}{2}M\eta_3^2(t)\dot{a}(t) + \frac{\beta}{2}J\eta_4^2(t)\dot{b}(t) \\
&= \beta\eta_3(t)\Big(-Kw_{Lx}(L_1,t) + Kw_{Rx}(L_1,t) + u(t) + d_1(t) + M\frac{\alpha}{\beta}\dot{w}_0(t)\Big)a(t) \\
&\quad + \beta\eta_3(t)\Big(-EI\varphi_{Lx}(L_1,t) + EI\varphi_{Rx}(L_1,t) + \tau(t) + d_2(t) + J\frac{\alpha}{\beta}\dot{\varphi}_0(t)\Big)b(t) \\
&\quad + \beta M\eta_3^2(t)\frac{w_0(t)\dot{w}_0(t)}{k_1^2 - w_0^2(t)} + \beta J\eta_4^2(t)\frac{\varphi_0(t)\dot{\varphi}_0(t)}{k_2^2 - \varphi_0^2(t)} + \beta k_1 w_0(t)\dot{w}_0(t) \\
&\quad + \beta k_2\varphi_0(t)\dot{\varphi}_0(t). \qquad (11.17)
\end{aligned}
$$

We define

$$
\begin{aligned}
\mu_1(t) &= k_3\eta_3(t)+\bar{D}_1\mathrm{sgn}(\eta_3(t)) - A(t) + M\frac{\alpha}{\beta}\dot{w}_0(t) \\
&\quad + \frac{k_5\eta_3(t) + k_1 w_0(t) + A(t)}{a(t)} + \frac{M\eta_3(t)}{a(t)}\frac{w_0(t)\dot{w}_0(t)}{k_1^2 - w_0^2(t)} \qquad (11.18)
\end{aligned}
$$

$$
\begin{aligned}
\mu_2(t) &= k_4\eta_4(t)+\bar{D}_2\mathrm{sgn}(\eta_4(t)) - B(t) + J\frac{\alpha}{\beta}\dot{\varphi}_0(t) \\
&\quad + \frac{k_6\eta_4(t) + k_2\varphi_0(t) + B(t)}{b(t)} + \frac{J\eta_4(t)}{b(t)}\frac{\varphi_0(t)\dot{\varphi}_0(t)}{k_2^2 - \varphi_0^2(t)} \qquad (11.19)
\end{aligned}
$$

where $k_3, k_4, k_5, k_6 > 0$, $A(t) = Kw_{Lx}(L_1,t) - Kw_{Rx}(L_1,t)$, and $B(t) = EI\varphi_{Lx}(L_1,t) - EI\varphi_{Rx}(L_1,t)$.

Then we have

$$
\begin{aligned}
&\dot{V}_1(t) \le \beta\eta_3(t) \\
&\Big(-k_3\eta_3(t) + \mu_1(t) - \frac{k_5\eta_3(t) + k_1 w_0(t) + A(t)}{a(t)} - \frac{M\eta_3(t)}{a(t)}\frac{w_0(t)\dot{w}_0(t)}{k_1^2 - w_0^2(t)} + u(t)\Big) \\
&\quad a(t) + \beta\eta_4(t)(-k_4\eta_4(t) + \mu_2(t) \\
&\quad - \frac{k_6\eta_4(t) + k_2\varphi_0(t) + B(t)}{b(t)} - \frac{J\eta_4(t)}{b(t)}\frac{\varphi_0(t)\dot{\varphi}_0(t)}{k_2^2 - \varphi_0^2(t)} + \tau(t)\Big)b(t) \\
&\quad + \beta M\eta_3^2(t)\frac{w_0(t)\dot{w}_0(t)}{k_1^2 - w_0^2(t)} + \beta J\eta_4^2(t)\frac{\varphi_0(t)\dot{\varphi}_0(t)}{k_2^2 - \varphi_0^2(t)} \\
&\quad + \beta k_1 w_0(t)\dot{w}_0(t) + \beta k_2\varphi_0(t)\dot{\varphi}_0(t) \\
&= -\beta k_3\eta_3^2(t)a(t) + \beta\eta_3(t)(\mu_1(t) + u(t))a(t) \\
&\quad - \beta k_4\eta_4^2(t)b(t) + \beta\eta_4(t)(\mu_2(t) + \tau(t))b(t) + D \qquad (11.20)
\end{aligned}
$$

where

$$
\begin{aligned}
D &= -\beta\eta_3(t)(k_5\eta_3(t) + k_1 w_0(t)) - \beta\eta_3(t)A(t) - \beta\eta_4(t)(k_6\eta_4(t) + k_2\varphi_0(t)) \\
&\quad - \beta\eta_4(t)B(t) + \beta k_1 w_0(t)\dot{w}_0(t) + \beta k_2\varphi_0(t)\dot{\varphi}_0(t) \\
&= -k_5\beta\eta_3^2(t) - \alpha k_1 w_0^2(t) - \beta\eta_3(t)A(t) - k_6\beta\eta_4^2(t) - \alpha k_2\varphi_0^2(t) - \beta\eta_4(t)B(t).
\end{aligned}
\tag{11.21}
$$

Design

$$u(t) = -\mu_1(t) \tag{11.22}$$

$$\tau(t) = -\mu_2(t) \tag{11.23}$$

and thus from inequality (11.20) we have

$$\dot{V}_1(t) \le -\beta k_3\eta_3^2(t)a(t) - \beta k_4\eta_4^2(t)b(t) + D. \tag{11.24}$$

Theorem 11.1 *Under Assumptions* 11.1 *and* 11.2, *robust control laws in Eqs.* (11.22) *and* (11.23) *can ensure the boundedness of the closed-loop signals of the overhead crane bridge system in Eqs.* (11.2)–(11.9) *subject to disturbances. The closed-loop system is exponentially stable with output constraints of the middle rigid body, which means that* $\lim_{t\to\infty} w(x,t) = \lim_{t\to\infty} \varphi(x,t) = 0$ *holds for* $\forall x \in [0, L]$, *and* $|w_0(t)| < k_1$ and $|\varphi_0(t)| < k_2$ *hold for* $\forall t \in [0, \infty)$.

Proof The system stability can be analyzed by designing the following Lyapunov function

$$V(t) = V_1(t) + V_2(t) + V_3(t) + V_4(t) \tag{11.25}$$

where

$$
\begin{aligned}
V_2(t) &= \frac{\beta}{2}\rho\int_0^{L_1}\dot{w}_L^2(x,t)\mathrm{d}x + \frac{\beta}{2}K\int_0^{L_1}(\varphi_L(x,t) - w_{Lx}(x,t))^2\mathrm{d}x + \frac{\beta}{2}I_p\int_0^{L_1}\dot{\varphi}_L^2(x,t)\mathrm{d}x \\
&\quad + \frac{\beta}{2}EI\int_0^{L_1}\varphi_{Lx}^2(x,t)\mathrm{d}x + \frac{\beta}{2}\rho\int_{L_1}^{L}\dot{w}_R^2(x,t)\mathrm{d}x \\
&\quad + \frac{\beta}{2}K\int_{L_1}^{L}(\varphi_R(x,t) - w_{Rx}(x,t))^2\mathrm{d}x \\
&\quad + \frac{\beta}{2}I_p\int_{L_1}^{L}\dot{\varphi}_R^2(x,t)\mathrm{d}x + \frac{\beta}{2}EI\int_{L_1}^{L}\varphi_{Rx}^2(x,t)\mathrm{d}x
\end{aligned}
\tag{11.26}
$$

$$V_3(t) = \alpha\rho \int_0^{L_1} \dot{w}_L(x,t) w_L(x,t)\mathrm{dx} + \alpha I_p \int_0^{L_1} \dot{\varphi}_L(x,t)\varphi_L(x,t)\mathrm{dx} + \alpha\rho \int_{L_1}^{L} \dot{w}_R(x,t) w_R(x,t)\mathrm{dx} + \alpha I_p \int_{L_1}^{L} \dot{\varphi}_R(x,t)\varphi_R(x,t)\mathrm{dx} \tag{11.27}$$

$$V_4(t) = \frac{\alpha\gamma_1}{2}\int_0^{L_1} w_L^2(x,t)\mathrm{dx} + \frac{\alpha\gamma_2}{2}\int_0^{L_1} \varphi_L^2(x,t)\mathrm{dx} + \frac{\alpha\gamma_1}{2}\int_{L_1}^{L} w_R^2(x,t)\mathrm{dx} + \frac{\alpha\gamma_2}{2}\int_{L_1}^{L} \varphi_R^2(x,t)\mathrm{dx}. \tag{11.28}$$

To verify the positive definiteness of $V(t)$, we define

$$\begin{aligned}\chi(t) = & \int_0^{L_1} \dot{w}_L^2(x,t)\mathrm{dx} + \int_0^{L_1} \dot{\varphi}_L^2(x,t)\mathrm{dx} + \int_0^{L_1} \varphi_{Lx}^2(x,t)\mathrm{dx} \\ & + \int_0^{L_1} w_{Lx}^2(x,t)\mathrm{dx} + \int_0^{L_1} \varphi_L^2(x,t)\mathrm{dx} + \int_0^{L_1} w_L^2(x,t)\mathrm{dx} \\ & + \int_{L_1}^{L} \dot{w}_R^2(x,t)\mathrm{dx} + \int_{L_1}^{L} \dot{\varphi}_R^2(x,t)\mathrm{dx} + \int_{L_1}^{L} \varphi_{Rx}^2(x,t)\mathrm{dx} \\ & + \int_{L_1}^{L} w_{Rx}^2(x,t)\mathrm{dx} + \int_{L_1}^{L} \varphi_R^2(x,t)\mathrm{dx} + \int_{L_1}^{L} w_R^2(x,t)\mathrm{dx}. \end{aligned} \tag{11.29}$$

From Eqs. (11.26) and (11.28) and Lemma 2.6 we can obtain

$$\begin{aligned} V_2(t) + V_4(t) \geq & \frac{\beta}{2}\rho \int_0^{L_1} \dot{w}_L^2(x,t)\mathrm{dx} + \frac{\alpha\gamma_1}{2}\int_0^{L_1} w_L^2(x,t)\mathrm{dx} + \frac{\beta}{2} I_p \int_0^{L_1} \dot{\varphi}_L^2(x,t)\mathrm{dx} \\ & + \frac{\beta}{4} EI \int_0^{L_1} \varphi_{Lx}^2(x,t)\mathrm{dx} + \int_0^{L_1} \mathbf{\Omega}_1^{\mathrm{T}}(x,t)\mathbf{\Xi}_1\mathbf{\Omega}_1(x,t)\mathrm{dx} \\ & + \frac{\beta}{2}\rho \int_{L_1}^{L} \dot{w}_R^2(x,t)\mathrm{dx} + \frac{\alpha\gamma_1}{2}\int_{L_1}^{L} w_R^2(x,t)\mathrm{dx} + \frac{\beta}{2} I_p \int_{L_1}^{L} \dot{\varphi}_R^2(x,t)\mathrm{dx} \end{aligned}$$

$$+\frac{\beta}{4}EI\int_{L_1}^{L}\varphi_{Rx}^2(x,t)\mathrm{dx}+\int_{L_1}^{L}\boldsymbol{\Omega}_2^{\mathrm{T}}(x,t)\boldsymbol{\Xi}_2\boldsymbol{\Omega}_2(x,t)\mathrm{dx} \tag{11.30}$$

in which $\boldsymbol{\Omega}_1(x,t)=\left[\varphi_L(x,t)\ \varphi_{Lx}(x,t)\right]^{\mathrm{T}}$, $\boldsymbol{\Omega}_2(x,t)=\left[\varphi_R(x,t)\ w_{Rx}(x,t)\right]^{\mathrm{T}}$, and $\boldsymbol{\Xi}_c=\begin{bmatrix}\frac{\beta EI}{16L_c^2}+\frac{\beta}{2}K+\frac{\alpha\gamma_2}{2} & -\frac{\beta}{2}K\\ -\frac{\beta}{2}K & \frac{\beta}{2}K\end{bmatrix}$ is a positive definite matrix for $c=1,2$.

Thus we have

$$V_2(t)+V_4(t)\geq\lambda_1\chi(t) \tag{11.31}$$

where $\lambda_1=\min\left\{\frac{\beta}{2}\rho,\frac{\beta}{2}I_p,\frac{\beta}{4}EI,\sigma_{\min}(\boldsymbol{\Xi}_1),\sigma_{\min}(\boldsymbol{\Xi}_2),\frac{\alpha\gamma_1}{2}\right\}>0$.

According to Eqs. (11.26) and (11.28) and Lemma 2.6, we achieve

$$\begin{aligned}V_2(t)+V_4(t)\leq&\frac{\beta}{2}\rho\int_0^{L_1}\dot{w}_L^2(x,t)\mathrm{dx}+\frac{\alpha\gamma_1}{2}\int_0^{L_1}w_L^2(x,t)\mathrm{dx}+\beta K\int_0^{L_1}w_{Lx}^2(x,t)\mathrm{dx}\\&+\frac{\beta}{2}I_p\int_0^{L_1}\dot{\varphi}_L^2(x,t)\mathrm{dx}+\frac{\beta}{2}EI\int_0^{L_1}\varphi_{Lx}^2(x,t)\mathrm{dx}\\&+\left(\beta K+\frac{\alpha\gamma_2}{2}\right)\int_0^{L_1}\varphi_L^2(x,t)\mathrm{dx}\\&+\frac{\beta}{2}\rho\int_{L_1}^{L}\dot{w}_R^2(x,t)\mathrm{dx}+\frac{\alpha\gamma_1}{2}\int_{L_1}^{L}w_R^2(x,t)\mathrm{dx}\\&+\beta K\int_{L_1}^{L}w_{Rx}^2(x,t)\mathrm{dx}+\frac{\beta}{2}I_p\int_{L_1}^{L}\dot{\varphi}_R^2(x,t)\mathrm{dx}\\&+\frac{\beta}{2}EI\int_{L_1}^{L}\varphi_{Rx}^2(x,t)\mathrm{dx}+\left(\beta K+\frac{\alpha\gamma_2}{2}\right)\int_{L_1}^{L}\varphi_R^2(x,t)\mathrm{dx}\\&\leq\lambda_2\chi(t)\end{aligned} \tag{11.32}$$

where $\lambda_2=\max\left\{\frac{1}{2}\beta\rho,\frac{\beta}{2}I_p,\frac{\beta}{2}EI,\beta K+\frac{\alpha\gamma_2}{2},\frac{\alpha\gamma_1}{2}\right\}>0$.

Using Eq. (11.27), we can get the following inequality

$$
\begin{aligned}
|V_3(t)| \le {} & \frac{\alpha\rho}{2}\int_0^{L_1} \dot{w}_L^2(x,t)\mathrm{dx} + \frac{\alpha\rho}{2}\int_0^{L_1} w_L^2(x,t)\mathrm{dx} + \frac{\alpha I_p}{2}\int_0^{L_1} \dot{\varphi}_L^2(x,t)\mathrm{dx} \\
& + \frac{\alpha I_p}{2}\int_0^{L_1} \varphi_L^2(x,t)\mathrm{dx} \\
& + \frac{\alpha\rho}{2}\int_{L_1}^{L} \dot{w}_R^2(x,t)\mathrm{dx} + \frac{\alpha\rho}{2}\int_{L_1}^{L} w_R^2(x,t)\mathrm{dx} + \frac{\alpha I_p}{2}\int_{L_1}^{L} \dot{\varphi}_R^2(x,t)\mathrm{dx} \\
& + \frac{\alpha I_p}{2}\int_{L_1}^{L} \varphi_R^2(x,t)\mathrm{dx} \\
\le {} & \lambda_3\chi(t)
\end{aligned}
\tag{11.33}
$$

in which $\lambda_3 = \max\left\{\frac{1}{2}\alpha I_p, \frac{1}{2}\alpha\rho\right\} > 0$.

If α and β are selected to meet $\lambda_1 > \lambda_3$, we have

$$
0 < (\lambda_1 - \lambda_3)\chi(t) \le V_2(t) + V_3(t) + V_4(t) \le (\lambda_2 + \lambda_3)\chi(t). \tag{11.34}
$$

Then we can further attain

$$
0 < \lambda_4(\chi(t) + V_1(t)) \le V(t) \le \lambda_5(\chi(t) + V_1(t)) \tag{11.35}
$$

where $\kappa_1 = \lambda_1 - \lambda_3 > 0$, $\kappa_2 = \lambda_2 + \lambda_3 > 0$, $\lambda_4 = \min\{\kappa_1, 1\} > 0$, and $\lambda_5 = \max\{\kappa_2, 1\} > 0$.

Differentiating Eq. (11.26) and substituting Eqs. (11.2)–(11.5) with respect to time lead to

$$
\begin{aligned}
\dot{V}_2(t) = {} & \beta K \dot{w}_L(L_1,t)(w_{Lx}(L_1,t) - \varphi_L(L_1,t)) + \beta EI\dot{\varphi}_L(L_1,t)\varphi_{Lx}(L_1,t) \\
& - \beta K \dot{w}_R(L_1,t)(w_{Rx}(L_1,t) - \varphi_R(L_1,t)) - \beta EI\dot{\varphi}_R(L_1,t)\varphi_{Rx}(L_1,t) \\
& - \beta\gamma_1\int_0^{L_1} \dot{w}_L^2(x,t)\mathrm{dx} - \beta\gamma_2\int_0^{L_1} \dot{\varphi}_L^2(x,t)\mathrm{dx} - \beta\gamma_1\int_{L_1}^{L} \dot{w}_R^2(x,t)\mathrm{dx} \\
& - \beta\gamma_2\int_{L_1}^{L} \dot{\varphi}_R^2(x,t)\mathrm{dx}.
\end{aligned}
\tag{11.36}
$$

Substituting Eqs. (11.6)–(11.9) into the derivative of Eq. (11.27) and using integration by parts, we have

$$\begin{aligned}
\dot{V}_3(t) = {} & \alpha K w_L(L_1,t) w_{Lx}(L_1,t) - \alpha K \int_0^{L_1} w_{Lx}^2(x,t)\mathrm{dx} - \alpha K w_L(L_1,t)\varphi_L(L_1,t) \\
& + \alpha\rho \int_0^{L_1} \dot{w}_L^2(x,t)\mathrm{dx} \\
& - \alpha\gamma_1 \int_0^{L_1} w_L(x,t)\dot{w}_L(x,t)\mathrm{dx} + \alpha EI \varphi_L(L_1,t)\varphi_{Lx}(L_1,t) \\
& - \alpha EI \int_0^{L_1} \varphi_{Lx}^2(x,t)\mathrm{dx} \\
& + 2\alpha K \int_0^{L_1} \varphi_L(x,t) w_{Lx}(x,t)\mathrm{dx} - \alpha K \int_0^{L_1} \varphi_L^2(x,t)\mathrm{dx} + \alpha I_p \int_0^{L_1} \dot{\varphi}_L^2(x,t)\mathrm{dx} \\
& - \alpha\gamma_2 \int_0^{L_1} \varphi_L(x,t)\dot{\varphi}_L(x,t)\mathrm{dx} - \alpha K w_R(L_1,t) w_{Rx}(L_1,t) \\
& - \alpha K \int_{L_1}^{L} w_{Rx}^2(x,t)\mathrm{dx} \\
& + \alpha K w_R(L_1,t)\varphi_R(L_1,t) + \alpha\rho \int_{L_1}^{L} \dot{w}_R^2(x,t)\mathrm{dx} \\
& - \alpha\gamma_1 \int_{L_1}^{L} w_R(x,t)\dot{w}_R(x,t)\mathrm{dx} \\
& - \alpha EI \varphi_R(L_1,t)\varphi_{Rx}(L_1,t) - \alpha EI \int_{L_1}^{L} \varphi_{Rx}^2(x,t)\mathrm{dx} \\
& + 2\alpha K \int_{L_1}^{L} \varphi_R(x,t) w_{Rx}(x,t)\mathrm{dx} \\
& - \alpha K \int_{L_1}^{L} \varphi_R^2(x,t)\mathrm{dx} + \alpha I_p \int_{L_1}^{L} \dot{\varphi}_R^2(x,t)\mathrm{dx} - \alpha\gamma_2 \int_{L_1}^{L} \varphi_R(x,t)\dot{\varphi}_R(x,t)\mathrm{dx}.
\end{aligned} \tag{11.37}$$

The time derivative of $\dot{V}_4(t)$ is supposed to be

$$\dot{V}_4(t) = \alpha\gamma_1 \int_0^{L_1} w_L(x,t)\dot{w}_L(x,t)\mathrm{dx} + \alpha\gamma_2 \int_0^{L_1} \varphi_L(x,t)\dot{\varphi}_L(x,t)\mathrm{dx}$$
$$+ \alpha\gamma_1 \int_{L_1}^{L} w_R(x,t)\dot{w}_R(x,t)\mathrm{dx} + \alpha\gamma_2 \int_{L_1}^{L} \varphi_R(x,t)\dot{\varphi}_R(x,t)\mathrm{dx}. \quad (11.38)$$

Combining inequality (11.24), Eqs. (11.36)–(11.38) and utilizing Lemma 2.4 yield

$$\begin{aligned}
\dot{V}(t) \le & -(\beta\gamma_1 - \alpha\rho) \int_0^{L_1} \dot{w}_L^2(x,t)\mathrm{d}x - \alpha K \int_0^{L_1} w_{Lx}^2(x,t)\mathrm{dx} \\
& - \left(\beta\gamma_2 - \alpha I_p\right) \int_0^{L_1} \dot{\varphi}_L^2(x,t)\mathrm{dx} - \alpha EI \int_0^{L_1} \varphi_{Lx}^2(x,t)\mathrm{dx} - \alpha K \int_0^{L_1} \varphi_L^2(x,t)\mathrm{dx} \\
& + 2\alpha K \int_0^{L_1} \varphi_L(x,t) w_{Lx}(x,t)\mathrm{dx} - (\beta\gamma_1 - \alpha\rho) \int_{L_1}^{L} \dot{w}_R^2(x,t)\mathrm{dx} \\
& - \alpha K \int_{L_1}^{L} w_{Rx}^2(x,t)\mathrm{dx} \\
& - \left(\beta\gamma_2 - \alpha I_p\right) \int_{L_1}^{L} \dot{\varphi}_R^2(x,t)\mathrm{dx} - \alpha EI \int_{L_1}^{L} \varphi_{Rx}^2(x,t)\mathrm{dx} - \alpha K \int_{L_1}^{L} \varphi_R^2(x,t)\mathrm{dx} \\
& + 2\alpha K \int_{L_1}^{L} \varphi_R(x,t) w_{Rx}(x,t)\mathrm{dx} - \beta k_3 \eta_3^2(t) a(t) - \beta k_4 \eta_4^2(t) b(t) \\
& - \alpha k_1 w_0^2(t) - \alpha k_2 \varphi_0^2(t) \\
\le & -(\beta\gamma_1 - \alpha\rho) \int_0^{L_1} \dot{w}_L^2(x,t)\mathrm{dx} - (\alpha K - 2\alpha K\delta_1) \int_0^{L_1} w_{Lx}^2(x,t)\mathrm{dx} \\
& - \left(\beta\gamma_2 - \alpha I_p\right) \int_0^{L_1} \dot{\varphi}_L^2(x,t)\mathrm{dx} - \left(\alpha K - \frac{2\alpha K}{\delta_1}\right) \int_0^{L_1} \varphi_L^2(x,t)\mathrm{dx} \\
& - \alpha EI \int_0^{L_1} \varphi_{Lx}^2(x,t)\mathrm{dx}
\end{aligned}$$

$$
\begin{aligned}
&-(\beta\gamma_1-\alpha\rho)\int_{L_1}^{L}\dot{w}_R^2(x,t)\mathrm{dx}-(\alpha K-2\alpha K\delta_2)\int_{L_1}^{L}w_{Rx}^2(x,t)\mathrm{dx}\\
&-\left(\beta\gamma_2-\alpha I_p\right)\int_{L_1}^{L}\dot{\varphi}_R^2(x,t)\mathrm{dx}-\left(\alpha K-\frac{2\alpha K}{\delta_2}\right)\int_{L_1}^{L}\varphi_R^2(x,t)\mathrm{dx}\\
&-\alpha EI\int_{L_1}^{L}\varphi_{Rx}^2(x,t)\mathrm{dx}\\
&-\beta k_3\eta_3^2(t)a(t)-\beta k_4\eta_4^2(t)b(t)-\alpha k_1 w_0^2(t)-\alpha k_2\varphi_0^2(t)
\end{aligned}\tag{11.39}
$$

in which $\delta_1 > 0$ and $\delta_2 > 0$.

Utilizing Lemma 2.6, we have

$$
\begin{aligned}
\dot{V}(t)\leq&-(\beta\gamma_1-\alpha\rho)\int_{0}^{L_1}\dot{w}_L^2(x,t)\mathrm{dx}-\zeta_1\int_{0}^{L_1}w_L^2(x,t)\mathrm{dx}\\
&-\left(\alpha K-2\alpha K\delta_1-4L_1^2\zeta_1\right)\int_{0}^{L_1}w_{Lx}^2(x,t)\mathrm{dx}\\
&-\left(\beta\gamma_2-\alpha I_p\right)\int_{0}^{L_1}\dot{\varphi}_L^2(x,t)\mathrm{dx}-\left(\alpha K+\frac{\alpha}{8L_1^2}EI-\frac{2\alpha K}{\delta_1}\right)\int_{0}^{L_1}\varphi_L^2(x,t)\mathrm{dx}\\
&-\frac{\alpha}{2}EI\int_{0}^{L_1}\varphi_{Lx}^2(x,t)\mathrm{dx}-(\beta\gamma_1-\alpha\rho)\int_{L_1}^{L}\dot{w}_R^2(x,t)\mathrm{dx}-\zeta_2\int_{L_1}^{L}w_R^2(x,t)\mathrm{dx}\\
&-\left(\alpha K-2\alpha K\delta_2-4L_2^2\zeta_2\right)\int_{L_1}^{L}w_{Rx}^2(x,t)\mathrm{dx}-\left(\beta\gamma_2-\alpha I_p\right)\int_{L_1}^{L}\dot{\varphi}_R^2(x,t)\mathrm{dx}\\
&-\left(\alpha K+\frac{\alpha}{8L_2^2}EI-\frac{2\alpha K}{\delta_2}\right)\int_{L_1}^{L}\varphi_R^2(x,t)\mathrm{dx}-\frac{\alpha}{2}EI\int_{L_1}^{L}\varphi_{Rx}^2(x,t)\mathrm{dx}\\
&-\beta k_3\eta_3^2(t)a(t)-\beta k_4\eta_4^2(t)b(t)-\alpha k_1 w_0^2(t)-\alpha k_2\varphi_0^2(t)
\end{aligned}\tag{11.40}
$$

where $\zeta_1 > 0$, $\zeta_2 > 0$, and δ_1, δ_2, ζ_1, ζ_2, α, β are selected to satisfy $\beta\gamma_1-\alpha\rho > 0$, $\beta\gamma_2-\alpha I_p > 0$, $\alpha K+\frac{\alpha}{8L_1^2}EI-\frac{2\alpha K}{\delta_1} > 0$, $\alpha K+\frac{\alpha}{8L_2^2}EI-\frac{2\alpha K}{\delta_2} > 0$,$\alpha K-2\alpha K\delta_1-4L_1^2\zeta_1 > 0$, $\alpha K-2\alpha K\delta_2-4L_2^2\zeta_2 > 0$.

Then $\dot{V}(t)$ can be rewritten as

$$\dot{V}(t) \leq -\lambda_6(\chi(t) + V_1(t)) \tag{11.41}$$

where

$$\lambda_6 = \min\Bigg\{\beta\gamma_1 - \alpha\rho, \zeta_1, \alpha K - 2\alpha K\delta_1 - 4L_1^2\zeta_1, \beta\gamma_2 - \alpha I_p, \alpha K + \frac{\alpha}{8L_1^2}EI - \frac{2\alpha K}{\delta_1},$$
$$\frac{\alpha}{2}EI,\ \zeta_2, \alpha K - 2\alpha K\delta_2 - 4L_2^2\zeta_2, \alpha K + \frac{\alpha}{8L_2^2}EI - \frac{2\alpha K}{\delta_2}, \frac{2\alpha}{\beta}, \frac{2k_3}{M}, \frac{2k_4}{J}\Bigg\}.$$

From inequalities (11.35) and (11.41), we have

$$\dot{V}(t) \leq -\lambda V(t) \tag{11.42}$$

where $\lambda = \frac{\lambda_6}{\lambda_5} > 0$.

Multiplying inequality (11.42) by $e^{\lambda t}$ yields

$$V(t) \leq V(0)e^{-\lambda t}. \tag{11.43}$$

Based on inequalities (11.43), (11.35), Lemmas 2.2 and 2.5, we have

$$\frac{1}{L_1}\varphi_L^2(x,t) \leq \frac{1}{\lambda_4}V(0)e^{-\lambda t}, \frac{1}{L_2}\varphi_R^2(x,t) \leq \frac{1}{\lambda_4}V(0)e^{-\lambda t}. \tag{11.44}$$

$$\frac{1}{L_1}w_L^2(x,t) \leq \frac{1}{\lambda_4}V(0)e^{-\lambda t}, \frac{1}{L_2}w_R^2(x,t) \leq \frac{1}{\lambda_4}V(0)e^{-\lambda t}. \tag{11.45}$$

$$\frac{\beta}{2}k_1w_0^2(t) \leq \frac{1}{\lambda_4}V(0)e^{-\lambda t}, \frac{\beta}{2}k_2\varphi_0^2(t) \leq \frac{1}{\lambda_4}V(0)e^{-\lambda t}. \tag{11.46}$$

Then it is obvious that

$$|w_L(x,t)|, |\varphi_L(x,t)| \leq \sqrt{\frac{L_1}{\lambda_4}V(0)e^{-\lambda t}} \tag{11.47}$$

$$|w_R(x,t)|, |\varphi_R(x,t)| \leq \sqrt{\frac{L_2}{\lambda_4}V(0)e^{-\lambda t}} \tag{11.48}$$

$$|w_0(t)| \leq \sqrt{\frac{2}{\beta k_1\lambda_4}V(0)e^{-\lambda t}}, |\varphi_0(t)| \leq \sqrt{\frac{2}{\beta k_2\lambda_5}V(0)e^{-\lambda t}}. \tag{11.49}$$

According to inequalities (11.47)–(11.49), it can be stated that if system initial conditions are bounded, then $V(0)$ is bounded, which means $w(x,t)$ and $\varphi(x,t)$ are all bounded for $\forall(x,t) \in [0, L] \times [0, \infty)$.

Besides, from inequalities (11.47)–(11.49), we have

$$\lim_{t\to\infty}|w(x,t)|, \lim_{t\to\infty}|\varphi(x,t)| = 0 \,\text{for}\, \forall x \in [0, L]. \tag{11.50}$$

Moreover, according to Lemma 2.3, we know that if the initial conditions in Assumption 11.2 hold, output constraints $|w_0(t)| \le k_1$ and $|\varphi_0(t)| \le k_2$ can always been guaranteed for $\forall t \in [0, \infty)$ under the proposed control scheme.

Therefore, the closed-loop system is exponentially stable under the proposed control with output constraints, and Theorem 11.1 has been completely proved.

11.4 Simulations

To verify the performance of the proposed control strategy, numerical simulations are done for the overhead crane bridge system in this chapter whose dynamic equations are given by Eqs. (11.2)–(11.9). The values of system parameters are chosen as: $\rho = 2\,\text{kg}\,\text{m}^{-1}$, $K = 0.5\,\text{N}$, $\gamma_1 = 2.5\,\text{kg/(ms)}$, $I_p = 2\,\text{kg}\,\text{m}^2$, $EI = 50\,\text{N}\,\text{m}^2$, $\gamma_2 = 2.5\,\text{kg/(ms)}$, $M = 60\,\text{kg}$, $J = 2\,\text{kg}\,\text{m}^2$, $L_1 = 2\,\text{m}$, $L_2 = 2\,\text{m}$, and $L = 4\,\text{m}$. Initial conditions of $w(x,t)$ and $\varphi(x,t)$ are respectively selected as $w(x,0) = \begin{cases} 0.03x, & \text{if } 0 \le x < L_1 \\ 0.03L_1, & \text{if } x = L_1 \\ -0.03\frac{L_1}{L_2}(x-L), & \text{if } L_1 < x \le L \end{cases}$, $\varphi(x,0) = 0$, $\dot{w}(x,0) = 0$, and $\dot{\varphi}(x,0) = 0$. External disturbances are presented as $d_1(t) = 0.05\sin(0.1t)$ and $d_2(t) = 5 \times 10^{-4}\sin(0.5t)$.

Control laws in Eqs. (11.22)–(11.23) are selected as $k_1 = 0.1$, $k_2 = 2 \times 10^{-5}$, $k_3 = 0.1$, $k_4 = 0.1$, $k_5 = 100$, $k_6 = 100$, $\alpha = 1$, $\beta = 10$, $\bar{D}_1 = 200$, $\bar{D}_2 = 10$, where k_1 and k_2 are also limited values of $w_0(t)$ and $\varphi_0(t)$, respectively.

To clearly compare the simulation results, Figs. 11.1 and 11.2 show the displace-

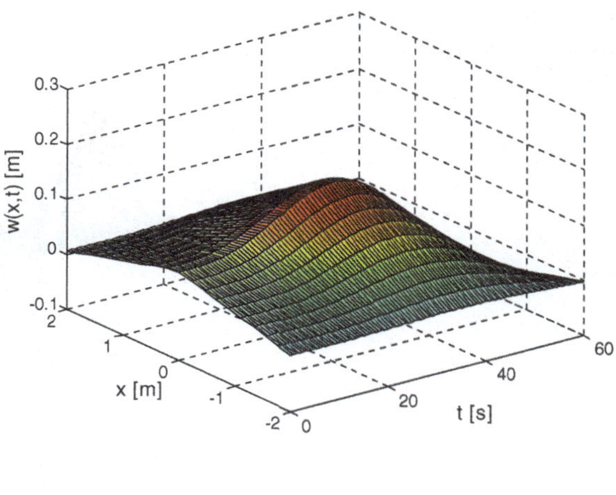

Fig. 11.1 Displacement $w(x,t)$ of the crane bridge in the open-loop system

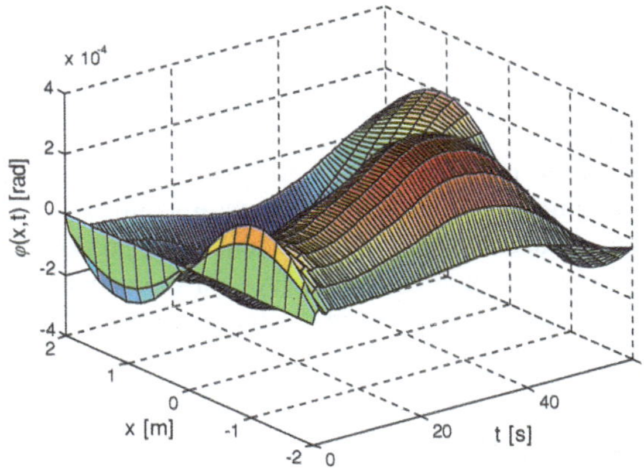

Fig. 11.2 Rotation $\varphi(x, t)$ of the crane bridge in the open-loop system

ment $w(x, t)$ and rotation $\varphi(x, t)$ of the crane bridge system without control. From Figs. 11.1 and 11.2, it is obvious that the system has serious vibration and rotation which need to be eliminated urgently.

With the actuated control in Eqs. (11.22) and (11.23), Figs. 11.3 and 11.4 present displacement $w(x, t)$ and rotation $\varphi(x, t)$ of the crane bridge. Furthermore, dynamic responses of the closed-loop system and open-loop system at several representative points, such as $w_L\left(\frac{L_1}{2}, t\right)$, $w(L_1, t)$, $w_R\left(L_1 + \frac{L_2}{2}, t\right)$, $\varphi_L\left(\frac{L_1}{2}, t\right)$, $\varphi(L_1, t)$, $\varphi_R\left(L_1 + \frac{L_2}{2}, t\right)$ are also as shown in Figs. 11.5, 11.6, 11.7, 11.8, 11.9, 11.10 to strengthen the persuasiveness of simulation results. Note from Figs. 11.3, 11.4, 11.5, 11.6, 11.7, 11.8, 11.9, 11.10 that the proposed control scheme has a significant effect on vibration elimination, and states of the closed-loop system can gradually tend to zero even in the presence of environmental disturbances. Moreover, according to Figs. 11.6 and 11.9, it is obvious that the control laws can always limit the displacement and rotation of the trolley into the given bounds, which means that

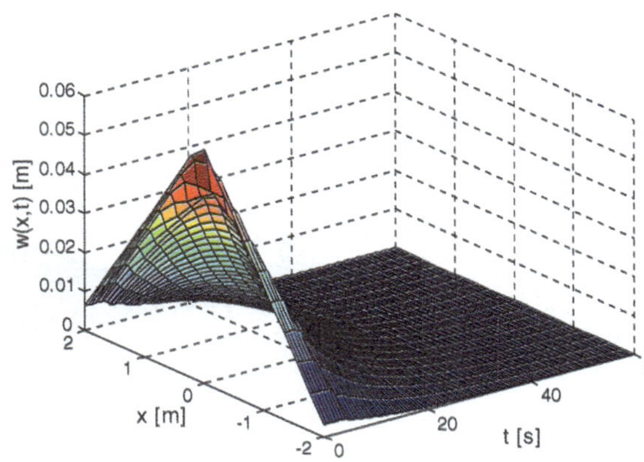

Fig. 11.3 Displacement $w(x, t)$ of the crane bridge under the proposed control laws

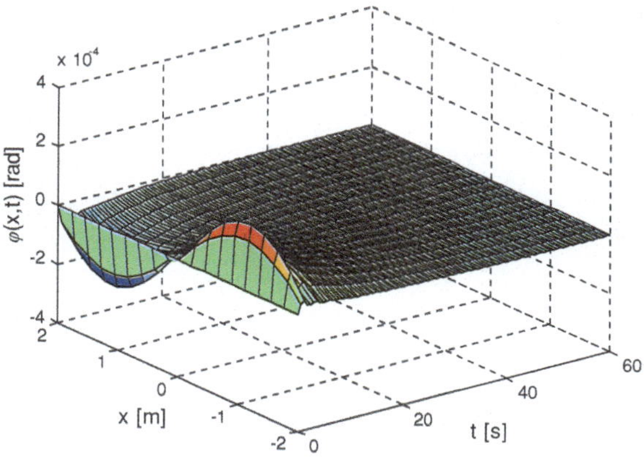

Fig. 11.4 Rotation $\varphi(x, t)$ of the crane bridge under the proposed control laws

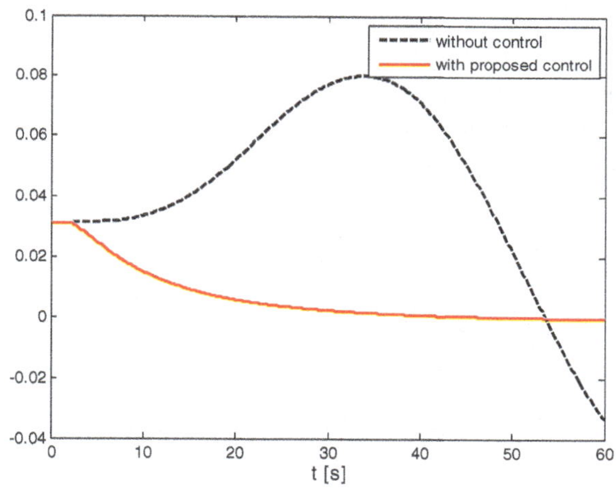

Fig. 11.5 Displacement $w_L\left(\frac{L_1}{2}, t\right)$ of the crane bridge under the proposed control laws

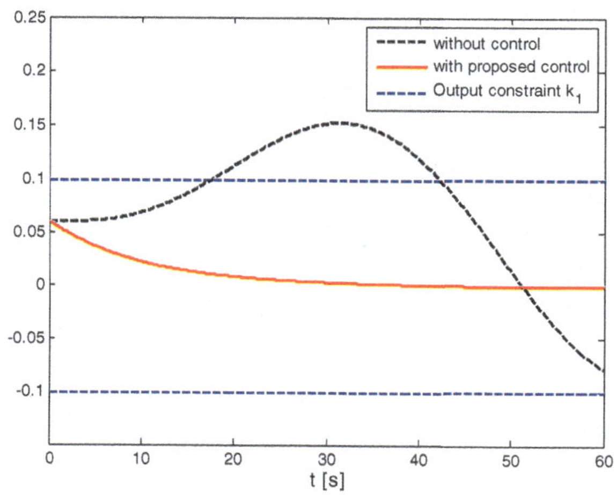

Fig. 11.6 Displacement $w_0(t)$ of the middle trolley under the proposed control laws

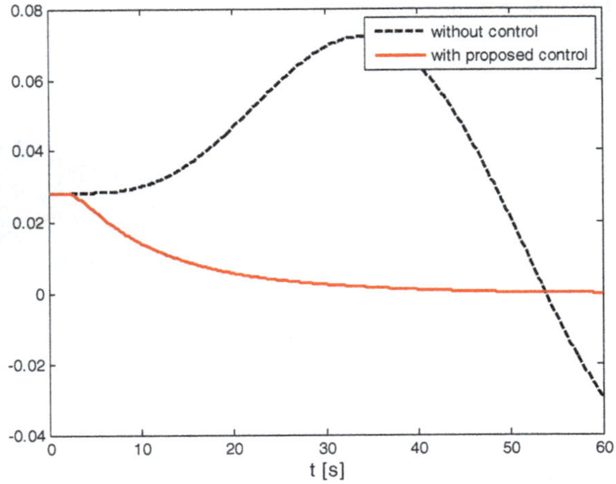

Fig. 11.7 Displacement $w_R\left(L_1 + \frac{L_2}{2}, t\right)$ of the crane bridge under the proposed control laws

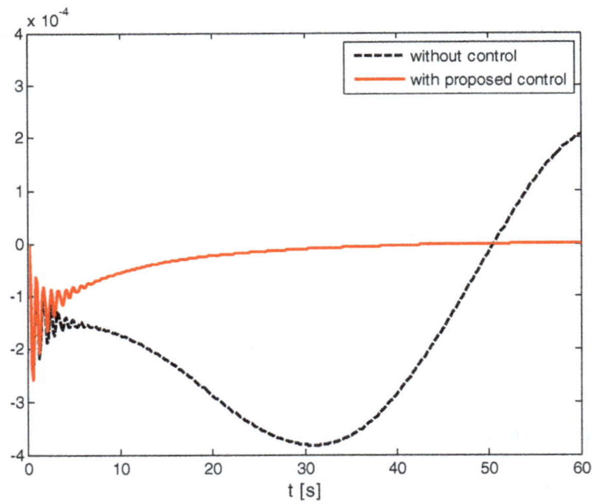

Fig. 11.8 Rotation $\varphi_L\left(\frac{L_1}{2}, t\right)$ of the crane bridge under the proposed control laws

output constraints $|w_0(t)| < k_1$ and $|\varphi_0(t)| < k_2$ can always be satisfied without any violation.

Figures 11.11 and 11.12 depict the control signals $u(t)$ and $\tau(t)$ flowing into the system.

Through the analysis of simulation results, it is noteworthy that states of the closed-loop system can be greatly restricted with the required output constraints of the trolley. Consequently, the simulation results in Figs. 11.1, 11.2, 11.3, 11.4, 11.5, 11.6, 11.7, 11.8, 11.9, 11.10, 11.11, 11.12 show the validity of the developed control strategy.

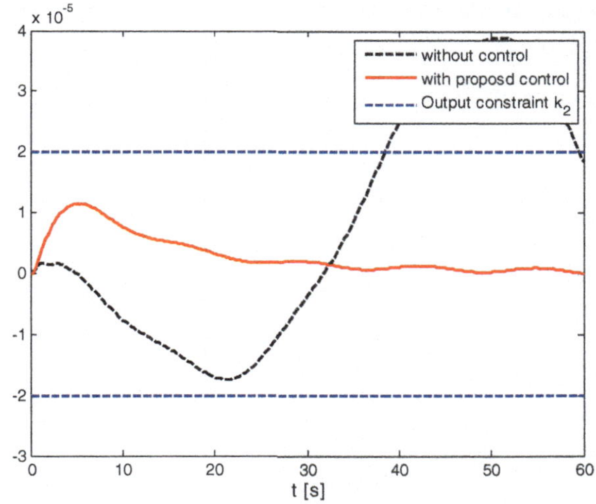

Fig. 11.9 Rotation $\varphi_0(t)$ of the middle trolley under the proposed control laws

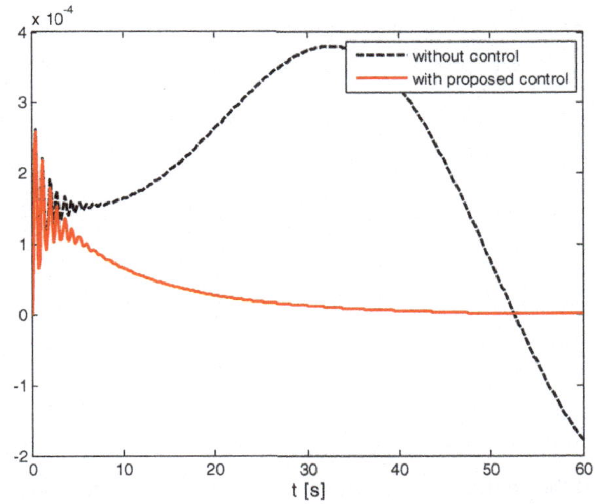

Fig. 11.10 Rotation $\varphi_R\left(L_1 + \frac{L_2}{2}, t\right)$ of the crane bridge under the proposed control laws

11.5 Conclusions

In this chapter, the PDE model has been used to depict the dynamics of the overhead crane bridge system which is abstracted as a Timoshenko beam with a middle attached rigid body, and a robust control scheme has been developed to realize system stabilization even in the presence of external perturbations and output constraints. Under the proposed control, the displacement and rotation of the system are eliminated even though there are unknown disturbances. In addition, the output constraints

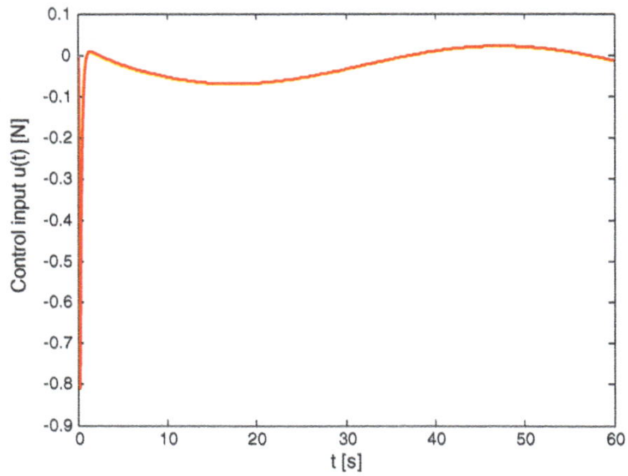

Fig. 11.11 Control force $u(t)$

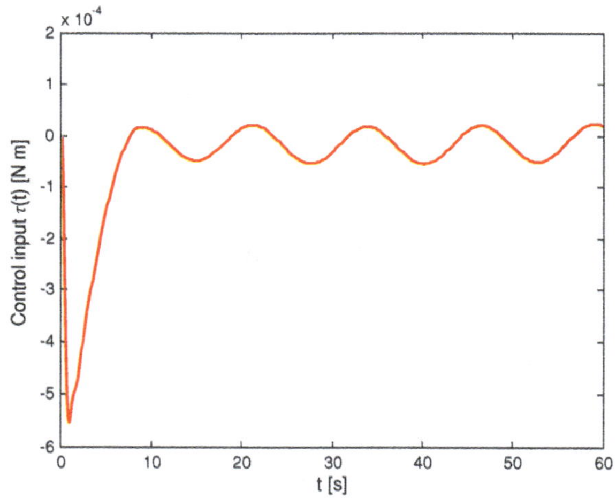

Fig. 11.12 Control torque $\tau(t)$

of the central trolley can always be ensured in the course of control. Simulations have conducted to verify the effectiveness of the proposed control.

References

1. W. He, S.S. Ge, Vibration control of a flexible beam with output constraint. IEEE Trans. Ind. Electron. **62**(8), 5023–5030 (2015)
2. K.P. Tee, B. Ren, S.S. Ge, Control of nonlinear systems with time-varying output constraints. Automatica **47**(11), 2511–2516 (2011)
3. H. Ouyang, Y. Lin, Adaptive fault-tolerant control for actuator failures: a switching strategy. Automatica **81**, 87–95 (2017)

Chapter 12
Vibration Control of Overhead Crane Bridge with Unknown Control Directions and Parametric Uncertainties

12.1 Introduction

In adaptive control researches, the signs of control gains are generally required to be known as priori information to simplify the control design. These signs are called control directions, which represent motion directions of the system under control [1, 2]. When the signs of control coefficients are uncertain, the system under control can be regarded a system with unknown control directions. As a result, it is essential to develop effective control methods to deal with the problem of uncertain control directions.

In this chapter, the vibration control is designed for an overhead crane bridge system based on a PDE model. With the infinite dimensional PDE model, an adaptive control approach is developed based on the Nussbaum function in order to settle out parametric uncertainties of the system with unknown control directions and disturbances. Under the proposed control method, states of the overhead crane bridge are globally bounded and finally converge to adjustable regions of zero. The stability of the closed-loop overhead crane bridge system is analyzed by employing Lyapunov's direct method. The effectiveness of the developed control strategy is verified through the simulation results.

12.2 Problem Formulation

The virtual work done by control signals and frictions is expressed by

$$W(t) = \theta_1 u(t) w_0(t) + \theta_2 \tau(t) \varphi_0(t) - \gamma_1 \int_0^{L_1} \dot{w}_L(x,t) w_L(x,t) \mathrm{d}x$$

X. Xing and J. Liu, *Vibration Control Methods of Mechanical Distributed Parameter Systems*, Springer Tracts in Mechanical Engineering,
https://doi.org/10.1007/978-981-16-1532-0_12

$$
- \gamma_1 \int_{L_1}^{L} \dot{w}_R(x,t) w_R(x,t) \mathrm{d}x - \gamma_2 \int_{0}^{L_1} \varphi_L(x,t) \dot{\varphi}_L(x,t) \mathrm{d}x
- \gamma_2 \int_{L_1}^{L} \varphi_R(x,t) \dot{\varphi}_R(x,t) \mathrm{d}x \tag{12.1}
$$

in which θ_1 and θ_2 are bounded and uncertain parameters with $\theta_i \neq 0, i = 1, 2$.

By using Hamilton principle in Chap. 2, governing equations of the overhead crane bridge can be achieved as

$$
\rho \ddot{w}_L(x,t) - K w_{Lxx}(x,t) + K \varphi_{Lx}(x,t) + \gamma_1 \dot{w}_L(x,t) = 0 \tag{12.2}
$$

$$
I_p \ddot{\varphi}_L(x,t) - EI \varphi_{Lxx}(x,t) - K w_{Lx}(x,t) + K \varphi_L(x,t) + \gamma_2 \dot{\varphi}_L(x,t) = 0 \tag{12.3}
$$

for $\forall (x,t) \in (0, L_1) \times [0, \infty)$, and

$$
\rho \ddot{w}_R(x,t) - K w_{Rxx}(x,t) + K \varphi_{Rx}(x,t) + \gamma_1 \dot{w}_R(x,t) = 0 \tag{12.4}
$$

$$
I_p \ddot{\varphi}_R(x,t) - EI \varphi_{Rxx}(x,t) - K w_{Rx}(x,t) + K \varphi_R(x,t) + \gamma_2 \dot{\varphi}_R(x,t) = 0 \tag{12.5}
$$

for $\forall (x,t) \in (L_1, L) \times [0, \infty)$, where $K = kGA$, and boundary conditions are

$$
M \ddot{w}_0(t) + K w_{Lx}(L_1,t) - K w_{Rx}(L_1,t) - d_1(t) - \theta_1 u(t) = 0 \tag{12.6}
$$

$$
J \ddot{\varphi}_0(t) + EI \varphi_{Lx}(L_1,t) - EI \varphi_{Rx}(L_1,t) - d_2(t) - \theta_2 \tau(t) = 0 \tag{12.7}
$$

$$
w_L(0,t) = w_R(L,t) = \varphi_L(0,t) = \varphi_R(L,t) = 0 \tag{12.8}
$$

$$
w_L(L_1,t) = w_R(L_1,t) = w_0(t), \varphi_L(L_1,t) = \varphi_R(L_1,t) = \varphi_0(t) \tag{12.9}
$$

for $\forall t \in [0, \infty)$, where θ_1 and θ_2 are unknown nonzero constants, and their signs, which represent control directions, are uncertain.

Remark 12.1 When $\theta_1 = \theta_2 = 1$, the system described by Eqs. (12.2)–(12.9) satisfies the structure specified by the so-called strict feedback condition. However, a broader class of systems are investigated recently in [3], in which the uncertainties of control directions have been considered. As we know, when partial fault happens to the actuator but not complete failure in practical engineering, the direction and amplitude of the control input signal are both unknown. In order to simulate the above

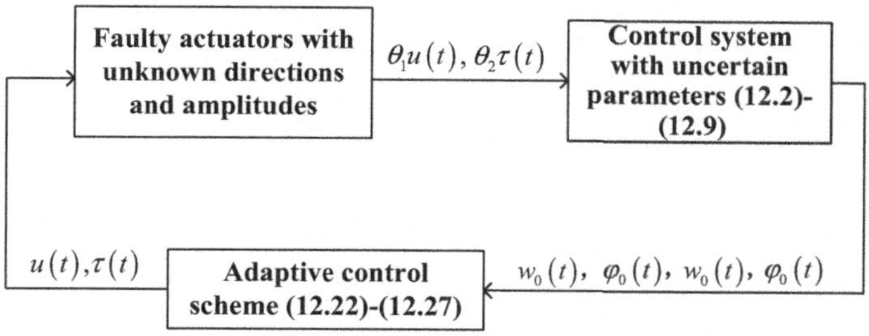

Fig. 12.1 Block diagram of the control system

conditions, uncertain constants θ_1 and θ_2 are introduced to the system in Eqs. (12.6) and (12.7), and the signs of θ_1 and θ_2, so-called control directions, are uncertain.

The block diagram of the control system is given in Fig. 12.1.

Assumption 12.1 Due to the limited disturbance energy, we assume that $d_1(t)$ and $d_2(t)$ are bounded, that is, there exist positive constants D_1 and D_2 satisfying $|d_1(t)| \leq D_1$ and $|d_2(t)| \leq D_2$.

12.3 Control Design and Stability Analysis

In this section, to eliminate the deflection and rotation of the crane bridge beam with a lumped mass in the middle, a control strategy is proposed to deal with both parametric uncertainties and unknown control directions. The control inputs are all applied on the middle trolley.

According to the definition in Definition 2.3, two Nussbaum functions are defined as

$$\mathcal{N}_1(k_1) = k_1 \cos(k_1) \tag{12.10}$$

$$\mathcal{N}_2(k_2) = k_2 \cos(k_2) \tag{12.11}$$

where k_1 and k_2 are two variables which will be designed later.

Four auxiliary variables are defined as

$$z_1(t) = w_0(t) \tag{12.12}$$

$$z_2(t) = \varphi_0(t) \tag{12.13}$$

$$z_3(t) = \dot{z}_1(t) + \eta_1(t) \tag{12.14}$$

$$z_4(t) = \dot{z}_2(t) + \eta_2(t) \tag{12.15}$$

where $\eta_1(t)$ and $\eta_2(t)$ are virtual control laws designed as

$$\eta_1(t) = \frac{\alpha}{\beta} z_1(t) = \frac{\alpha}{\beta} w_0(t) \tag{12.16}$$

$$\eta_2(t) = \frac{\alpha}{\beta} z_2(t) = \frac{\alpha}{\beta} \varphi_0(t) \tag{12.17}$$

where $\alpha > 0$ and $\beta > 0$.

A Lyapunov function is selected as

$$V_{11}(t) = \frac{\beta \bar{k}_1}{2} z_1^2(t) + \frac{\beta \bar{k}_2}{2} z_2^2(t) \tag{12.18}$$

where $\bar{k}_1 > 0$ and $\bar{k}_2 > 0$.

Differentiating Eq. (12.18) and substituting Eqs. (12.14) and (12.15), we have

$$\begin{aligned} \dot{V}_{11}(t) &= \beta \bar{k}_1 z_1(t) \left(z_3(t) - \frac{\alpha}{\beta} z_1(t) \right) + \beta \bar{k}_2 z_2(t) \left(z_4(t) - \frac{\alpha}{\beta} z_2(t) \right) \\ &= -\alpha \bar{k}_1 z_1^2(t) - \alpha \bar{k}_2 z_2^2(t) + \beta \bar{k}_1 z_1(t) z_3(t) + \beta \bar{k}_2 z_2(t) z_4(t). \end{aligned} \tag{12.19}$$

Then a new Lyapunov function is chosen on the basis of Eq. (12.18)

$$V_1(t) = V_{11}(t) + \frac{\beta}{2} M z_3^2(t) + \frac{\beta}{2} J z_4^2(t). \tag{12.20}$$

The derivative of Eq. (12.20) can be obtained from Eqs. (12.14)–(12.17) and (12.19)

$$\begin{aligned} \dot{V}_1(t) = &-\alpha \bar{k}_1 z_1^2(t) - \alpha \bar{k}_2 z_2^2(t) + \beta \bar{k}_1 z_1(t) z_3(t) + \beta \bar{k}_2 z_2(t) z_4(t) \\ &+ \beta z_3(t)(-K w_{Lx}(L_1, t) + K w_{Rx}(L_1, t) + \theta_1 u(t) + d_1(t) + M \dot{\eta}_1(t)) \\ &+ \beta z_4(t)(-EI \varphi_{Lx}(L_1, t) + EI \varphi_{Rx}(L_1, t) + \theta_2 \tau(t) + d_2(t) + J \dot{\eta}_2(t)). \end{aligned} \tag{12.21}$$

Considering uncertain system parameters and control directions in the practical reality, the adaptive control is developed as follows

$$u(t) = \mathcal{N}_1(k_1) \Big(\kappa_1 z_3(t) + \hat{M}(t) \dot{\eta}_1(t) + \bar{k}_1 z_1(t) + D_1 \operatorname{sgn}(z_3(t)) \Big) \tag{12.22}$$

$$\tau(t) = \mathcal{N}_2(k_2)\Big(\kappa_2 z_4(t) + \hat{J}(t)\dot{\eta}_2(t) + \bar{k}_2 z_2(t) + D_2 \mathrm{sgn}(z_4(t))\Big) \tag{12.23}$$

in which the adaptive laws are designed as

$$\dot{k}_1(t) = z_3(t)\Big(\kappa_1 z_3(t) + \hat{M}(t)\dot{\eta}_1(t) + \bar{k}_1 z_1(t) + D_1 \mathrm{sgn}(z_3(t))\Big) \tag{12.24}$$

$$\dot{k}_2(t) = z_4(t)\Big(\kappa_2 z_4(t) + \hat{J}(t)\dot{\eta}_2(t) + \bar{k}_2 z_2(t) + D_2 \mathrm{sgn}(z_4(t))\Big) \tag{12.25}$$

$$\dot{\hat{M}}(t) = \mu_1 \dot{\eta}_1(t) z_3(t) - \mu_2 \hat{M}(t) \tag{12.26}$$

$$\dot{\hat{J}}(t) = \mu_3 \dot{\eta}_2(t) z_4(t) - \mu_4 \hat{J}(t) \tag{12.27}$$

where $\kappa_1 > 0$, $\kappa_2 > 0$ and $\mu_i > 0$, $i = 1, 2, 3, 4$.

Theorem 12.1 If the PDE model of overhead crane bridge in Eqs. (12.2)–(12.9) satisfies Assumption 12.1 and system initial conditions are bounded, the closed-loop system is globally stable, and following properties of the crane bridge system with control laws in Eqs. (12.22)–(12.27) hold when the control direction and system parameters J and M are uncertain:

(1) The states of the closed-loop system are respectively uniformly bounded for $\forall (x, t) \in [0, L] \times [0, \infty)$;
(2) The undesirable deflection and rotation of the system can eventually converge to a given compact set which is defined as $\Omega_1 = \Big\{ w(x,t), \varphi(x,t) \in R \,|\, \lim_{t\to\infty} |w(x,t)|, \lim_{t\to\infty} |\varphi(x,t)| \le C, \forall x \in [0, L] \Big\}$ when $t \to \infty$, where $C = \min\left\{ \sqrt{\frac{L_1 \bar{D}_0}{\lambda_5}}, \sqrt{\frac{L_2 \bar{D}_0}{\lambda_5}}, \sqrt{\frac{2\bar{D}_0}{\beta \bar{k}_1 \lambda_5}}, \sqrt{\frac{2\bar{D}_0}{\beta \bar{k}_2 \lambda_5}} \right\}$, $\bar{D}_0$ and λ_5 are defined in the following proof process.

Proof The following Lyapunov function is constructed as

$$V(t) = V_1(t) + V_2(t) + V_3(t) + V_4(t) + V_5(t) \tag{12.28}$$

where

$$V_2(t) = \frac{\beta}{2}\rho \int_0^{L_1} \dot{w}_L^2(x,t)\mathrm{d}x + \frac{\beta}{2} I_p \int_0^{L_1} \dot{\varphi}_L^2(x,t)\mathrm{d}x + \frac{\beta}{2} EI \int_0^{L_1} \varphi_{Lx}^2(x,t)\mathrm{d}x$$
$$+ \frac{\beta}{2} K \int_0^{L_1} \left(\varphi_L(x,t) - w_{Lx}(x,t)\right)^2 \mathrm{d}x$$

$$+\frac{\beta}{2}\rho\int_{L_1}^{L}\dot{w}_R^2(x,t)\mathrm{d}x+\frac{\beta}{2}I_p\int_{L_1}^{L}\dot{\varphi}_R^2(x,t)\mathrm{d}x$$

$$+\frac{\beta}{2}EI\int_{L_1}^{L}\varphi_{Rx}^2(x,t)\mathrm{d}x+\frac{\beta}{2}K\int_{L_1}^{L}\left(\varphi_R(x,t)-w_{Rx}(x,t)\right)^2\mathrm{d}x \tag{12.29}$$

$$V_3(t)=\alpha\rho\int_0^{L_1}\dot{w}_L(x,t)w_L(x,t)\mathrm{d}x+\alpha I_p\int_0^{L_1}\dot{\varphi}_L(x,t)\varphi_L(x,t)\mathrm{d}x$$

$$+\alpha\rho\int_{L_1}^{L}\dot{w}_R(x,t)w_R(x,t)\mathrm{d}x+\alpha I_p\int_{L_1}^{L}\dot{\varphi}_R(x,t)\varphi_R(x,t)\mathrm{d}x \tag{12.30}$$

$$V_4(t)=\frac{\alpha\gamma_1}{2}\int_0^{L_1}w_L^2(x,t)\mathrm{d}x+\frac{\alpha\gamma_2}{2}\int_0^{L_1}\varphi_L^2(x,t)\mathrm{d}x+\frac{\alpha\gamma_1}{2}\int_{L_1}^{L}w_R^2(x,t)\mathrm{d}x$$

$$+\frac{\alpha\gamma_2}{2}\int_{L_1}^{L}\varphi_R^2(x,t)\mathrm{d}x \tag{12.31}$$

$$V_5(t)=\frac{\beta}{2\mu_1}\tilde{M}^2(t)+\frac{\beta}{2\mu_3}\tilde{J}^2(t) \tag{12.32}$$

where $\tilde{M}(t)=\hat{M}(t)-M$ and $\tilde{J}(t)=\hat{J}(t)-J$.

Define a positive function as

$$V_0(t)=\int_0^{L_1}\left(\dot{w}_L^2(x,t)+\dot{\varphi}_L^2(x,t)+\varphi_{Lx}^2(x,t)+w_{Lx}^2(x,t)+\varphi_L^2(x,t)+w_L^2(x,t)\right)\mathrm{d}x$$

$$+\int_{L_1}^{L}\left(\dot{w}_R^2(x,t)+\dot{\varphi}_R^2(x,t)+\varphi_{Rx}^2(x,t)+w_{Rx}^2(x,t)+\varphi_R^2(x,t)+w_R^2(x,t)\right)\mathrm{d}x. \tag{12.33}$$

Using Lemma 2.6, it yields

$$\int_0^{L_1}w_L^2(x,t)\mathrm{d}x\le 4L_1^2\int_0^{L_1}w_{Lx}^2(x,t)\mathrm{d}x,\ \int_0^{L_1}\varphi_L^2(x,t)\mathrm{d}x\le 4L_1^2\int_0^{L_1}\varphi_{Lx}^2(x,t)\mathrm{d}x \tag{12.34}$$

$$\int_{L_1}^{L} w_R^2(x,t)\mathrm{d}x \le 4L_2^2 \int_{L_1}^{L} w_{Rx}^2(x,t)\mathrm{d}x, \int_{L_1}^{L} \varphi_R^2(x,t)\mathrm{d}x \le 4L_2^2 \int_{L_1}^{L} \varphi_{Rx}^2(x,t)\mathrm{d}x. \tag{12.35}$$

Using Lemmas 2.2, 2.5, 2.6 and inequalities (12.34) and (12.35), the following inequality can be obtained

$$\begin{aligned} V_2(t)+V_4(t) \ge{}& \frac{\beta}{2}\rho \int_0^{L_1} \dot{w}_L^2(x,t)\mathrm{d}x + \frac{\beta}{2} I_p \int_0^{L_1} \dot{\varphi}_L^2(x,t)\mathrm{d}x + \frac{\beta}{4} EI \int_0^{L_1} \varphi_{Lx}^2(x,t)\mathrm{d}x \\ &+ \int_0^{L_1} \mathbf{\Phi}^{\mathrm{T}}(x,t)\mathbf{\Delta}_1\mathbf{\Phi}(x,t)\mathrm{d}x + \frac{\beta}{2}\rho \int_{L_1}^{L} \dot{w}_R^2(x,t)\mathrm{d}x \\ &+ \frac{\beta}{2} I_p \int_{L_1}^{L} \dot{\varphi}_R^2(x,t)\mathrm{d}x \\ &+ \frac{\beta}{4} EI \int_{L_1}^{L} \varphi_{Rx}^2(x,t)\mathrm{d}x + \int_{L_1}^{L} \mathbf{\Psi}^{\mathrm{T}}(x,t)\mathbf{\Delta}_2\mathbf{\Psi}(x,t)\mathrm{d}x \\ &+ \frac{\alpha\gamma_1}{2} \int_0^{L_1} w_L^2(x,t)\mathrm{d}x + \frac{\alpha\gamma_1}{2} \int_{L_1}^{L} w_R^2(x,t)\mathrm{d}x \end{aligned} \tag{12.36}$$

where $\mathbf{\Phi}(x,t) = \left[\varphi_L(x,t)\ w_{Lx}(x,t)\right]^{\mathrm{T}}$, $\mathbf{\Psi}(x,t) = \left[\varphi_R(x,t)\ w_{Rx}(x,t)\right]^{\mathrm{T}}$, and $\mathbf{\Delta}_i \in R^{2\times 2}$, $i = 1, 2$, is a positive matrix defined as $\mathbf{\Delta}_i = \begin{bmatrix} \frac{\beta EI}{16L_i^2} + \frac{\beta}{2}K + \frac{\alpha\gamma_2}{2} & -\frac{\beta}{2}K \\ -\frac{\beta}{2}K & \frac{\beta}{2}K \end{bmatrix} > 0.$

From Eq. (12.33) and inequality (12.36), it is obvious that

$$V_2(t)+V_4(t) \ge \lambda_1 V_0(t) \tag{12.37}$$

where $\lambda_1 = \min\left\{\frac{\beta}{2}\rho, \frac{\beta}{2}I_p, \frac{\beta}{4}EI, \sigma_{\min}(\mathbf{\Delta}_1), \sigma_{\min}(\mathbf{\Delta}_2), \frac{\alpha\gamma_1}{2}\right\} > 0.$

Similarly, we have

$$V_2(t)+V_4(t) \le \frac{\beta}{2}\rho \int_0^{L_1} \dot{w}_L^2(x,t)\mathrm{d}x + \frac{\beta}{2} I_p \int_0^{L_1} \dot{\varphi}_L^2(x,t)\mathrm{d}x + \frac{\beta}{2} EI \int_0^{L_1} \varphi_{Lx}^2(x,t)\mathrm{d}x$$

$$
\begin{aligned}
&+\left(\beta K+\frac{\alpha\gamma_2}{2}\right)\int\limits_0^{L_1}\varphi_L^2(x,t)\mathrm{d}x+\beta K\int\limits_0^{L_1}w_{Lx}^2(x,t)\mathrm{d}x\\
&+\frac{\beta}{2}\rho\int\limits_{L_1}^{L}\dot{w}_R^2(x,t)\mathrm{d}x\\
&+\frac{\beta}{2}I_p\int\limits_{L_1}^{L}\dot{\varphi}_R^2(x,t)\mathrm{d}x+\frac{\beta}{2}EI\int\limits_{L_1}^{L}\varphi_{Rx}^2(x,t)\mathrm{d}x\\
&+\left(\beta K+\frac{\alpha\gamma_2}{2}\right)\int\limits_{L_1}^{L}\varphi_R^2(x,t)\mathrm{d}x\\
&+\beta K\int\limits_{L_1}^{L}w_{Rx}^2(x,t)\mathrm{d}x+\frac{\alpha\gamma_1}{2}\int\limits_0^{L_1}w_L^2(x,t)\mathrm{d}x+\frac{\alpha\gamma_1}{2}\int\limits_{L_1}^{L}w_R^2(x,t)\mathrm{d}x\\
&\leq\lambda_2V_0(t)
\end{aligned}
\tag{12.38}
$$

where $\lambda_2=\max\left\{\frac{1}{2}\beta\rho,\frac{\beta}{2}I_p,\frac{\beta}{2}EI,\beta K+\frac{\alpha\gamma_2}{2},\frac{\alpha\gamma_1}{2}\right\}>0$.

According to Eq. (12.30), we know

$$
\begin{aligned}
|V_3(t)|&\leq\frac{\alpha\rho}{2}\int\limits_0^{L_1}\dot{w}_L^2(x,t)\mathrm{d}x+\frac{\alpha\rho}{2}\int\limits_0^{L_1}w_L^2(x,t)\mathrm{d}x+\frac{\alpha I_p}{2}\int\limits_0^{L_1}\dot{\varphi}_L^2(x,t)\mathrm{d}x\\
&+\frac{\alpha I_p}{2}\int\limits_0^{L_1}\varphi_L^2(x,t)\mathrm{d}x+\frac{\alpha\rho}{2}\int\limits_{L_1}^{L}\dot{w}_R^2(x,t)\mathrm{d}x+\frac{\alpha\rho}{2}\int\limits_{L_1}^{L}w_R^2(x,t)\mathrm{d}x\\
&+\frac{\alpha I_p}{2}\int\limits_{L_1}^{L}\dot{\varphi}_R^2(x,t)\mathrm{d}x+\frac{\alpha I_p}{2}\int\limits_{L_1}^{L}\varphi_R^2(x,t)\mathrm{d}x\\
&\leq\lambda_3V_0(t)
\end{aligned}
\tag{12.39}
$$

in which $\lambda_3=\frac{1}{2}\alpha\max\left\{I_p,\rho\right\}>0$.

Let $\lambda_1>\lambda_3$ by choosing proper control coefficients α and β, and then we have

$$
0<(\lambda_1-\lambda_3)V_0(t)\leq V_2(t)+V_3(t)+V_4(t)\leq(\lambda_2+\lambda_3)V_0(t). \tag{12.40}
$$

Define $a_1=\lambda_1-\lambda_3>0$ and $a_2=\lambda_2+\lambda_3>0$

$$
\lambda_5(V_0(t)+V_1(t)+V_5(t))\leq V(t)\leq\lambda_4(V_0(t)+V_1(t)+V_5(t)) \tag{12.41}
$$

where $\lambda_4 = \max\{a_2, 1\}$ and $\lambda_5 = \min\{a_1, 1\}$.

With Eqs. (12.22)–(12.25), the time derivative of $V_1(t)$ can be written as

$$\begin{aligned}
\dot{V}_1(t) \le\ & \beta z_3(t)(-Kw_{Lx}(L_1,t) + Kw_{Rx}(L_1,t)) \\
& + \beta z_4(t)(-EI\varphi_{Lx}(L_1,t) + EI\varphi_{Rx}(L_1,t)) \\
& + \beta(\theta_1\mathcal{N}_1(k_1)+1)z_3(t)\Big(\kappa_1 z_3(t) + \hat{M}(t)\dot{\eta}_1(t) + \bar{k}_1 z_1(t) + D_1\mathrm{sgn}(z_3(t))\Big) \\
& - \alpha\bar{k}_1 z_1^2(t) - \beta\kappa_1 z_3^2(t) - \alpha\bar{k}_2 z_2^2(t) - \beta\kappa_2 z_4^2(t) \\
& + \beta(\theta_2\mathcal{N}_2(k_2)+1)z_4(t)\Big(\kappa_2 z_4(t) + \hat{J}(t)\dot{\eta}_2(t) + \bar{k}_2 z_2(t) + D_2\mathrm{sgn}(z_4(t))\Big) \\
& - \beta z_3(t)\tilde{M}(t)\dot{\eta}_1(t) - \beta z_4(t)\tilde{J}(t)\dot{\eta}_2(t) \\
=\ & \beta z_3(t)(-Kw_{Lx}(L_1,t) + Kw_{Rx}(L_1,t)) \\
& + \beta z_4(t)(-EI\varphi_{Lx}(L_1,t) + EI\varphi_{Rx}(L_1,t)) \\
& - \alpha\bar{k}_1 z_1^2(t) - \beta\kappa_1 z_3^2(t) - \alpha\bar{k}_2 z_2^2(t) - \beta\kappa_2 z_4^2(t) \\
& + \beta(\theta_1\mathcal{N}_1(k_1)+1)\dot{k}_1(t) + \beta(\theta_2\mathcal{N}_2(k_2)+1)\dot{k}_2(t) \\
& - \beta z_3(t)\tilde{M}(t)\dot{\eta}_1(t) - \beta z_4(t)\tilde{J}(t)\dot{\eta}_2(t).
\end{aligned} \tag{12.42}$$

Differentiating Eq. (12.29) and substituting Eqs. (12.2)–(12.9), we have

$$\begin{aligned}
\dot{V}_2(t) =\ & \beta K\dot{w}_L(L_1,t)(w_{Lx}(L_1,t) - \varphi_L(L_1,t)) + \beta EI\dot{\varphi}_L(L_1,t)\varphi_{Lx}(L_1,t) \\
& - \beta K\dot{w}_R(L_1,t)(w_{Rx}(L_1,t) - \varphi_R(L_1,t)) - \beta EI\dot{\varphi}_R(L_1,t)\varphi_{Rx}(L_1,t) \\
& - \beta\gamma_1\int_0^{L_1}\dot{w}_L^2(x,t)\mathrm{d}x - \beta\gamma_2\int_0^{L_1}\dot{\varphi}_L^2(x,t)\mathrm{d}x \\
& - \beta\gamma_1\int_{L_1}^{L}\dot{w}_R^2(x,t)\mathrm{d}x - \beta\gamma_2\int_{L_1}^{L}\dot{\varphi}_R^2(x,t)\mathrm{d}x.
\end{aligned} \tag{12.43}$$

After a series of calculation, we have

$$\begin{aligned}
\dot{V}_3(t) =\ & \alpha EI\varphi_L(L_1,t)\varphi_{Lx}(L_1,t) - \alpha EI\int_0^{L_1}\varphi_{Lx}^2(x,t)\mathrm{d}x \\
& + 2\alpha K\int_0^{L_1}\varphi_L(x,t)w_{Lx}(x,t)\mathrm{d}x \\
& - \alpha K\int_0^{L_1}\varphi_L^2(x,t)\mathrm{d}x + \alpha I_p\int_0^{L_1}\dot{\varphi}_L^2(x,t)\mathrm{d}x - \alpha\gamma_2\int_0^{L_1}\varphi_L(x,t)\dot{\varphi}_L(x,t)\mathrm{d}x
\end{aligned}$$

$$
\begin{aligned}
&- \alpha EI \varphi_R(L_1, t)\varphi_{Rx}(L_1, t) - \alpha EI \int_{L_1}^{L} \varphi_{Rx}^2(x, t)\mathrm{d}x \\
&+ 2\alpha K \int_{L_1}^{L} \varphi_R(x, t) w_{Rx}(x, t)\mathrm{d}x \\
&- \alpha K \int_{L_1}^{L} \varphi_R^2(x, t)\mathrm{d}x + \alpha I_p \int_{L_1}^{L} \dot{\varphi}_R^2(x, t)\mathrm{d}x - \alpha\gamma_2 \int_{L_1}^{L} \varphi_R(x, t)\dot{\varphi}_R(x, t)\mathrm{d}x \\
&+ \alpha K w_L(L_1, t) w_{Lx}(L_1, t) - \alpha K \int_0^{L_1} w_{Lx}^2(x, t)\mathrm{d}x + \alpha\rho \int_0^{L_1} \dot{w}_L^2(x, t)\mathrm{d}x \\
&- \alpha\gamma_1 \int_0^{L_1} w_L(x, t)\dot{w}_L(x, t)\mathrm{d}x - \alpha K w_R(L_1, t) w_{Rx}(L_1, t) \\
&- \alpha K \int_{L_1}^{L} w_{Rx}^2(x, t)\mathrm{d}x + \alpha\rho \int_{L_1}^{L} \dot{w}_R^2(x, t)\mathrm{d}x - \alpha\gamma_1 \int_{L_1}^{L} w_R(x, t)\dot{w}_R(x, t)\mathrm{d}x.
\end{aligned} \tag{12.44}
$$

The derivative of Eq. (12.31) is attained as

$$
\begin{aligned}
\dot{V}_4(t) = &\alpha\gamma_1 \int_0^{L_1} w_L(x, t)\dot{w}_L(x, t)\mathrm{d}x + \alpha\gamma_2 \int_0^{L_1} \varphi_L(x, t)\dot{\varphi}_L(x, t)\mathrm{d}x \\
&+ \alpha\gamma_1 \int_{L_1}^{L} w_R(x, t)\dot{w}_R(x, t)\mathrm{d}x + \alpha\gamma_2 \int_{L_1}^{L} \varphi_R(x, t)\dot{\varphi}_R(x, t)\mathrm{d}x.
\end{aligned} \tag{12.45}
$$

After taking the derivative of $V_5(t)$, we obtain

$$
\begin{aligned}
\dot{V}_5(t) = &\frac{\beta}{\mu_1}\tilde{M}(t)\Big(\mu_1\dot{\eta}_1(t)z_3(t) - \mu_2\hat{M}(t)\Big) \\
&+ \frac{\beta}{\mu_3}\tilde{J}(t)\Big(\mu_3\dot{\eta}_2(t)z_4(t) - \mu_4\hat{J}(t)\Big) \\
\le &\beta\tilde{M}(t)\dot{\eta}_1(t)z_3(t) - \frac{\beta\mu_2}{2\mu_1}\tilde{M}^2(t) + \frac{\beta\mu_2}{2\mu_1}M^2 \\
&+ \beta\tilde{J}(t)\dot{\eta}_2(t)z_4(t) - \frac{\beta\mu_4}{2\mu_3}\tilde{J}^2(t) + \frac{\beta\mu_4}{2\mu_3}J^2.
\end{aligned} \tag{12.46}
$$

Combing Eqs. (12.43)–(12.45) and inequalities (12.46) and (12.42), we have

$$
\begin{aligned}
\dot{V}(t) \le & -(\beta\gamma_1-\alpha\rho)\int_0^{L_1}\dot{w}_L^2(x,t)\mathrm{d}x-\left(\beta\gamma_2-\alpha I_p\right)\int_0^{L_1}\dot{\varphi}_L^2(x,t)\mathrm{d}x\\
&-(\beta\gamma_1-\alpha\rho)\int_{L_1}^{L}\dot{w}_R^2(x,t)\mathrm{d}x-\left(\beta\gamma_2-\alpha I_p\right)\int_{L_1}^{L}\dot{\varphi}_R^2(x,t)\mathrm{d}x\\
&-\alpha EI\int_0^{L_1}\varphi_{Lx}^2(x,t)\mathrm{d}x+2\alpha K\int_0^{L_1}\varphi_L(x,t)w_{Lx}(x,t)\mathrm{d}x-\alpha K\int_0^{L_1}\varphi_L^2(x,t)\mathrm{d}x\\
&-\alpha EI\int_{L_1}^{L}\varphi_{Rx}^2(x,t)\mathrm{d}x+2\alpha K\int_{L_1}^{L}\varphi_R(x,t)w_{Rx}(x,t)\mathrm{d}x-\alpha K\int_{L_1}^{L}\varphi_R^2(x,t)\mathrm{d}x\\
&-\alpha K\int_0^{L_1}w_{Lx}^2(x,t)\mathrm{d}x-\alpha K\int_{L_1}^{L}w_{Rx}^2(x,t)\mathrm{d}x-\alpha\bar{k}_1z_1^2(t)\\
&-\beta\kappa_1z_3^2(t)-\alpha\bar{k}_2z_2^2(t)\\
&-\beta\kappa_2z_4^2(t)+\beta(\theta_1\mathcal{N}_1(k_1)+1)\dot{k}_1(t)+\beta(\theta_2\mathcal{N}_2(k_2)+1)\dot{k}_2(t)\\
&-\frac{\beta\mu_2}{2\mu_1}\tilde{M}^2(t)+\frac{\beta\mu_2}{2\mu_1}M^2-\frac{\beta\mu_4}{2\mu_3}\tilde{J}^2(t)+\frac{\beta\mu_4}{2\mu_3}J^2.
\end{aligned}
\tag{12.47}
$$

Substituting inequalities (12.34) and (12.35) into inequality (12.47) and using Lemmas 2.2, 2.4, 2.5, the following inequality can be achieved

$$
\begin{aligned}
\dot{V}(t) \le & -(\beta\gamma_1-\alpha\rho)\int_0^{L_1}\dot{w}_L^2(x,t)\mathrm{d}x-\left(\beta\gamma_2-\alpha I_p\right)\int_0^{L_1}\dot{\varphi}_L^2(x,t)\mathrm{d}x\\
&-(\beta\gamma_1-\alpha\rho)\int_{L_1}^{L}\dot{w}_R^2(x,t)\mathrm{d}x-\left(\beta\gamma_2-\alpha I_p\right)\int_{L_1}^{L}\dot{\varphi}_R^2(x,t)\mathrm{d}x\\
&-\frac{\alpha}{2}EI\int_0^{L_1}\varphi_{Lx}^2(x,t)\mathrm{d}x-\left(\alpha K+\frac{\alpha}{8L_1^2}EI-\frac{2\alpha K}{\varepsilon_1}\right)\int_0^{L_1}\varphi_L^2(x,t)\mathrm{d}x\\
&-\frac{\alpha}{2}EI\int_{L_1}^{L}\varphi_{Rx}^2(x,t)\mathrm{d}x-\left(\alpha K+\frac{\alpha}{8L_2^2}EI-\frac{2\alpha K}{\varepsilon_2}\right)\int_{L_1}^{L}\varphi_R^2(x,t)\mathrm{d}x\\
&-\xi_1\int_0^{L_1}w_L^2(x,t)\mathrm{d}x
\end{aligned}
$$

$$
\begin{aligned}
&- \xi_2 \int\limits_{L_1}^{L} w_R^2(x,t)\mathrm{d}x - \left(\alpha K - 2\alpha K\varepsilon_1 - 4L_1^2\xi_1\right) \int\limits_0^{L_1} w_{Lx}^2(x,t)\mathrm{d}x \\
&- \left(\alpha K - 2\alpha K\varepsilon_2 - 4L_2^2\xi_2\right) \int\limits_{L_1}^{L} w_{Rx}^2(x,t)\mathrm{d}x \\
&- \alpha\bar{k}_1 z_1^2(t) - \beta\kappa_1 z_3^2(t) - \alpha\bar{k}_2 z_2^2(t) - \beta\kappa_2 z_4^2(t) \\
&+ \beta(\theta_1\mathcal{N}_1(k_1)+1)\dot{k}_1(t) + \beta(\theta_2\mathcal{N}_2(k_2)+1)\dot{k}_2(t) \\
&- \frac{\beta\mu_2}{2\mu_1}\tilde{M}^2(t) + \frac{\beta\mu_2}{2\mu_1}M^2 - \frac{\beta\mu_4}{2\mu_3}\tilde{J}^2(t) + \frac{\beta\mu_4}{2\mu_3}J^2
\end{aligned}
\tag{12.48}
$$

where $\varepsilon_1 > 0$ and $\varepsilon_2 > 0, \xi_1 > 0, \xi_2 > 0$, and $\alpha, \beta, \varepsilon_1, \varepsilon_2, \xi_1, \xi_2$ are selected to satisfy $\beta\gamma_1 - \alpha\rho > 0, \beta\gamma_2 - \alpha I_p > 0, \alpha K + \frac{\alpha}{8L_1^2}EI - \frac{2\alpha K}{\varepsilon_1} > 0, \alpha K + \frac{\alpha}{8L_2^2}EI - \frac{2\alpha K}{\varepsilon_2} > 0,$ $\alpha K - 2\alpha K\varepsilon_1 - 4L_1^2\xi_1 > 0, \alpha K - 2\alpha K\varepsilon_2 - 4L_2^2\xi_2 > 0.$

Then from inequality (12.48), we have

$$
\dot{V}(t) \le -\lambda_6(V_0(t) + V_1(t) + V_5(t)) + D \tag{12.49}
$$

where $D = \frac{\beta\mu_2}{2\mu_1}M^2 + \frac{\beta\mu_4}{2\mu_3}J^2 + \beta(\theta_1\mathcal{N}_1(k_1)+1)\dot{k}_1(t) + \beta(\theta_2\mathcal{N}_2(k_2)+1)\dot{k}_2(t)$ and

$$
\begin{aligned}
\lambda_6 = \min\Bigg\{ &\beta\gamma_1 - \alpha\rho, \beta\gamma_2 - \alpha I_p, \frac{\alpha}{2}EI, \alpha K + \frac{\alpha}{8L_1^2}EI - \frac{2\alpha K}{\varepsilon_1}, \\
&\alpha K + \frac{\alpha}{8L_2^2}EI - \frac{2\alpha K}{\varepsilon_2}, \alpha K - 2\alpha K\varepsilon_1 - 4L_1^2\xi_1, \\
&\alpha K - 2\alpha K\varepsilon_2 - 4L_2^2\xi_2, \xi_1, \xi_2, \frac{2\kappa_1}{M}, \frac{2\kappa_2}{J}, \frac{2\alpha}{\beta}, \mu_2, \mu_4 \Bigg\} > 0.
\end{aligned}
$$

According to inequality (12.41), we have

$$
\dot{V}(t) \le -\lambda V(t) + D \tag{12.50}
$$

where $\lambda = \frac{\lambda_6}{\lambda_4} > 0$.

Multiplying inequality (12.50) by $e^{\lambda t}$ and integrating yield

$$
V(t) \le V(0)e^{-\lambda t} + \bar{D} \tag{12.51}
$$

where

$$
\bar{D} = \frac{\beta}{2\lambda}\left(\frac{\mu_2}{\mu_1}M^2 + \frac{\mu_4}{\mu_3}J^2\right) + e^{-\lambda t}\int\limits_0^t \beta e^{\lambda s}\left[(\theta_1\mathcal{N}_1(k_1)+1)\dot{k}_1(s) + (\theta_2\mathcal{N}_2(k_2)+1)\dot{k}_2(s)\right]\mathrm{d}s.
$$

From Lemma 2.7, we can know that $V(t)$, $k_1(t)$, $k_2(t)$ and $\bar{D}$ are bounded. Let $|\bar{D}| \leq \bar{D}_0$, where $\bar{D}_0$ is a positive constant. Then using Lemmas 2.2 and 2.5 we have

$$\frac{1}{L_1}\varphi_L^2(x,t) \leq \int_0^{L_1} \varphi_{Lx}^2(x,t)\mathrm{d}x \leq \frac{1}{\lambda_5}V(t) \leq \frac{1}{\lambda_5}\left(V(0)e^{-\lambda t}+\bar{D}_0\right) \tag{12.52}$$

$$\frac{\beta\bar{k}_2}{2}\varphi_0^2(t) \leq \frac{1}{\lambda_5}V(t) \leq \frac{1}{\lambda_5}\left(V(0)e^{-\lambda t}+\bar{D}_0\right) \tag{12.53}$$

$$\frac{1}{L_2}\varphi_R^2(x,t) \leq \int_{L_1}^{L} \varphi_{Rx}^2(x,t)\mathrm{d}x \leq \frac{1}{\lambda_5}V(t) \leq \frac{1}{\lambda_5}\left(V(0)e^{-\lambda t}+\bar{D}_0\right) \tag{12.54}$$

$$\frac{1}{L_1}w_L^2(x,t) \leq \int_0^{L_1} w_{Lx}^2(x,t)\mathrm{d}x \leq \frac{1}{\lambda_5}V(t) \leq \frac{1}{\lambda_5}\left(V(0)e^{-\lambda t}+\bar{D}_0\right) \tag{12.55}$$

$$\frac{\beta\bar{k}_1}{2}w_0^2(t) \leq \frac{1}{\lambda_5}V(t) \leq \frac{1}{\lambda_5}\left(V(0)e^{-\lambda t}+\bar{D}_0\right) \tag{12.56}$$

$$\frac{1}{L_2}w_R^2(x,t) \leq \int_{L_1}^{L} w_{Rx}^2(x,t)\mathrm{d}x \leq \frac{1}{\lambda_5}V(t) \leq \frac{1}{\lambda_5}\left(V(0)e^{-\lambda t}+\bar{D}_0\right). \tag{12.57}$$

Hence, when $V(t) \leq V(0)e^{-\lambda t}+\bar{D}$ holds, we have

$$|\varphi_L(x,t)|, |w_L(x,t)| \leq \sqrt{\frac{L_1}{\lambda_5}\left(V(0)e^{-\lambda t}+\bar{D}_0\right)} \tag{12.58}$$

$$|\varphi_R(x,t)|, |w_R(x,t)| \leq \sqrt{\frac{L_2}{\lambda_5}\left(V(0)e^{-\lambda t}+\bar{D}_0\right)} \tag{12.59}$$

$$|w_0(t)| \leq \sqrt{\frac{2}{\beta\bar{k}_1\lambda_5}\left(V(0)e^{-\lambda t}+\bar{D}_0\right)} \tag{12.60}$$

$$|\varphi_0(t)| \leq \sqrt{\frac{2}{\beta\bar{k}_2\lambda_5}\left(V(0)e^{-\lambda t}+\bar{D}_0\right)} \tag{12.61}$$

and

$$\lim_{t\to\infty}|\varphi_L(x,t)|, \lim_{t\to\infty}|w_L(x,t)| \leq \sqrt{\frac{L_1\bar{D}_0}{\lambda_5}} \tag{12.62}$$

$$\lim_{t\to\infty}\left|\varphi_R(x,t)\right|, \lim_{t\to\infty}\left|w_R(x,t)\right| \le \sqrt{\frac{L_2\bar{D}_0}{\lambda_5}} \tag{12.63}$$

$$\lim_{t\to\infty}|w_0(t)| \le \sqrt{\frac{2\bar{D}_0}{\beta\bar{k}_1\lambda_5}}, \lim_{t\to\infty}|\varphi_0(t)| \le \sqrt{\frac{2\bar{D}_0}{\beta\bar{k}_2\lambda_5}}. \tag{12.64}$$

Therefore, if the boundedness of system initial conditions can be guaranteed, states of the closed-loop system are bounded subject to the developed control laws in Eqs. (12.22)–(12.27), and can eventually converge to small regions of zero by adjusting corresponding control parameters. Then the Theorem 12.1 is completely proved.

12.4 Simulations

In this section, several numerical simulations are implemented for evaluating the performance of control strategy in Eqs. (12.22)–(12.27).

System parameters are provided as $L = 4\,\text{m}$, $L_1 = 2\,\text{m}$, $L_2 = 2\,\text{m}$, $\rho = 2\,\text{kg/m}$, $\gamma_1 = 2.5\,\text{kg/m}$, $\gamma_2 = 2.5\,\text{kg/m}$, $M = 60\,\text{kg}$, $EI = 50\,\text{Nm}^2$, $I_p = 2\,\text{kgm}^2$, $K = 0.5\,\text{N}$, $J = 1\,\text{kgm}^2$. Other control parameters are given as: $\alpha = 1$, $\beta = 10$, $\bar{k}_1 = 10$, $\bar{k}_2 = 100$, $\kappa_1 = 300$, $\kappa_2 = 1000$, $\mu_1 = 10$, $\mu_2 = 0.01$, μ_3=10, $\mu_4 = 0.01$, $D_1 = 1$, and $D_2 = 1$. System initial states are set as $\dot{w}(x,0) = 0$, $\dot{\varphi}(x,0) = 0$, $w(x,0) = \begin{cases} 0.3x, & \text{if } 0 \le x < L_1 \\ 0.3L_1, & \text{if } x = L_1 \\ -0.3\frac{L_1}{L_2}(x-L), & \text{if } L_1 < x \le L \end{cases}$, $\varphi(x,0) = 0$. External disturbances acting on the overhead crane bridge system are $d_1(t) = \sin 0.5t$ and $d_2(t) = 5\times 10^{-4}\sin 0.5t$. In the simulation, θ_1 and θ_2 can be set artificially to simulate the problem of unknown control directions and verify the effectiveness of the proposed control scheme. Let θ_1 and θ_2 vary from $\theta_1 = 1$ and $\theta_2 = 1$ to $\theta_1 = -0.5$ and $\theta_2 = 0.5$ at $t = 10$ s to simulate the dynamic change of control input signals with uncertain directions and amplitudes. With these parameters, dynamic responses of the system are conducted.

Figures 12.2, 12.3 and 12.4 depict the flexural and rotational vibrations $w(x,t)$, $\varphi(x,t)$, $w_0(t)$ and $\varphi_0(t)$ of the crane bridge system respectively without control. It is observed that the system shows large oscillatory response without control inputs, which needs to be further suppressed to ensure the system performance.

The dynamic responses of states $w_L\left(\frac{L_1}{2},t\right)$, $\varphi_L\left(\frac{L_1}{2},t\right)$, $w_0(t)$, $\varphi_0(t)$, $w_R\left(L_1+\frac{L_2}{2},t\right)$, $\varphi_R\left(L_1+\frac{L_2}{2},t\right)$ in the closed-loop system are presented in Figs. 12.5, 12.6, 12.7, 12.8 and 12.9. Through observation, it is obvious that although the proposed control laws only act on the trolley, every point of the crane bridge decays rapidly to a small desirable range of zero, and the vibration of the trolley can be reduced.

Control input force $u(t)$ and control torque $\tau(t)$ are presented in Figs. 12.10 and

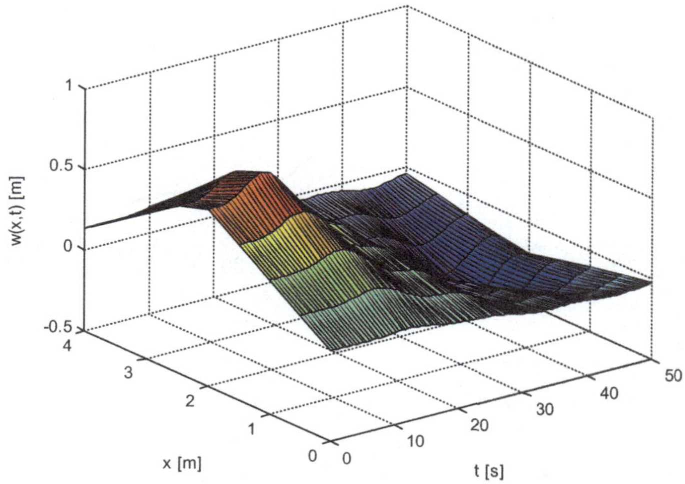

Fig. 12.2 Bending displacement $w(x, t)$ of the crane bridge without control

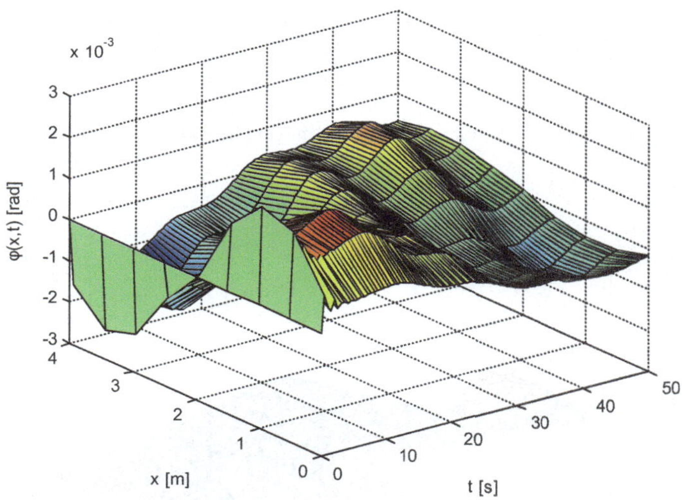

Fig. 12.3 Rotation $\varphi(x, t)$ of the crane bridge without control

12.11, respectively. It is shown that at $t = 10$ s, an obvious change occurs to the control force $u(t)$, which is caused by the variation of control directions.

From Figs. 12.2, 12.3, 12.4, 12.5, 12.6, 12.7, 12.8, 12.9, 12.10 and 12.11, the states of the crane bridge system can converge to acceptable neighborhoods of zero, and the undesirable vibration and rotation of the system can be greatly restricted by utilizing the proposed control scheme, which indicates the validity of the adaptive control laws in Eqs. (12.22)–(12.27).

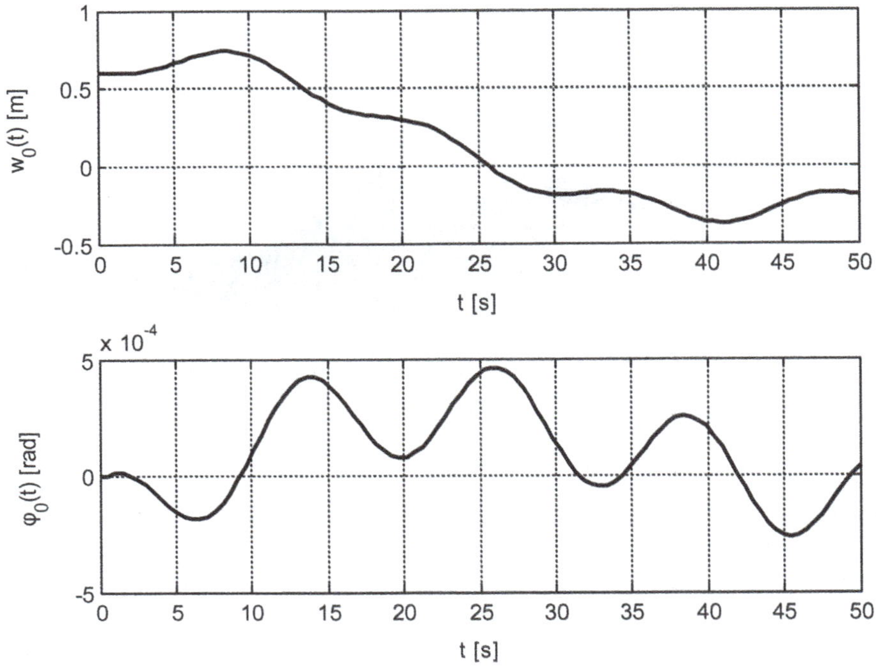

Fig. 12.4 Displacement $w_0(t)$ and rotation $\varphi_0(t)$ of the middle rigid body without control

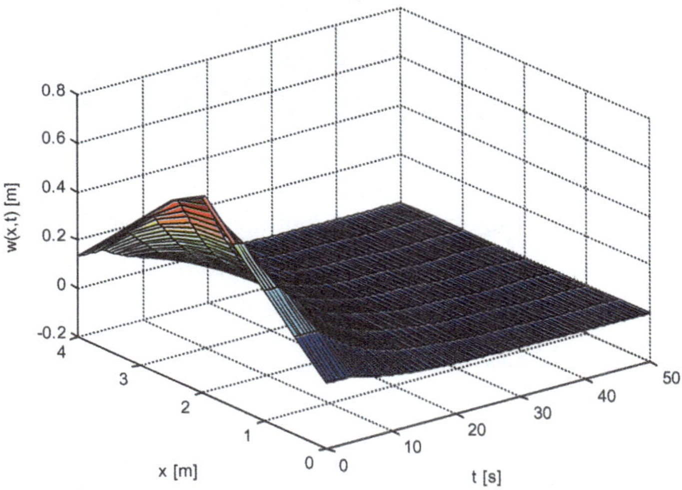

Fig. 12.5 Bending displacement $w(x, t)$ of the crane bridge with the proposed control

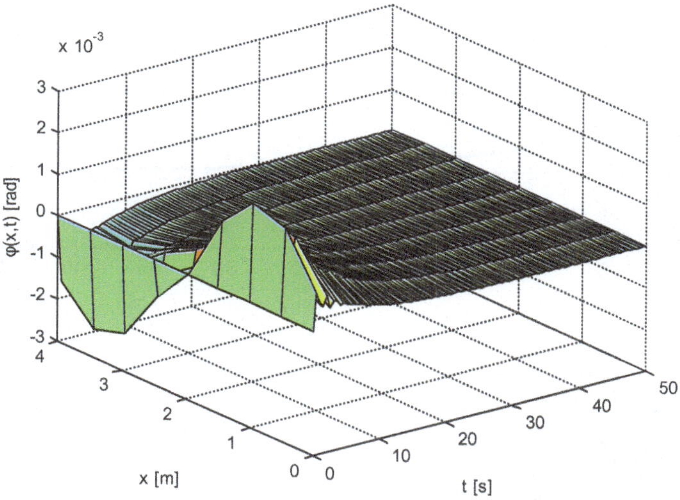

Fig. 12.6 Rotation $\varphi(x, t)$ of the crane bridge with the proposed control

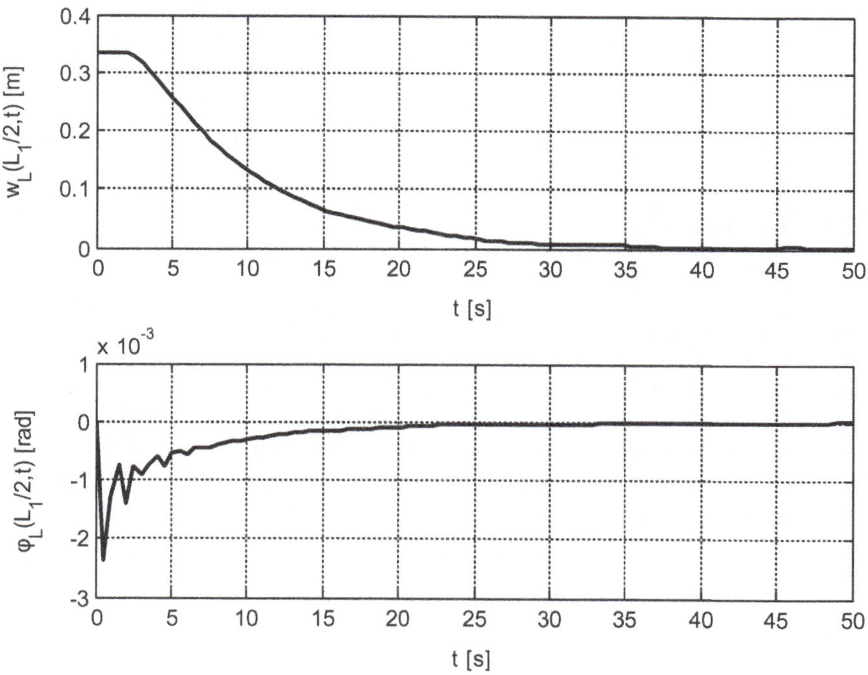

Fig. 12.7 Bending displacement $w_L\left(\frac{L_1}{2}, t\right)$ and rotation $\varphi_L\left(\frac{L_1}{2}, t\right)$ of the crane bridge with the proposed control

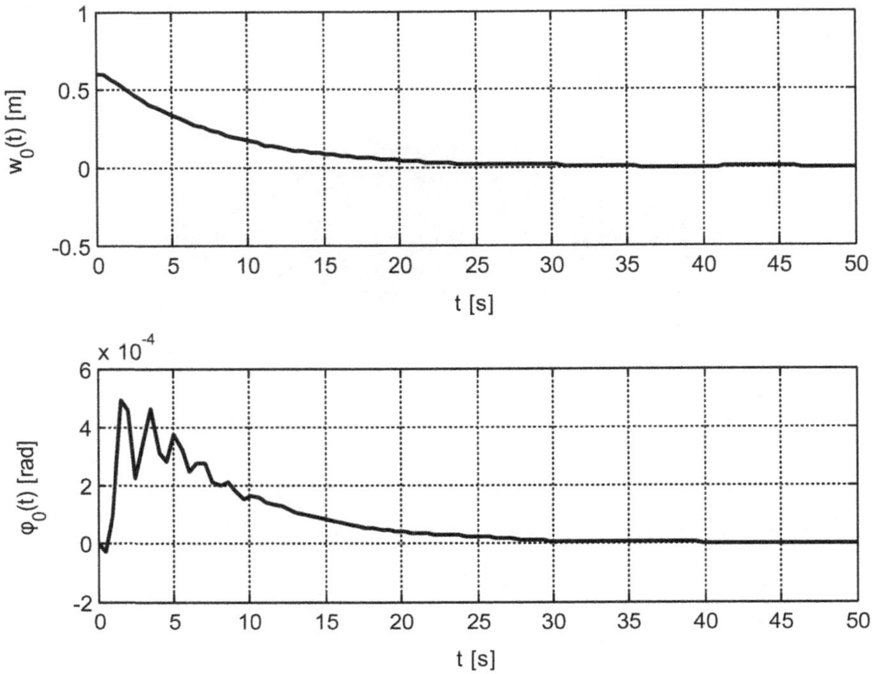

Fig. 12.8 Displacement $w_0(t)$ and rotation $\varphi_0(t)$ of the middle rigid body with the proposed control

12.5 Conclusions

In this chapter, the vibration control problem of the disturbed overhead crane bridge system, which is regarded as a Timoshenko beam equipped with a rigid body representing an attached trolley, has been settled out. Considering the parametric uncertainties and unknown control directions, an adaptive control scheme has been developed without needing to exactly know the system parameters and control directions. Lyapunov's direct method is employed to analyze the system stability under the designed control laws. Simulation results has been conducted to verify the effectiveness of the system.

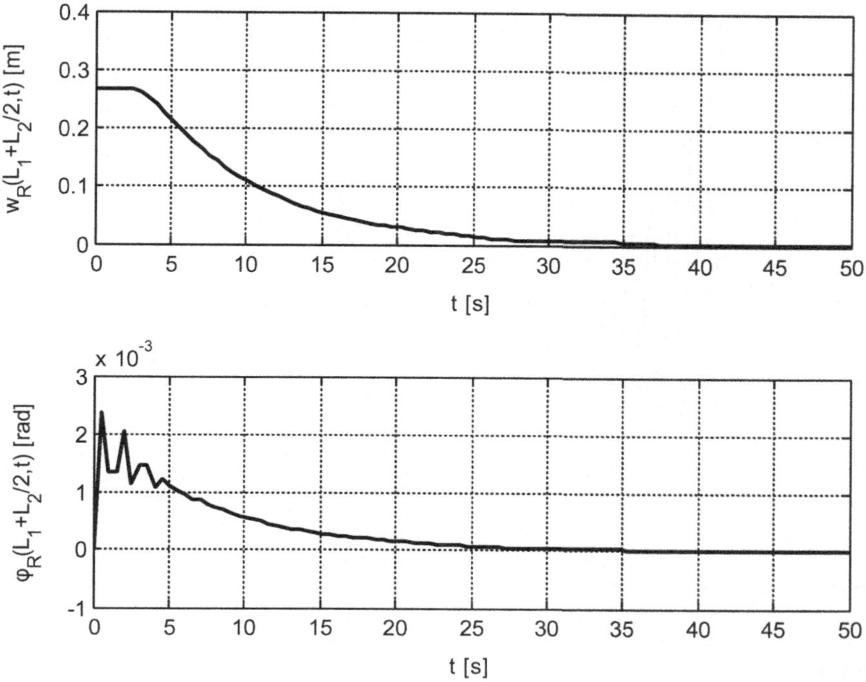

Fig. 12.9 Bending displacement $w_R\left(L_1 + \frac{L_2}{2}, t\right)$ and rotation $\varphi_R\left(L_1 + \frac{L_2}{2}, t\right)$ of the crane bridge with the proposed control

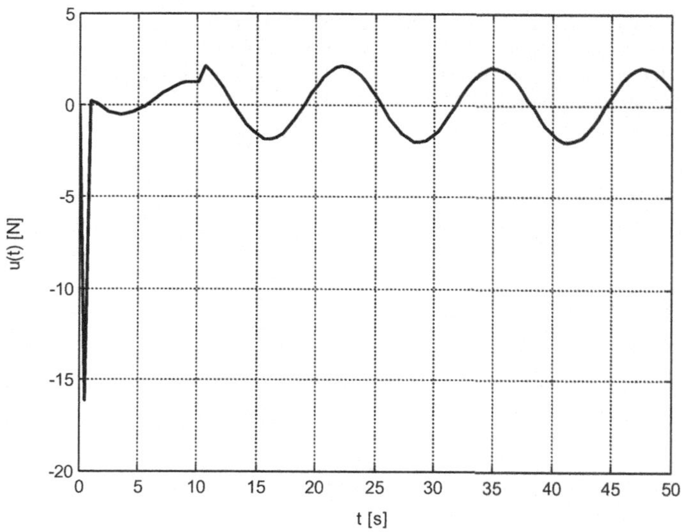

Fig. 12.10 Control input $u(t)$

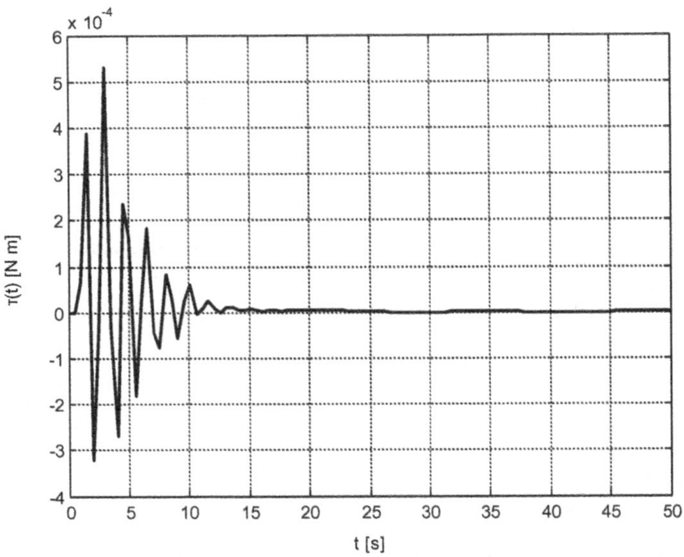

Fig. 12.11 Control input $\tau(t)$

References

1. S.S. Ge, C. Yang, T.H. Lee, Adaptive robust control of a class of nonlinear strict-feedback discrete-time systems with unknown control directions. Syst. Control Lett. **57**(11), 888–895 (2008)
2. C. Yang, S.S. Ge, T.H. Lee, Output feedback adaptive control of a class of nonlinear discrete-time systems with unknown control directions. Automatica **45**(1), 270–276 (2009)
3. X. Ye, J. Jiang, Adaptive nonlinear design without a priori knowledge of control directions. IEEE Trans. Autom. Control **43**(11), 1617–1621 (1998)

Conclusion

The book devotes to the PDE modeling and distributed and boundary controller design of several kinds of distributed parameter systems. This book is divided into 12 chapters and each chapter is dedicated to a control problem. The research results in this book are summarized in each chapter, and verified by numerical simulation results. The major research results are presented as following.

In Chap. 3, based on the PDE model of the Euler-Bernoulli beam which is subject to unknown parametric and boundary modeling uncertainties, the LMI-based vibration control strategy is carried out using RBF neural networks for stabilization the closed-loop system. Stability analysis is provided by Lyapunov's direct method. Simulation results are analyzed to verify the effectiveness of the proposed control strategy.

In Chap. 4, the vibration control problem of the Euler-Bernoulli beam disturbed by unknown distributed disturbance is investigated. Based on the PDE model with uncertain parameters, disturbance observer is developed and vibration control laws are given by using LMI techniques to stabilize the system. Exponential stability analysis of the beam is provided by Lyapunov's direct method. The proposed control law is proved to be effective by numerical simulations.

In Chap. 5, exponentially stable boundary and distributed observers are designed to estimate the external disturbances of the flexible string system which is described by PDEs. Based on the proposed observers, boundary and distributed control laws are given by using LMI techniques to suppress the vibration, with which the exponential stability of the closed-loop system is guaranteed. The proposed observers and control laws are verified to be effective by numerical simulations.

In Chap. 6, the PDE model of the vehicle-mounted flexible manipulator system is derived by Hamilton principle. Then basic control laws are discussed and the asymptotic stability analyses of the closed-loop system are provided by Lyapunov's direct method and LaSalle's invariance principle. Numerical simulations are given to demonstrate the effectiveness of the proposed controllers.

X. Xing and J. Liu, *Vibration Control Methods of Mechanical Distributed Parameter Systems*, Springer Tracts in Mechanical Engineering,
https://doi.org/10.1007/978-981-16-1532-0

In Chap. 7, the proposed RBF neural network-based adaptive fault-tolerant control is proposed for the vehicle-mounted flexible manipulator system and the asymptotic stability of closed-loop system is analyzed. The results of numerical simulations are given to verify the effectiveness of the controller.

In Chap. 8, a switching fault-tolerant control is developed using the BLF function to guarantee the state constraints of the vehicle-mounted flexible manipulator system under the condition of actuator failures. The proof of the asymptotic stability of the fault-free system is provided. The validity of proposed control is demonstrated by numerical simulations.

In Chap. 9, the state-feedback control laws which can guarantee the prescribed performance are designed for the vehicle-mounted flexible manipulator system and the asymptotic stability of the closed-loop system is proved. Simulation results are finally given to validate the proposed control method.

In Chap. 10, the PDE modeling problem of the overhead crane bridge is investigated and the nonlinear coupling between the bridge and the trolley is considered in the established model. Based on the PDE model, control laws are designed to suppress the vibration and rotation of the system and numerical simulation results are given.

In Chap. 11, a robust control scheme is addressed for stabilizing the overhead crane bridge system and ensuring output constraints of the trolley using the BLF. With the proposed control, closed-loop signals of the disturbed closed-loop system are bounded and the closed-loop system is exponentially stable with output constraints of the middle trolley. A simulation example is given for demonstrating the validity of the designed control method.

In Chap. 12, the PDE control strategy is developed to eliminate the vibration and rotation of the overhead crane bridge system subject to uncertain control directions, parametric uncertainties and environmental disturbances by employing Nussbaum functions. Under the proposed control, the undesirable deflection and rotation of the system can eventually converge to given compacts.

To summarize, this book covers the modeling and controller design of several common mechanical distributed parameter systems. The book is primarily intended for researchers and engineers in the control field. It can also serve as a complementary reading on modeling and control of mechanical systems at the post-graduate level.

The manufacturer's authorised representative in the EU is Springer Nature Customer Service Centre GmbH, Europaplatz 3, 69115 Heidelberg, Germany. If you have any concerns regarding our products, please contact ProductSafety@springernature.com

Printed and bound by CPI Group (UK) Ltd, Croydon, CR0 4YY
30/09/2026
02225289-0004